Lecture Notes in Computational Science and Engineering 25

Research and Technology Organization of NATO:
Educational Notes, RTO-EN-021

Springer
Berlin
Heidelberg
New York
Hong Kong
London
Milan
Paris
Tokyo

Timothy J. Barth
Herman Deconinck
Editors

Error Estimation and Adaptive Discretization Methods in Computational Fluid Dynamics

With 143 Figures and 24 Tables

Springer

Editors

Timothy J. Barth

NASA Ames Research Center
NAS Division
Moffett Field, CA 94035, USA
e-mail: barth@nas.nasa.gov

Herman Deconinck

von Karman Institute for Fluid Dynamics
Waterloosesteenweg, 72
1640 Sint-Genesius-Rode, Belgium
e-mail: deconinck@vki.ac.be

Cataloging-in-Publication Data applied for

Die Deutsche Bibliothek - CIP-Einheitsaufnahme

Error estimation and adaptive discretization methods in computational fluid dynamics /
Timothy J. Barth ; Herman Deconinck ed.. - Berlin ; Heidelberg ; New York ; Hong Kong ;
London ; Milan ; Paris ; Tokyo : Springer, 2003
(Lecture notes in computational science and engineering ; 25)
ISBN 3-540-43758-4

Mathematics Subject Classification (2000):
65L60, 76M10, 76M12, 49M29

ISSN 1439-7358
ISBN 3-540-43758-4 Springer-Verlag Berlin Heidelberg New York

Springer-Verlag Berlin Heidelberg New York
a member of BertelsmannSpringer Science + Business Media GmbH
http://www.springer.de

Printed in Germany

Cover Design: Friedhelm Steinen-Broo, Estudio Calamar, Spain
Cover production: *design & production*
Typeset by the authors using a Springer TEX macro package

Printed on acid-free paper SPIN: 10555421 46/3142/LK - 5 4 3 2 1 0

Preface

As computational fluid dynamics (CFD) is applied to ever more demanding fluid flow problems, the ability to compute numerical fluid flow solutions to a user specified tolerance as well as the ability to quantify the accuracy of an existing numerical solution are seen as essential ingredients in robust numerical simulation. Although the task of accurate error estimation for the nonlinear equations of CFD seems a daunting problem, considerable effort has centered on this challenge in recent years with notable progress being made by the use of advanced error estimation techniques and adaptive discretization methods.

To address this important topic, a special course was jointly organized by the NATO Research and Technology Office (RTO), the von Karman Institute for Fluid Dynamics, and the NASA Ames Research Center. The NATO RTO sponsored course entitled "Error Estimation and Solution Adaptive Discretization in CFD" was held September 10-14, 2002 at the NASA Ames Research Center and October 15-19, 2002 at the von Karman Institute in Belgium. During the special course, a series of comprehensive lectures by leading experts discussed recent advances and technical progress in the area of numerical error estimation and adaptive discretization methods with specific emphasis on computational fluid dynamics. The lecture notes provided in this volume are derived from the special course material. The volume consists of 6 articles prepared by the special course lecturers. The articles cover a wide range of topics related to *a posteriori* error estimation and adaptive meshing that should be useful to those readers with a modest background in numerical methods and computational fluid dynamics.

The organizers gratefully acknowledge the special course lecturers for their considerable time and effort in preparing the lecture notes for publication. The organizers also acknowledge the support of the NATO RTO, the von Karman Institute, and the NASA Ames Research Center for sponsoring and holding the special course. Additional thanks is given to the NATO RTO and Springer-Verlag for their enthusiastic support of the volume and attention to detail.

NASA Ames Research Center, USA, *Timothy Barth*
Von Karman Institute for Fluid Dynamics, Belgium *Herman Deconinck*
July 2002

Table of Contents

Adaptive Mesh Generation

Marshall Bern

Xerox Palo Alto Research Center
3333 Coyote Hill Rd., Palo Alto, CA 94304, USA
bern@parc.xerox.com

Abstract. These notes cover topics in mesh generation from a computational geometry perspective. This perspective means emphasis on difficiult domain geometry, unstructured triangular and tetrahedral meshes, and provable bounds on quality and complexity. We concentrate on new results, especially those results applicable to solution-adaptive methods in computational fluid dynamics.

Contents

1 Introduction

We start with an old story of geometric discretization gone awry. Imagine we discretize a cylinder as shown in Figure 1(a). The picture shows a stack of $n = 4$ congruent prisms, each with $m = 6$ triangulated rectangular sides. As m and n go to infinity, the sum of the areas of the rectangles approaches the surface area of the cylinder just as it should. If $n \gg m$, then the rectangles are long and skinny and the two triangles in each rectangle are *sharp*, meaning having an angle near 0°.

Now let us discretize the cylinder as shown in Figure 1(b). This picture shows a stack of n "antiprisms", a solid in which the top and bottom m-gons are rotated with respect to each other by a turn of π/m radians. Now as m and n go to infinity, we find something surprising, called the Schwarz paradox. The sum of the areas of the triangles depends upon exactly how fast m and n go to infinity. If $n/m^2 \to 0$ then the sum converges as it should to the area of the cylinder. But if $n/m^2 \to \infty$, then the sum diverges, and if n/m^2 is somewhere in between, then the sum could converge to an incorrect finite number. The calculation can be found in [93] or [55]. Now $n \gg m$ implies not only that the triangles are sharp, but even worse, that they are *flat*, meaning having an angle near 180°.

Notice that any fixed bound on the maximum angle of a triangle less than 180° prevents the "paradox". For example, as a triangle with maximum angle 170° shrinks to be much smaller than the radius of the cylinder, it must flatten onto the cylinder's surface. Thus a triangulation in which the overall maximum angle is at most 170° and the longest edge length is small necessarily approximates the surface area of the cylinder.

Also notice that it is not the flat angles themselves that are the problem; it is the combination of the flat angles and the directions of the triangles. For if we triangulate the cylinder as shown in Figure 1(c), we do not encounter the paradox, no matter how flat the triangles.

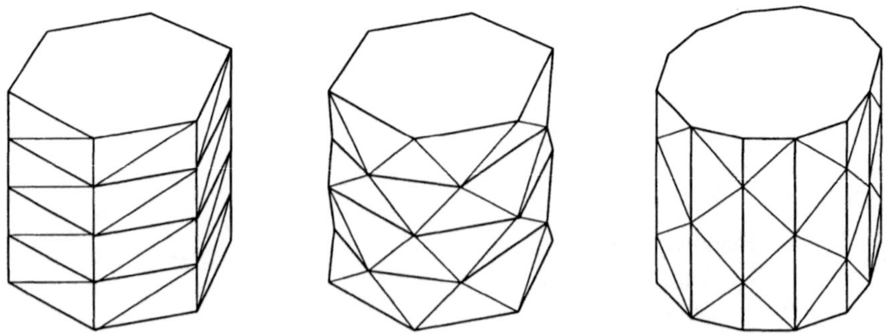

Fig. 1. A cylinder discretized in three different ways: (a) sharp horizontal triangles, (b) flat horizontal triangles, (c) flat vertical triangles.

With this story to whet our appetite, we now give an overview of the material in this survey article. We start with a discussion of element shape (Section 2). We then give a short course on Delaunay triangulation in Sections 3 and 4. (Some people think that computational geometry *is* the study of Delaunay triangulation!) We consider point placement methods next, first point placement dictated by domain geometry (Section 5) and then solution-adaptive refinement (Section 6). We continue with mesh simplification in Section 7. Finally we look at hierarchical meshing such as Cartesian meshes and subdivision surfaces (Section 8) and data compression of meshes (Section 9). The author of this article has co-authored two previous surveys of mesh generation [21,27]. This article concentrates on new results, and its overlap with the previous two should be less than 50%. For the most up-to-date information on mesh generation, we recommend the Web [62,78,89,104].

2 Element Shape

The cylinder example of Figure 1 should motivate us to be at least a little bit careful about the shapes of elements used in discretization of geometry. For purposes of discussion, let us define the *aspect ratio* of an element to be the ratio of its maximum to its minimum width, where width refers to the distance between parallel supporting hyperplanes. There are many other roughly equivalent definitions of aspect ratio.

In general, elements of large aspect ratio are bad. Large aspect ratios lead to poorly conditioned matrices, worsening the speed and accuracy of the linear solver. Speed degrades before accuracy; a triangular mesh with a rather mild sharpest angle (say 10°) can be noticeably slower to converge than a triangular mesh with a minimum angle of 45°.

Moreover, even assuming that the solver gives an exact answer, large aspect ratios may give unacceptable interpolation error. Here it is useful to distinguish between sharp and flat elements as in the cylinder example. Early results [38] showed convergence of finite element methods as triangular elements shrink, assuming that all angles are bounded away from 0°. In a frequently cited paper, Babuška and Aziz [9] showed convergence assuming only that angles are bounded away from 180°, a weaker condition. There have been a number of more recent analyses of interpolation error as well, some of which [28,95] include information about the finite element solution into triangle quality estimators. Bank and Smith [15] gave a not-just-asymptotic analysis of interpolation error and suggested a quality measure that only slightly favors sharp triangles over flat triangles: the area of a triangle divided by the sum of squared edge lengths.

In three dimensions, the equivalent of a sharp triangle is a tetrahedron with a small solid angle and the equivalent of a flat triangle is a tetrahedron with a small dihedral angle. Thus the generalization of Babuška-Aziz shows convergence assuming that dihedrals are bounded away from 0°, even if some solid angle is small (as in the 1-3 tetrahedron in Figure 16).

2.1 Anisotropy

Sometimes elements of large aspect ratio are good. If the solution to the differential equation is *anisotropic*, meaning that its second derivative varies greatly with direction, then properly aligned high-aspect-ratio elements give a very efficient mesh. The cylinder serves as a simple example: relatively few sharp or flat vertical triangles (that is, $m \gg n$, even $n = 1$) are needed for a good approximation of the surface area. The ideal aspect ratio of a triangle is the square root of the ratio of the larger to smaller eigenvalue of the Hessian [95]. For triangular meshes, it does not make much difference whether long skinny elements have large angles as well as small angles, but if the aspect ratio exceeds the ideal then large angles are much worse than small [95].

Fluid flow problems, especially Navier-Stokes simulation, are strongly anisotropic. For example, in aerodynamic simulations ideal aspect ratios may reach 10,000 along the surface of the aircraft. Quadrilateral and hexahedral meshes have an advantage in accuracy over triangular and tetrahedral meshes for finite volume and finite difference formulations of these problems, because such meshes allow faces of elements in the boundary layers to be either almost parallel or almost orthogonal to the surface. (There is still some debate over exactly how advantageous are quad and hex meshes [12].) However, it may be difficult or impossible to mesh complicated domains with hex boundary layers.

Hybrid meshes with a mix of element shapes are a practical solution. In two dimensions, one can use quad boundary layers along with triangles away from the boundary. In three dimensions, hex boundary layers require some other sorts of elements, such as prisms or pyramids, in order to transition to tetrahedra. A more esthetic sort of three-dimensional hybrid mesh uses triangular prisms in its boundary layers and tetrahedra away from the boundary. Hybrid meshing is gaining popularity in fluid dynamics research [116] and in industry [33,52], but has so far been neglected by the computational geometry community.

Simulations with shock fronts—for example, supersonic air flow over a wing—are also strongly anisotropic. In this case, the locations and directions for high-aspect-ratio elements cannot be predicted in advance. The need for general solution-adaptivity—local remeshing based on an initial solution—now favors triangles and tetrahedra over quads and hexes. As we shall see in Section 6, triangles and tetrahedra can be refined and derefined (coarsened) much more easily than quads, hexes, or prisms. Hybrid meshes should be a workable solution even with shock fronts, however, because hex or prism surface layers should not be as hard to refine as unstructured hex meshes.

2.2 M-Matrix

Element shape affects another property of the linear system besides condition number. A triangular mesh with well-shaped elements gives a symmetric M-matrix—positive definite with negative off-diagonal entries—for a finite element formulation of an equation with a Laplacian operator. M-matrices are exactly those matrices that satisfy a discrete maximum principle; this desirable property rules out oscillation of the numerical method. In this case, "well-shaped" has a precise meaning: the two angles opposite each interior edge of the mesh should sum to at most 180° [17,39].

This requirement implies that no quadrilaterals are "reversed" (Section 3), so the triangulation must be the Delaunay or constrained Delaunay triangulation. Depending on the boundary conditions associated with the differential equation, an M-matrix may also require that the single angle opposite a boundary edge should measure at most 90°. This requirement goes beyond Delaunay, but it is not hard to satisfy this requirement for domains without internal boundaries: simply split outwards-facing obtuse angles by dropping perpendiculars to the boundary, flip back to a new Delaunay triangulation, and repeat until there are no reversed quadrilaterals and no outwards-facing obtuse angles.

In three dimensions, an unstructured tetrahedral mesh gives an M-matrix for a finite element formulation if and only if, for each edge e' in the mesh, the sum $\sum_e |e| \cot \theta_e$ is nonnegative, where the sum is over all edges e that are opposite to e' in tetrahedra of the mesh, and where $|e|$ denotes the length of e and θ_e the dihedral angle at e [121]. All such sums will be nonnegative if all dihedrals in the mesh are nonobtuse, but this simplified condition is more restrictive than necessary. The condition for obtaining an M-matrix is fairly demanding for a mesh generator (or maybe just difficult to work with); there is no mesh generation algorithm for polyhedral domains that is guaranteed to produce an M-matrix.

For a finite volume formulation using Voronoi control volumes, a sufficient condition for obtaining an M-matrix is that each triangular face and each tetrahedral element be *self-centered*, meaning that it contains its circumcenter [92]. Again nonobtuse dihedrals suffices but is overly restrictive. And as in the finite element case, there is no mesh generation algorithm guaranteed to produce an M-matrix.

3 Voronoi Diagrams and Delaunay Triangulations

Our first topic is mesh generation based on Delaunay triangulation. We start by reviewing the basic geometric constructions. Bear with us if the initial material seems familiar, as we shall soon come to some constructions not so well-known outside computational geometry. In later sections, we shall see how to apply these constructions in mesh generation.

Let $S = \{ s_1, s_2, \ldots, s_n \}$ be a set of distinct points, called *sites*, in the Euclidean plane. The *Voronoi cell* $V(s_i)$ for site s_i is the set of all points at least as close to s_i as to any other site:

$$V(s_i) = \bigcap_{j \neq i} \{ p \in \mathbb{R}^2 \mid d(p, s_i) \leq d(p, s_j) \} .$$

The region of the plane at least as close to s_i as to s_j is a half-plane, bounded by the perpendicular bisector of line segment $s_i s_j$. Since $V(s_i)$ is the intersection of half-planes, it is a convex (possibly unbounded) polygon. The *Voronoi diagram* of S is the subdivision of the plane into Voronoi cells and shared edges and vertices.

A *planar straight-line graph* consists of a set of vertices in the plane, along with line segments representing graph edges between pairs of vertices; edges meet only

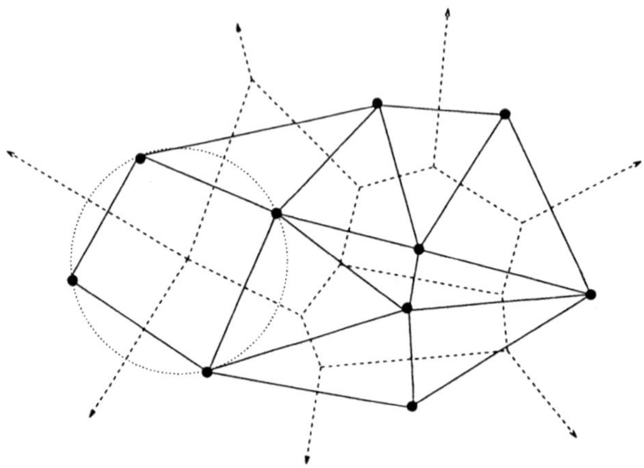

Fig. 2. The Voronoi diagram and Delaunay triangulation of points in the plane. This case includes four cocircular points, and hence the Delaunay triangulation must be completed to form a proper triangulation.

at shared endpoints. The *Delaunay triangulation* of S, denoted $DT(S)$, is the planar straight-line graph with vertex set S and an edge between s_i and s_j if and only if $V(s_i)$ and $V(s_j)$ meet at an edge in the Voronoi diagram of S. The Delaunay triangulation is thus the *planar dual* of the Voronoi diagram, with a Delaunay vertex corresponding to a Voronoi face and vice versa. If every vertex in the Voronoi diagram is the meeting point of exactly three Voronoi cells, then the Delaunay triangulation is indeed a triangulation of the convex hull of S. But if four or more Voronoi cells meet at a common point, then the Delaunay triangulation includes a face with four or more sides as shown in Figure 2. This is a degenerate case, occurring only when four or more points of S lie on a common circle, and can be easily *completed* by triangulating the untriangulated face.

We can define the Delaunay triangulation directly, without going through the Voronoi diagram, by the *empty circumcircle condition*: a triangle $s_i s_j s_k$ is a face of $DT(S)$ if and only if the circle through s_i, s_j, and s_k neither passes through nor encloses any other sites. We can also state an equivalent empty circle condition for edges of $DT(S)$: $s_i s_j$ is an edge of $DT(S)$ if and only if there is some circle through s_i and s_j that neither passes through nor encloses any other sites.

Figure 3 shows the two possible triangulations of four points forming a convex quadrilateral. Notice that either both circumcircles are empty or both are not. An equivalent condition to both circumcircles being empty is that the angles at the vertices of the quadrilateral, not incident to the diagonal, sum to less than 180°. Another equivalent condition is that the Delaunay triangulation is the triangulation that maximizes the minimum angle. In fact, this property holds for arbitrary planar point sets: $DT(S)$ maximizes the minimum angle over all triangulations of S. The Delaunay triangulation actually has a number of global optimality properties, including some rather exotic ones. For example, when used for piecewise-linear

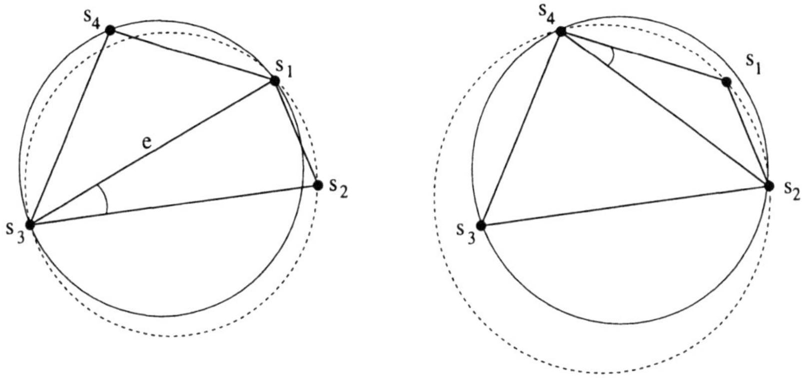

Fig. 3. (a) The Delaunay triangulation of four points forming a convex quadrilateral. (b) The reversed triangulation has smaller minimum angle (marked).

interpolation, it maximizes the total "roughness' of the surface [94], meaning the integral of the gradient squared. For another example, if we scale all z-coordinates downwards relative to x and y, at some point before the surface completely flattens out, the Delaunay triangulation gives the least-area piecewise-linear interpolating surface.

Voronoi diagrams and Delaunay triangulations generalize to higher dimensions in a straightforward way. In the absence of degeneracies, a Delaunay triangulation in $\mathbb{R}^3$ is a *tetrahedralization* (sometimes "tetrahedrization") of the convex hull of the sites. Each tetrahedron has an empty circumsphere, meaning that the sphere neither passes through nor encloses any other sites.

3.1 Lifting

It is often useful to view a Delaunay triangulation in $\mathbb{R}^d$ as a projection of a convex polytope in $\mathbb{R}^{d+1}$. Consider *lifting* points in $\mathbb{R}^2$ by projecting them vertically onto the paraboloid Λ in $\mathbb{R}^3$ given by

$$\Lambda = \{ (x, y, z) \mid z = x^2 + y^2 \}.$$

Let $\hat{S}$ be the image of S under this mapping; site $s_i = (x_i, y_i)$ maps to point $\hat{s_i} = (x_i, y_i, z_i = x_i^2 + y_i^2)$. Consider the *lower convex hull* of $\hat{S}$, that is, the part of the convex hull of $\hat{S}$ consisting of faces lying in planes that separate $\hat{S}$ from a point arbitrarily far down the z-axis, denoted $(0, 0, -\infty)$. The Delaunay triangulation $DT(S)$ is the projection of the lower convex hull of $\hat{S}$ onto the xy-plane. See Figure 4.

3.2 Power Diagrams and Weighted Delaunay Triangulations

Voronoi diagrams and Delaunay triangulations have been generalized in numerous ways: line segment or polygonal sites, multiplicatively or additively weighted sites,

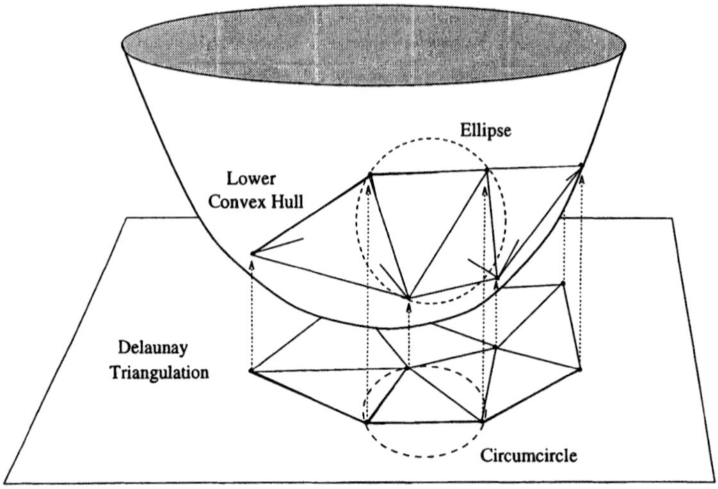

Fig. 4. The lower convex hull of lifted sites projects to the Delaunay triangulation. A plane supporting a face of the lower hull cuts the paraboloid in an ellipse that projects to a circumcircle.

and so forth [8]. Most of these generalizations, however, lose the elegance of the original constructions. For example, imagine that each site s_i is endowed with an additive weight w_i, such that the distance to s_i is the usual Euclidean distance plus w_i. Under this definition, the region closer to s_i than to s_j is bounded by the locus of points (x, y) such that $((x - x_i)^2 + (y - y_i)^2)^{1/2} + w_i = ((x - x_j)^2 + (y - y_j)^2)^{1/2} + w_j$, a hyperbola. Hence, in an additively weighted Voronoi diagram, cells have curved sides.

There is one generalization of Voronoi diagrams and Delaunay triangulations, however, that preserves almost all of the mathematics. This generalization is not very well known to the finite element community, although it is starting to find uses in mesh generation. Define the *power distance* of a point to site s_i to be the square of the Euclidean distance minus the weight w_i. Now the territory closer to $s_i = (x_i, y_i)$ than to $s_j = (x_j, y_j)$ is bounded by the locus of points (x, y) such that $(x - x_i)^2 + (y - y_i)^2 - w_i = (x - x_j)^2 + (y - y_j)^2 - w_j$. This time the x^2 and y^2 terms cancel, and the equation defines a line,

$$2(x_i - x_j)x + 2(y_i - y_j)y + ((x_j - x_i)^2 + (y_j - y_i)^2 + w_i - w_j) = 0. \quad (1)$$

Hence cells in the *power diagram* are—like Voronoi cells—convex, possibly unbounded, polygons. A power cell, however, may be empty.

We can regard the power diagram as a sort of Voronoi diagram for circular sites C_i, with the radius of site C_i being $\sqrt{w_i}$. For a point p outside the circle, the power distance is the square of the length of a tangent from the circle to p. For a point p inside the circle, the power distance is the square of half the length of the shortest chord through p, only negated. (An interesting geometric fact: the product of the lengths of the two pieces of a chord through p is always the same.) In the case of overlapping circles, the line dividing the power cells given by equation (1) contains the mutual chord. A circle completely enclosed by another circle has an empty cell. In Figure 5 we can see the mutual chord for circles C_i and C_j and an empty cell for circle C_m.

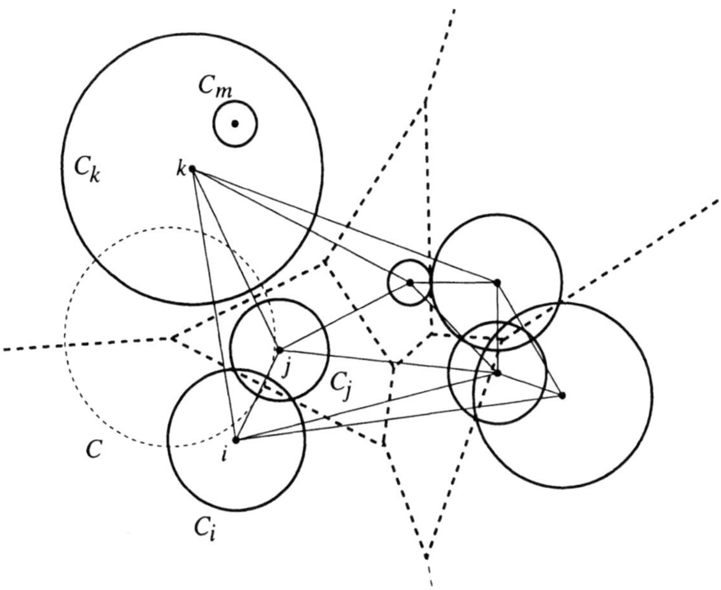

Fig. 5. A power diagram (dashed), along with its dual weighted Delaunay triangulation. The dashed circle C is the orthogonal circle for triangle ijk.

Following Edelsbrunner [45], we shall call the dual of the power diagram the *weighted Delaunay triangulation* of the set of weighted sites. In general, due to circles with empty cells, the weighted Delaunay triangulation is a graph on a subset of the sites S, but in the special case that all weights are the same, the weighted Delaunay triangulation is exactly the Delaunay triangulation. Unless weights and site locations conspire so that some point has equal power distance to four sites, the weighted Delaunay triangulation is indeed a triangulation. The analogue of the empty circumcircle condition for the weighted Delaunay triangulation is the *empty orthogonal circle condition*: the triangle for circular sites C_i, C_j, and C_k appears in the triangulation if and only if the unique circle C that crosses each of C_i, C_j, and C_k orthogonally crosses no other circles "more deeply" than orthogonal.

The lifting map described above generalizes to weighted Delaunay triangulations. The weighted Delaunay triangulation is the projection of the lower convex hull of points lifted by the map that takes site $s_i = (x_i, y_i)$ to $(x_i, y_i, x_i^2 + y_i^2 - w_i)$. The lifting map is perhaps even more natural for weighted than for unweighted Delaunay triangulations, because it now defines a bijection between weighted points in two dimensions and point sets in three dimensions, whereas it previously mapped

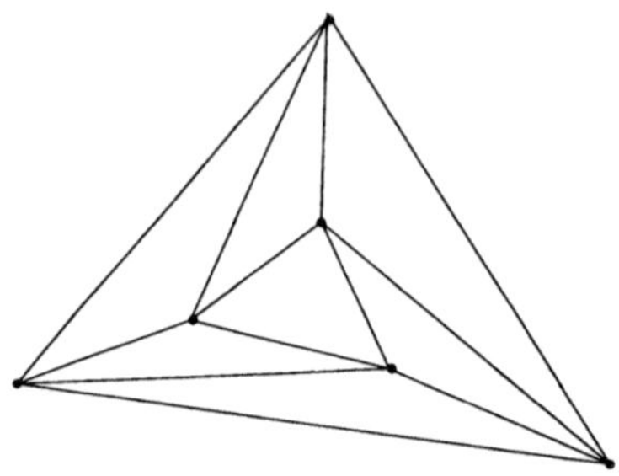

Fig. 6. This triangulation is not the projection of the lower half of a convex polytope, because any set of vertex elevations must make some edge a mountain (or flat) rather than a valley.

only to special sets of points. Hence weighted Delaunay triangulations are exactly the projections of (lower halves of) convex polytopes. This does not give all triangulations, only the so-called regular triangulations, as shown by the counterexample in Figure 6.

3.3 Algorithms

We now turn to the subject of algorithms for computing Delaunay triangulations. We shall look at two different algorithms, called *edge flipping* and *randomized incremental insertion*.

The edge flipping algorithm, due to Lawson [68], starts from any triangulation of S and then locally optimizes each edge. Let e be an internal (non-convex-hull) edge and Q_e be the triangulated quadrilateral formed by the triangles sharing e. Quadrilateral Q_e is *reversed* if it is triangulated the wrong way, meaning that each triangle's circumcircle contains the opposite vertex as in Figure 3(b). If Q_e is reversed, we "flip" it by exchanging e for the other diagonal.

compute an initial triangulation of S
place all internal edges onto a queue
while the queue is not empty **do**
  remove the first edge e
  if quadrilateral Q_e is reversed **then**
    flip it and add the outside edges of Q_e to the queue **endif**
endwhile

An initial triangulation can be computed by a sweep-line algorithm. This algorithm adds the points of S by x-coordinate order, to the left of an imaginary vertical line sweeping across the data. Upon each addition, the algorithm walks around the convex hull of the already-added points, starting from the rightmost previous point and adding edges until the slope reverses, as shown in Figure 7. The following theorem [44] guarantees the success of edge flipping: a triangulation in which no quadrilateral is reversed must be (a completion of) the Delaunay triangulation. Although

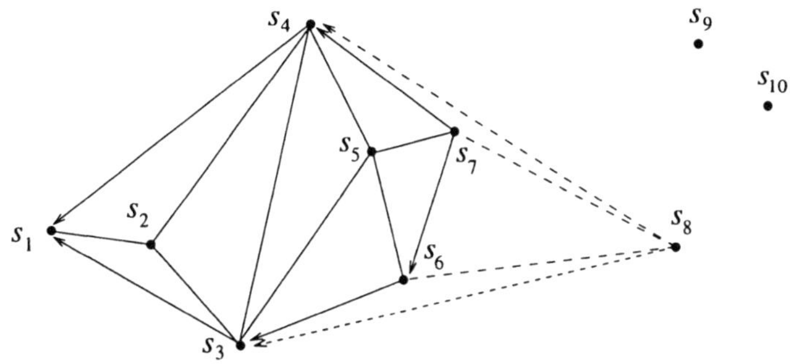

Fig. 7. A sweep algorithm for computing an initial triangulation

the worst-case running time of the edge flipping algorithm is $O(n^2)$ rather than the optimal $O(n \log n)$, it performs reasonably well in practice.

Randomized incremental insertion, due to Guibas et al. [56], adds points one at a time in random order, and after each addition flips edges to restore a Delaunay triangulation. The algorithm starts with a bounding box or bounding triangle that contains all the points of S. At the end all the triangles incident to corners of the bounding polygon can be thrown away to give $DT(S)$.

form an initial bounding box B_∞ containing all points
split the bounding box into two triangles
for $i = 1 \ldots n$ **do**
    find the triangle Δ that contains s_i
    split Δ into three by adding s_i
    place the outside edges of Δ onto a queue
    while the queue is not empty **do**
        remove the first edge e
        if quadrilateral Q_e is reversed **then**
            flip it and add the outside edges of Q_e to the queue **endif**
    endwhile
endfor

The novelty in this algorithm is the point location step in the fourth line of the pseudocode. The suggested solution for this step is quite elegant: the *history* of all triangles ever produced by the algorithm defines a data structure for answering point location queries. We keep the history in a DAG search structure (like a tree, but a node may have more than one parent), with one node for each triangle. Each time we split or flip a triangle, we add new nodes corresponding to the new triangles and make them child nodes of the old triangles, as shown in Figure 8. Notice that all new triangles are incident to the newly inserted vertex, so the flipping does not propagate very far, on the average less than six triangles (the maximum average degree of a planar graph). The fact that S is inserted in random order (rather than say left-to-right) is important to keep the expected height of the DAG for any input to $O(\log n)$ and hence the expected running time of the algorithm to the optimal $O(n \log n)$.

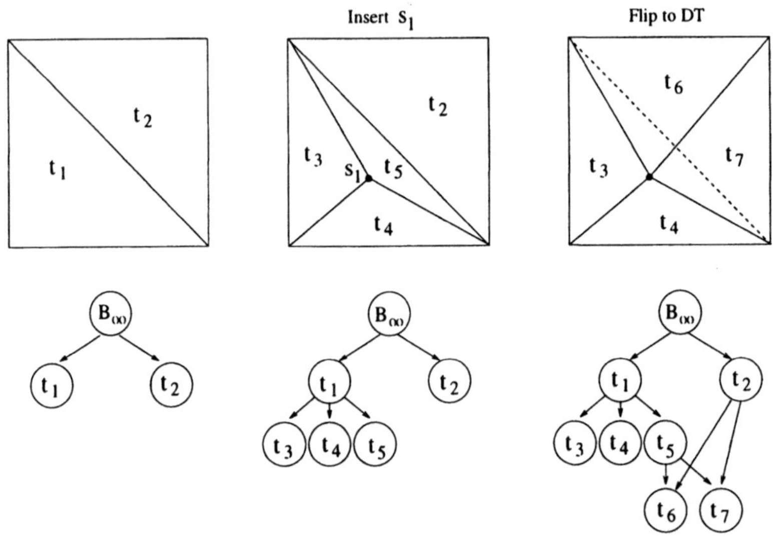

Fig. 8. Randomized incremental insertion uses a history DAG to locate inserted points.

Randomized incremental insertion generalizes to three and higher dimensions, but there are some subtleties. The generalization of edge flipping exchanges the two possible triangulations of five points in convex position, as shown in Figure 9. The two sides of the flip are no longer so symmetric: one side consists of three tetrahedra sharing an edge and the other consists of two tetrahedra sharing a triangle; the flipping part of the algorithm must consider both cases. In Section 5.5 we shall take a more general view of flipping, and flips will include splitting a triangle into three or a tetrahedron into four by adding an interior vertex.

We call a flip a *Delaunay flip* if, after the flip, the triangulation of the five points satisfies the empty sphere condition—no circumsphere encloses a point. In three dimensions, it is no longer true that any triangulation can be transformed into the Delaunay triangulation by a sequence of Delaunay flips; Joe [64] gave a simple example of a tetrahedralization that is not the Delauany triangulation and yet has no Delaunay flip. The triangulations produced by randomized incremental insertion, however, are all nicely behaved (in particular they are all regular triangulations) and the algorithm cannot get stuck.

Finally we say something about computing large 3D Delaunay triangulations. Randomized incremental insertion works quite well for computing the 3D Delaunay triangulation of up to 100,000 points. We obtained the best results [4] with a non-random, strategic order: the first 1000 points are random from all over the mesh in order to sketch out the history DAG, but the rest of the points are kept in some locality-preserving order so as to cut down on the amount of memory paging. For larger point sets, we must be smarter still. Inspired by a parallel algorithm [83], Amenta (personal communication) has proposed a divide-and-conquer approach that repeatedly cuts the points into slightly-overlapping subsets, until the subsets

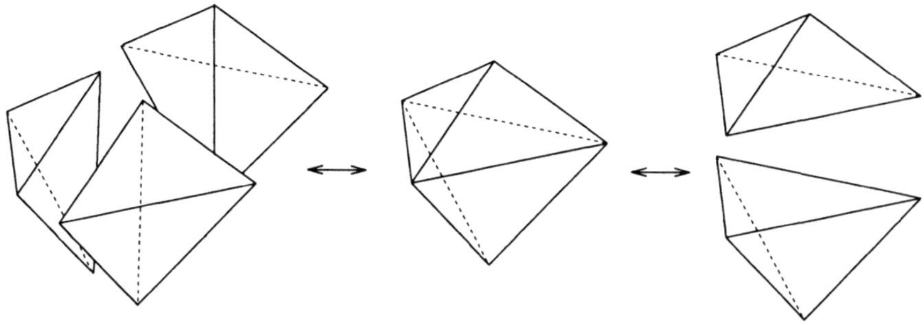

Fig. 9. In three dimensions, an edge flip exchanges three tetrahedra sharing an edge for two tetrahedra sharing a triangle, or vice versa.

are small enough to fit into the computer memory. The lifting map can be used to make the divisions. We divide the lifted points in $\mathbb{R}^4$ with a vertical hyperplane P, and think of the lifted points as lying on a paraboloid centered on P. Only the lifted points that appear on the convex hull when projected horizontally onto P need be included in both halves. We add right-half points to the left half (and vice versa) by increasing distance from P; when the projection of the remaining part of the paraboloid lies interior to the convex hull we can stop the sweep.

4 Constrained and Conforming Delaunay Triangulation

In mesh generation, we typically must conform to a domain boundary and not just to a set of vertices. Figure 10 shows that simply taking the Delaunay triangulation of the vertices of a polygonal domain does not suffice. We are now getting into the central problems of mesh generation: conforming to domain geometry, defining initial elements of appropriate sizes and shapes, and adapting a mesh based on a solution. These problems are respectively the topics Sections 4, 5, and 6; of course there is a certain amount of overlap among the three topics.

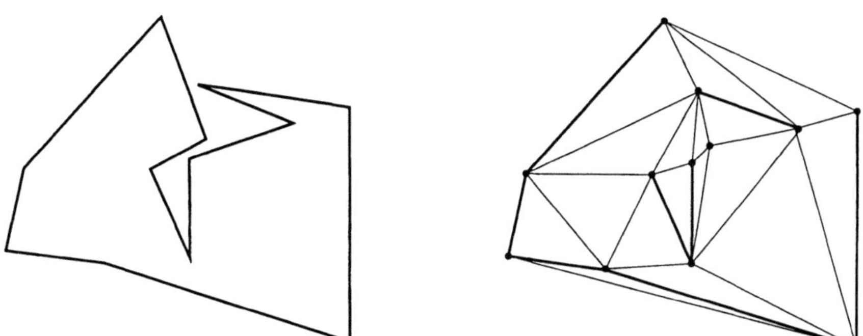

Fig. 10. The Delaunay triangulation of a polygon's vertices may not include all edges.

4.1 Two Dimensions

Let us start with two-dimensional domains. How do we define the Delaunay triangulation of a planar straight-line graph G?

One possible answer is to line the edges of G with closely-spaced vertices, so that the Delaunay triangulation of the augmented vertex set—the extra vertices are often called *Steiner points*—includes all the edges of G. This approach is called *conforming Delaunay triangulation*.

We shall sketch an algorithm due to Edelsbrunner and Tan [49] that uses only $O(n^3)$ Steiner points, regardless of the input geometry. A polynomial-size, geometry independent bound is a theoretical triumph, but such a bound is probably of no great practical importance for this problem because we shall eventually need a geometry-dependent number of interior Steiner points. Indeed in Section 5.1, we shall see another algorithm (Shewchuk's version of Delaunay refinement) that solves conforming Delaunay triangulation as a preliminary stage to interior point placement.

Edelsbrunner and Tan's algorithm places Steiner points in two stages, the "blocking" and "propagation" phases. The blocking phase computes a set of $O(n)$ non-overlapping disks whose union is connected and touches all the vertices of G. A Steiner point is added wherever a disk crosses an edge or meets another disk, $O(n^2)$ Steiner points total. The propagation phase treats the segments of input edges that are not yet covered by empty circles. This phase takes any not-yet-covered segment ab and attempts to add its diameter circle. If the diameter circle encloses another vertex c visible to ab, the orthogonal projection of c onto ab is added as a new vertex, forming two shorter segments. (We say that c is *visible* to ab if c can be connected to some point on ab by a line segment that does not cross any other vertices or edges of G.) Because the blocking phase has cut to all the vertices of G, new points cannot propagate in a spiral (which would necessarily enclose a vertex) to hit the same edge twice.

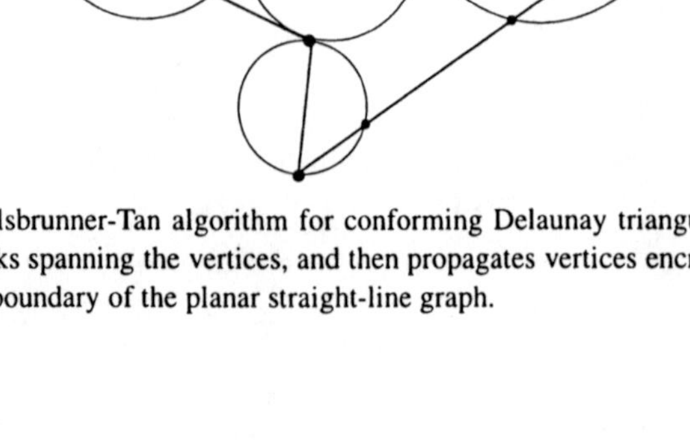

Fig. 11. The Edelsbrunner-Tan algorithm for conforming Delaunay triangulation first computes a set of disks spanning the vertices, and then propagates vertices encroaching on input edges out to the boundary of the planar straight-line graph.

Constrained Delaunay triangulation offers a more intrinsic solution to meshing a polygonal domain than does conforming Delaunay triangulation. The edges of the input domain are forced into a constrained Delaunay triangulation; all other edges arise naturally.

The *constrained Delaunay triangulation* $CDT(G)$ contains all vertices and edges of G. In addition, $CDT(G)$ contains the edge $\{a, b\}$ between two vertices of G, if and only if a is visible to b, and some circle passes through a and b, but no other vertices of G, and encloses no vertex c visible to segment ab. If G consists only of isolated vertices, this definition reduces to the Delaunay triangulation. Figure 12 depicts the constrained Delaunay triangulations of a planar straight-line graph and a simple polygon; the circles of two edges are shown. The polygon is a special case of a planar straight-line graph, but here we triangulate only the interior of the polygon, rather than all the way out to the convex hull.

The flipping algorithm for ordinary Delaunay triangulation can be adapted to compute the constrained Delaunay triangulation. As in Section 3.3, an initial triangulation can be computed by a sweep-line algorithm, and then edge flips can be performed in arbitrary order until no "flippable" quadrilateral is reversed. In the constrained case, a quadrilateral whose diagonal is one of the input edges is not flippable. The sweep-line algorithm is slightly more complicated in the constrained case. Whereas previously the algorithm kept a sorted list of all points on the current convex hull to the left of the sweep-line, the algorithm must now maintain such a list for each interval along the sweep-line between two input line segments. A sorted list or dictionary data structure can be used to determine which interval contains a newly encountered vertex.

4.2 Three Dimensions

A natural generalization of a planar-straight line graph to three dimensions is called a *piecewise-linear complex* [106]. A piecewise-linear complex X consists of a set of vertices, line segments, and facets. The facets can be arbitrary planar polygons, possibly with holes. Line segments can only intersect at shared endpoints (which

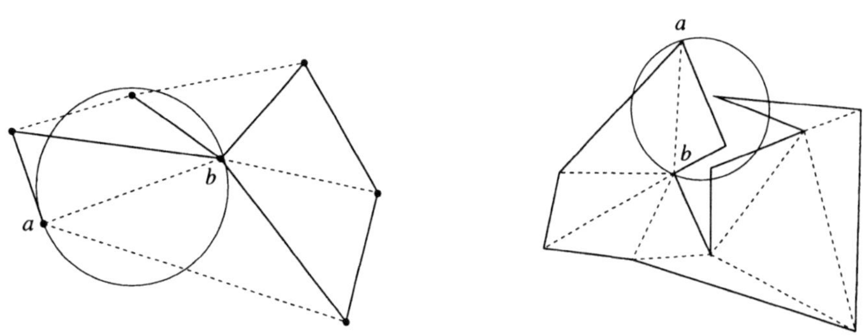

Fig. 12. Constrained Delaunay triangulations of a planar straight-line graph and a polygon.

must be vertices of X), and facets can only intersect at shared boundary segments (which must be segments of X). A line segment can intersect a facet only if it is a boundary segment of the facet, or if one of its endpoints is a vertex of the facet. If a line segment intersects a facet at a point in the segment's relative interior, then the entire segment must lie in the facet.

Given a piecewise-linear complex X, the conforming Delaunay triangulation problem in three dimensions asks for a set of Steiner points such that the Delaunay triangulation of the vertices, input plus Steiner points, contains all the segments and facets of X (possibly subdivided). Some mesh generation programs [51,117,118] take an approach that computes the Delaunay triangulation of the vertices, finds missing input edges and facets, and then adds Steiner points where input segments cross Delaunay balls or faces and where input facets cross Delaunay edges. This algorithm has not been analyzed. Murphy et al. [86] generalize a planar algorithm due to Saalfeld [103]; this algorithm covers the vertices, segments, and facets of X with small tangent balls, and then places Steiner points where the Voronoi diagram of the ball centers crosses X. This algorithm is at least guaranteed to terminate, but it has no simple-to-state bound on the number of Steiner points, not even a bound that depends on the geometry as well as on the combinatorial complexity of X.

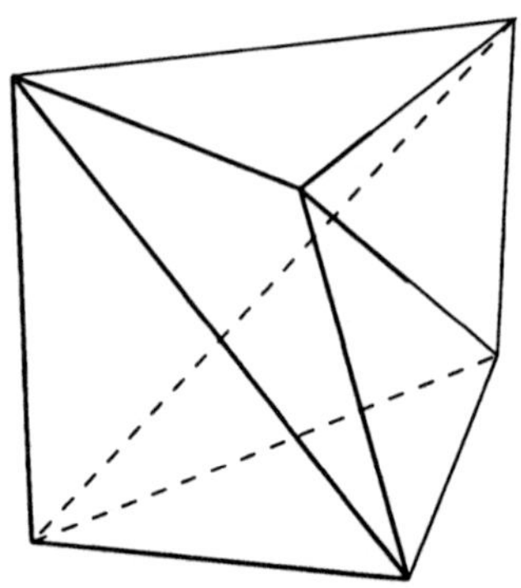

Fig. 13. Schönhardt's polyhedron is a triangular prism, twisted so that its three rectangular sides bend inwards and form a spiral of reflex edges. This polyhedron cannot be tetrahedralized without Steiner points.

There is no exact analogue of constrained Delaunay triangulation in three dimensions. One fundamental obstacle is that for some inputs triangulation is impossible without Steiner points. Figure 13 shows a classic example due to Schönhardt [105] of a polyhedron that cannot be triangulated without Steiner points. To form this polyhedron, start with a triangular prism and then twist the top relative to the bottom so that all three rectangular faces bend inwards. For this to work, the diagonals of the rectangular faces must be pairwise disjoint, forming a sort of a spiral. Now none of the three bottom vertices can see all of the top triangle, so we have no fourth vertex available for a tetrahedron containing this triangle.

Shewchuk [106] devised the closest thing to a 3D constrained Delaunay triangulation. He calls a segment of X *strongly Delaunay* if it has a circumsphere that neither encloses nor passes through any other vertex of X. He calls a simplex (line, triangle, tetrahedron) *constrained Delaunay* if it has a circumsphere that encloses no vertex of X visible to any point in the relative interior of the simplex. He shows that if each of X's segments is strongly Delaunay, then X has a triangulation in which each simplex is constrained Delaunay. In the non-degenerate case of no five cospherical vertices, the constrained Delaunay triangulation of X is unique.

Schönhardt's polyhedron provides a good example. This polyhedron can obviously be inscribed in a sphere. The edges of the top and bottom triangles are strongly Delaunay right from the start. By pushing two of the top vertices into the sphere we can also make any two of the three reflex edges strongly Delaunay, but it is not possible to make all three strongly Delaunay. We can subdivide the third reflex edge into two strongly Delaunay segments, but now the polyhedron becomes tetrahedralizable, as indeed it must for Shewchuk's result to hold. At this point we can apply Shewchuk's result, without adding any further Steiner points. This is the practical advantage of constrained over conforming Delaunay triangulation: fewer Steiner points are needed and in particular no Steiner points interior to facets. In Section 5 we shall see some 3D constrained Delaunay meshes. An interesting theoretical question [49,106] is whether there is a polynomial bound (or even a simple-to-state bound depending upon geometry) on the number of Steiner points needed to make all segments of X strongly Delaunay.

Shewchuk proved his result by showing that each point in the convex hull of X lies in some constrained Delaunay tetrahedron, and that together the constrained Delaunay tetrahedra fill the convex hull without overlapping. For the first part, Shewchuk showed that it is possible to grow a circumsphere of any constrained Delaunay simplex σ of X (edge of X or constrained Delaunay triangle on a facet of X) into a circumsphere of a constrained Delaunay tetrahedron. The key to this proof is an acyclicity result: facets of X bounded by strongly Delaunay segments cannot overlap in a cycle to block completely σ's visibility of vertices (Figure 14).

In order to compute 3D constrained Delaunay triangulations, Shewchuk [106,107] recommends the following steps: subdivide segments until they are all strongly Delaunay, compute the unconstrained Delaunay triangulation of the vertices, and then recover each facet in turn by deleting all the tetrahedra it crosses and retriangulating the holes on either side. More recently, Shewchuk [109] gave sweep algorithms with asymptotically faster running times for computing constrained Delaunay triangulations. Shewchuk's 3D constrained Delaunay triangulation also generalizes to $\mathbb{R}^d$ for $d > 3$ [106]. Here the requirement is that each *ridge*, that is, $(d-2)$-dimensional face, of the piecewise-linear complex be strongly Delaunay.

5 Point Placement

In this section, we discuss methods for placing points in order to achieve a high-quality initial mesh. There is a close connection between the topic of this section

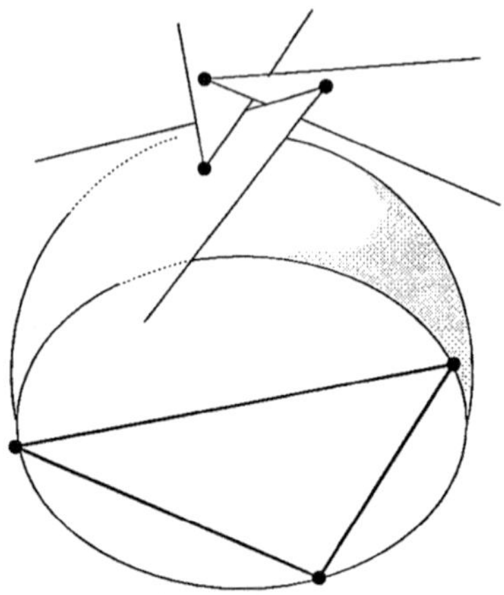

Fig. 14. Shewchuk's argument rules out the case that a cycle of facets blocks a growing circumsphere from finding a visible vertex.

Fig. 15. (a) A mesh computed by Ruppert's algorithm. (b) Detail of the right side of Lake Superior. (J. Ruppert)

(point placement) and that of the next section (mesh refinement). In this section, we cover techniques driven by domain geometry and *a priori* knowledge rather than by *a posteriori* error estimates, but much of the discussion, especially the material on mesh improvement (Sections 5.4, 5.5, and 5.6), applies to remeshing as well.

5.1 Delaunay Refinement

Delaunay refinement starts from a conforming or constrained Delaunay triangulation and then iteratively adds new Steiner points in order to improve the sizes and shapes of elements.

We first explain an algorithm for planar domains due to Ruppert [102], who extended work of Chew [36]. The first step of Ruppert's mesh generator cuts off all vertices of the domain Ω at which the interior angle measures less than 45°. The cutting line at such a vertex v should not introduce a new small feature to Ω; it is best to cut off an isosceles triangle whose base is about halfway from v to its closest visible neighbor. If v has degree greater than two, as might be the case in a *multiple domain* (a domain with internal boundaries such as general planar straight-line graph), then the bases of the isosceles triangles around v should match up so that no isosceles triangle receives a Steiner point on one of its legs. Shewchuk [108] recently gave a better algorithm for handling domains with sharp angles.

Next the algorithm computes a Delaunay triangulation of the modified domain; this Delaunay triangulation may be either constrained or conforming. (Figure 15, however, was actually computed started from the unconstrained Delaunay triangulation of input vertices, as in Shewchuk's 3D algorithm given below.) Ruppert's algorithm then goes through the loop given below. The last line of the loop repairs a constrained Delaunay triangulation after the addition of a new Steiner point c. To accomplish this step, there is no need to recompute the entire triangulation, but only to repair the triangulation around c as in randomized incremental insertion. The removed old triangles are exactly those with circumcircles containing c, which can be found by searching outwards from the triangle that contains c, and the new triangles that replace the removed triangles are all incident to the new vertex c.

while there exists a triangle t with an angle smaller than 20° **do**
    let c be the center of t's circumcircle
    if c lies within or on the diameter semicircle of a boundary edge e **then**
        add the midpoint m of e
    else add c **endif**
    recompute the Delaunay (or constrained Delaunay) triangulation
endwhile

The loop is guaranteed to halt with all angles larger than 20°. After the loop finishes, the cut-off isosceles triangles are returned to the domain, and the mesh is complete.

Ruppert's algorithm comes with a strong theoretical guarantee: for non-multiple domains it produces a triangulation in which no new angle (angle not already present in the domain boundary) measures less than 20° and in which the number of triangles is no more than a constant times the minimum number in any such *no-small-angle triangulation* of the same domain. Shewchuk showed that a similar theorem holds for multiple domains [108], only now the sharpest new angle may be as small as about one-third of the minimum input angle. The number of triangles needed is a function of geometry and not just of the number of vertices, as can be seen by considering the simple example of a long skinny rectangular domain. Such a guarantee was first proved by Bern et al. [24] for a quadtree mesh generator. In fact, it was the opportunity to prove such guarantees that attracted the attention of computational geometers to mesh generation.

To prove the bound on the number of triangles, Ruppert shows that each triangle in the final mesh is within a constant factor of the *local feature size* at its vertices. The local feature size at point $p \in \Omega$ is defined to be the radius of the smallest circle centered at p that touches two nonadjacent edges of the boundary; this is a *spacing function* intrinsic to the domain. In later sections, we shall see externally defined spacing functions.

Shewchuk [107] extended Delaunay refinement to three dimensions using his definition of 3D constrained Delaunay triangulation. The input domain is assumed to be a "facet-bounded" piecewise-linear complex, meaning that it has a watertight outer boundary; the input is also assumed to have no overly sharp diheral angles. The first step computes the unconstrained Delaunay triangulation of the vertices. Then the algorithm goes into a loop, as in Ruppert's 2D algorithm, this time adding new vertices according to three different rules. (1) If the diametral sphere of an input segment encloses or passes through a vertex not on the segment, the segment is subdivided by adding its midpoint, and midpoint subdivision continues until subsegment diametral spheres are empty, (2) If the equatorial sphere of a triangular boundary face encloses or passes through a vertex not on the face, the face is split by adding its circumcenter. Rule (1), however, takes precedence over rule (2), so that if the circumcenter would encroach upon an input segment (that is, land in its diametral sphere) then we split that segment first. (3) If a tetrahedron is *badly shaped* we add the center of its circumsphere. Rules (1) and (2) take precedence over rule (3), so we may have to split input segments and faces before adding the tetrahedron's circumcenter.

The meaning of the term "badly shaped" deserves more attention. Following [45] we can classify all the badly shaped tetrahedra into five types of skinny tetrahedra (Figure 16) and four types of flat tetrahedra (Figure 17). A previous, somewhat simpler, classification [20], divided tetrahedra by no-small- and no-large-angle criteria for each of solid and dihedral angles. In either classification, a *sliver* is a flat tetrahedron whose projection onto a plane that passes near all its vertices is fairly square.

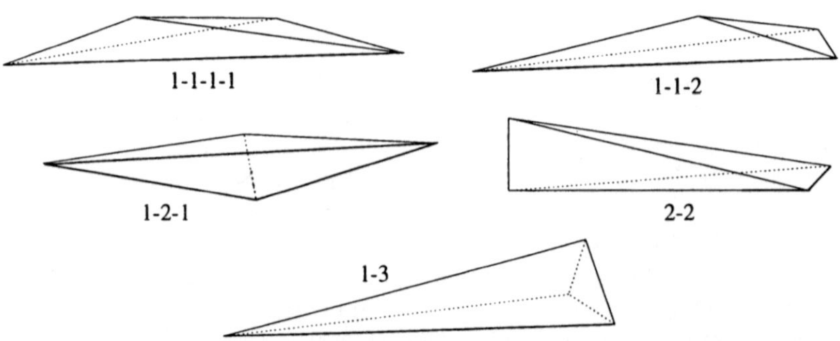

Fig. 16. Five fuzzy types of skinny tetrahedra. Vertices are close to colinear; in a 1-1-1-1 tetrahedron the vertices are all well-separated along the line, in a 2-2 tetrahedron the vertices fall into two pairs, and so forth.

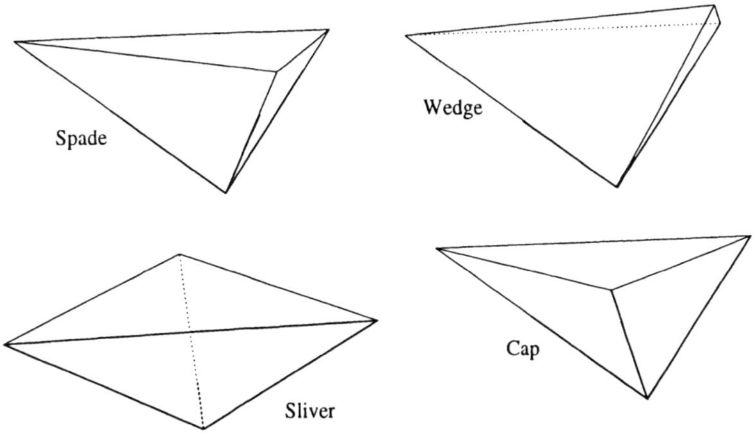

Fig. 17. Four fuzzy types of flat tetrahedra, classified according to the projection of the vertices onto the plane. The sliver is the unique type of badly shaped tetrahedron with a good circumradius to shortest edge ratio.

This bad tetrahedron is special: it is the only type of bad tetrahedron that has a small ratio of circumsphere radius (circumradius) to shortest edge.

Shewchuk proved that his mesh generator can eliminate all types of bad tetrahedra except slivers. When the circumcenter of a sufficiently bad tetrahedron of any other type is added, the worst circumradius to shortest-edge ratio in the entire mesh necessarily improves because the new vertex reduces the circumradius without creating any new short edges. (This observation was also used in earlier 3D Delaunay refinement algorithms by Miller et al. [84] and Talmor [114].) One can attempt to remove slivers using Delaunay refinement—Shewchuk gives some encouraging empirical results—but there is no guarantee that the algorithm will halt.

Recently, Edelsbrunner and Guoy [46] proposed a more selective form of Delaunay refinement called sink insertion. This method applies to both 2D and 3D Delaunay triangulation; for continuity we explain it in 3D. Call a Delaunay tetrahedron that contains its own circumcenter *self-centered* (see for example [20]) and the circumcenter of such a tetrahedron a *sink*. The name refers to the fact that self-centered tetrahedra are sinks (vertices without out-edges) in the directed graph in which each tetrahedron points to the neighboring tetrahedron in the direction of its circumcenter. Moreover, sinks locally maximize the distance to the closest vertex, and also locally maximize the error in interpolating a spherical quadratic function by linear basis functions over the tetrahedral elements. Sink insertion follows a path in the directed graph from a badly shaped tetrahedron to a self-centered tetrahedron and then adds the associated sink. Sink insertion gives the same theoretical guarantee as Delaunay refinement (no bad tetrahedra except slivers); experimentally it appears to be much faster and more parallelizable, yet equally successful at removing bad tetrahedra [46,58]. See Figure 25 below.

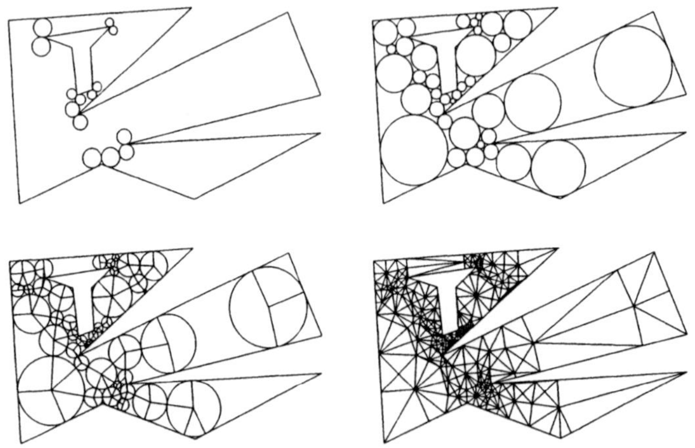

Fig. 18. An algorithm for nonobtuse triangulation of a polygonal domain: (1) place disks at reflex vertices and holes, (2) pack disks arbitrarily until each gap has three or four sides, (3) connect centers to break the domain into small polygons, and (4) triangulate small polygons using circumcenters of gaps.

5.2 Disks, Balls, and Bubbles

You may have already noticed that computational geometers are crazy about circles and spheres. In this section, we really indulge ourselves and look at point placement methods based on these lovely objects.

We start with hard-edged disks in the plane. *Disk packings* in which each gap between disks has exactly three sides are, in a very real and rigorous way, discrete analogues of conformal mappings (see for example [113]). A conformal mapping is the ultimate in "provably good" meshing: a conformal mapping of a fine square grid onto a planar domain gives all interior angles tending to 90° as elements shrink. Conformal mappings and three-sided-gap disk packings, however, are too global to be of much use in adaptive mesh generation; there are only three degree of freedom—choice of conformal center and orientation—in a Riemann mapping from a planar domain without holes to a disk. Nevertheless, it makes an interesting theoretical problem to optimize some geometric criterion over these three degrees of freedom [23].

If we relax the requirements so that gaps can have either three or four sides, we gain a lot of flexibility while retaining some control over the angles in the mesh. Figure 18 illustrates an algorithm [26] for nonobtuse triangulation of a polygon. This algorithm is not intended to give usable finite element meshes, but rather it proves an interesting theorem: any n-vertex polygon with holes can be triangulated with only $O(n)$ nonobtuse triangles. This theorem shows that *no-large-angle triangulation* is quite different from no-small-angle triangulation, which as mentioned above requires a number of triangles dependent upon the geometry of the domain rather than just n.

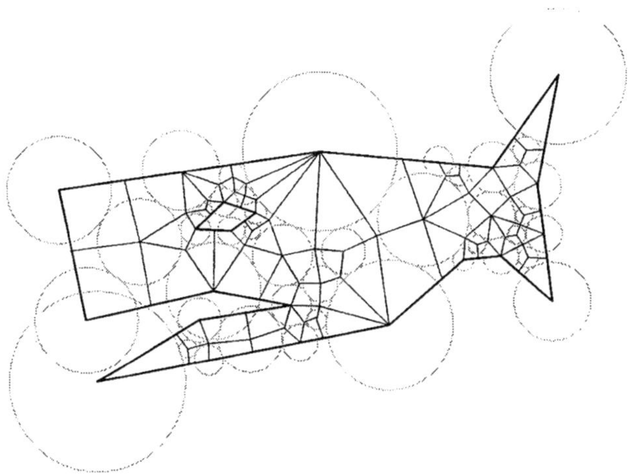

Fig. 19. A quadrilateral mesh, produced by disk-packing, in which each element is a Voronoi cell.

Disk-packing with three- and four-sided gaps can also be used for quadrilateral meshing. Bern and Eppstein [25] gave $O(n)$-size quad meshes for four different theoretical meshing problems: all elements are Voronoi cells (Figure 19), all elements are kites, all elements have two opposite right angles (implying that they have cocircular vertices), and finally all elements have maximum angle at most 120°.

Disk packing with three- and four-sided gaps has not found its way into practical mesh generation algorithms. If the disk sizes in the algorithm of Figure 18 could be graded nicely so that no disk is adjacent to one more than twice or three times its size, then this algorithm could give an elegant solution to nonobtuse meshing. Similarly, the results on quadrilateral meshing could be relevant to automatic multiblock meshing, but to our knowledge no-one has ever pursued this possibility.

A still more relaxed sort of disk packing—and in three dimensions ball packing—has been used by Miller, Talmor, Teng, and their collaborators [70,84,114]. Say we have a *spacing function* $f(p)$ defined on points p of the domain Ω. Let $B(p,r)$ denote the ball with center p and radius r. We call a set of balls $\mathcal{P}$ a *β-packing* for some $\beta > 0$ if (1) no two balls of $\mathcal{P}$ overlap, (2) the radius of a ball of $\mathcal{P}$ with center p is $f(p)$, and (3) for each point $q \in \Omega$, $B(q, \beta f(p))$ intersects a ball of $\mathcal{P}$. Thus the sizes of balls in a β-packing grade according to the spacing function.

The algorithm for creating a β-packing on Ω places balls centered on the vertices of Ω first, then on the segments, then on the facets, and finally in the interior. A simple way of ensuring non-overlap is to generate balls on facets and interior randomly, and then discard them if they overlap previously placed balls. A grid or quadtree can be used to keep track of uncovered regions in order to accelerate this process. Talmor [114] shows that the Delaunay triangulation of the ball centers of a β-packing generated in this way conforms to the domain. In addition, in $\mathbb{R}^2$ the Delaunay triangulation has no new small angles and optimal size up to constant factors just like Ruppert's meshes. In $\mathbb{R}^3$ the size remains optimal, but the quality guarantee is somewhat weaker: the mesh contains no bad tetrahedra except slivers. Figure 26 below shows some 2D meshes in which the vertices were placed by β-packing.

Finally *bubble systems*, introduced by Shimada [110,111], use methods from mesh smoothing (Section 5.4) inside the point placement process. Balls are now allowed to overlap each other a bit; neighboring balls exert forces on each other—typically repulsive at short range and attractive at long range—in order to space themselves appropriately. Sometimes "birth" and "death" processes are used to create vertices in overly sparse regions and destroy vertices in overly crowded regions, and damping forces are used to ensure convergence.

We give a few more details on one particular bubble algorithm due to Bossen and Heckbert [29]. Bossen and Heckbert use a force whose dependence upon distance d is $(1-d^4)\cdot\exp(-d^4)$, and for efficiency they define neighboring balls to be those that are connected by the current Delaunay triangulation of ball centers. They delete balls that are too crowded and subdivide edges that are too long, with both measurements taken relative to an externally defined spacing function. The Delaunay triangulation is repaired at each iteration.

5.3 Anisotropy Again

As mentioned in Section 2, anisotropy is often a desirable feature in meshes for fluid dynamics. Boundary layers in the flow demand anisotropic elements surrounding domain boundaries, for example, long skinny elements over an airfoil. A popular approach to point placement for resolving boundary layers is the *advancing front* method [79,91]. With a few recent exceptions [71], computational geometers have not worked on advancing front, so we will not say much about this approach.

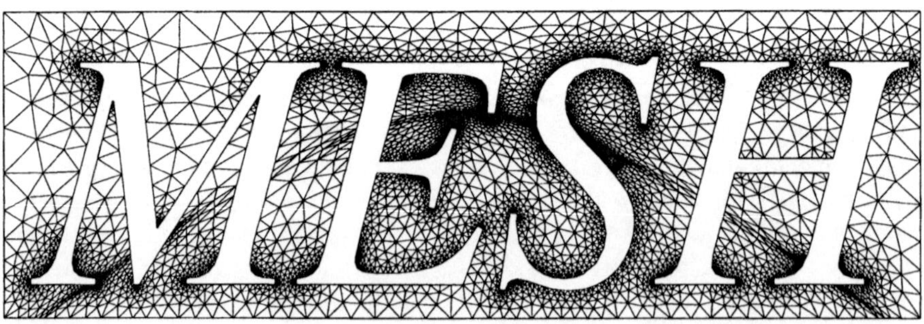

Fig. 20. A mesh computed by Bossen and Heckbert's bubble-based anisotropic mesh generator. (F. Bossen and P. Heckbert)

A more generic approach to anisotropy, which we shall call *stretching*, was first introduced by Mavriplis [81], Castro-Díaz [31], and others. Stretching replaces the usual scalar spacing function with a matrix-valued function that controls the desired local element size and orientation. We can picture this function as an "ellipse field",

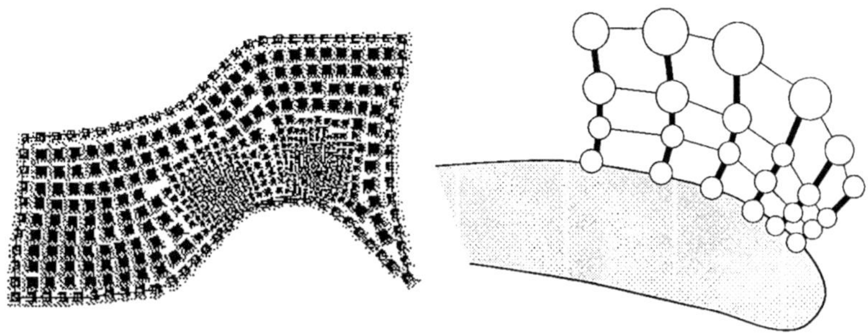

Fig. 21. Either (a) square packings (IBM - Japan) or (b) bubble packings with a mix of stiff and loose connections may be able to produce anisotropy without flat angles.

giving the linearly transformed unit ball at each position. Edge lengths are now measured relative to the local unit ball. Delaunay triangulation is replaced by a modified Delaunay triangulation in which the linear transformation associated with one of its points is applied to a quadrilateral before testing whether it is reversed or not. We must be a little bit careful about exactly which linear transformation to apply, so that we do not fall into an infinite flipping loop (personal communication with S. Teng). Bossen and Heckbert included stretching in their two-dimensional bubble mesher, obtaining esthetically pleasing results (Figure 20). Yamakawa and Shimada [122] incorporated stretching into a three-dimensional bubble mesher.

One possible drawback of stretching is that it can produce flat angles as well as the desired sharp angles. Recall that some numerical methods, such as finite difference and finite volume formulations, prefer anisotropic meshes in which skinny elements have their shortest edges in the direction of high second derivative. It should be possible to modify bubble methods to produce such meshes by building anisotropy into the simulated physical forces, rather than just into the length measurements. Shimada (personal communication) recommends rectangular or stretched square bubbles, which would tend to form stacks normal to the surface. It may also be possible to use stacks of stiffly connected spherical bubbles in order to transition more gracefully to isotropic regions. See Figure 21 for illustrations of these two ideas.

5.4 Mesh Smoothing

This section and the next two discuss mesh improvement methods, ways to improve element shape and size that can be applied as postprocessing steps or—as we saw in bubble meshing—at the same time as point placement. Mesh smoothing adjusts the locations of mesh vertices in order to improve element shapes and overall mesh quality [3,5,30,50,90]. In mesh smoothing, the topology of the mesh remains invariant, thus preserving important features such as the nonzero pattern of the linear system.

In bubble meshing, the smoothing procedure uses a spacing function or the radii of the bubbles to determine ideal distances between vertices, but more often smooth-

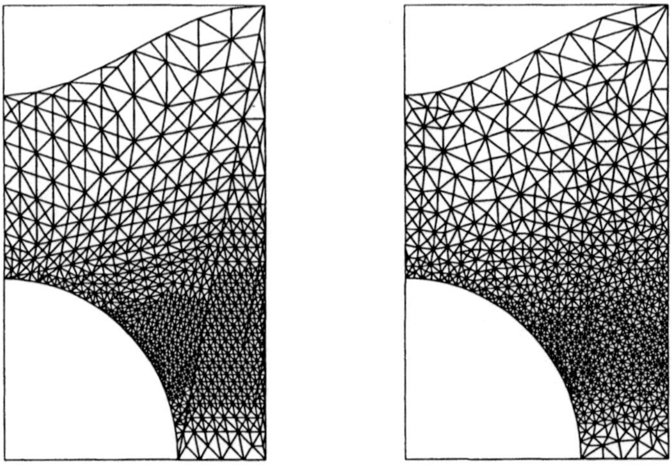

Fig. 22. (a) A mesh resulting from bisection refinement without smoothing. (b) The same mesh after local optimization-based smoothing. (L. Freitag)

ing proceeds without such outside information. Laplacian smoothing [50,76] is the most commonly used technique. This method sweeps over the entire mesh several times, repeatedly moving each vertex, not constrained to lie on a boundary feature, to the centroid (arithmetic average) of the vertices adjacent to it. Variations weight each adjacent vertex by the total area of the elements around it, or use the centroid of the incident elements rather than the centroid of the neighboring vertices [120]. Laplacian smoothing is computationally inexpensive and fairly effective, but it does not guarantee improvement in element quality, and it can even invert an element, unless the algorithm performs an explicit check before moving a vertex.

Another class of smoothing algorithms uses global or local optimization techniques to determine new vertex locations. Global techniques simultaneously adjust all unconstrained vertices; such an approach involves an optimization problem as large as the number of unconstrained vertices, and consequently is computationally expensive [30,90]. Local techniques adjust vertices one by one—or an independent set of vertices in parallel [53]—resulting in a cost more comparable to Laplacian smoothing. Many quality measures, including maximum angle and area divided by sum of squared edge lengths, can be optimized by techniques related to linear programming [3]. Figure 22 shows the results of a local optimization-based smoothing algorithm developed by Freitag et al. [53]; smoothing was applied after each refinement step. Overall, the global minimum angle has improved from 11.3° to 21.7° and the average minimum element angle from 35.7° to 41.1°.

5.5 Flipping Revisited

The flip moves used in the algorithms for computing Delaunay triangulations also make useful mesh improvement steps. As before, 2D flipping exchanges the diag-

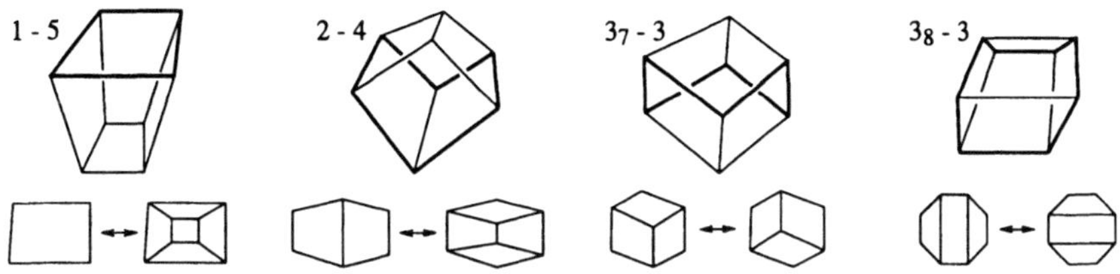

Fig. 23. Flips for quad meshes correspond to exchanging the upper and lower facets of a 3D cube, just as flips for triangle meshes correspond to exchanging the upper and lower facets of tetrahedron.

onals of a triangulated quadrilateral as in Figure 3, and 3D flipping exchanges two tetrahedra for three as in Figure 9, only now the criterion for making the exchange need not be the Delaunay empty sphere test. Flipping can instead be used to regularize vertex degrees, minimize the maximum angle, or improve almost any other quality measure of triangles or tetrahedra. For quality measures optimized by the Delaunay triangulation, flipping computes a true global optimum, but for other criteria it computes only a local optimum.

There is a nice way to view flipping that unifies it with the splitting move used in the randomized incremental insertion algorithm. Using the lifting map, we can view the two different ways to triangulate a convex quadrilateral as projections of the lower and upper convex hulls of the lifted points, whose entire convex hull is simply a tetrahedron. Similarly, the split move exchanges the lower and upper convex hulls of a tetrahedron, only this time the lower hull consists of a single triangle and the upper hull consists of three triangles (or vice versa), because back on the plane one point lies inside the triangle formed by the other three. Thus the split is sometimes called a 1-3 flip.

The 3D flip moves (Figure 9) are now seen to be the exchange of lower and upper convex hulls of a four-dimensional simplex. The split move, which subdivides a single tetrahedron into four by adding an interior point, is the only other such exchange; it is a 1-4 flip.

Although it does not seem to be well-known to the meshing community, there are canonical sets of flip moves for quad and hex meshes as well. The flip moves for quad meshes exchange the lower and upper hulls of a three-dimensional cubical polytope, as shown in Figure 23. There are 1-5 and 2-4 flips and two different kinds of 3-3 flips. We have denoted the two kinds of 3-3 flips as 3_7-3 and 3_8-3, depending upon the number of vertices appearing in the "left hand side" of the flip. Notice that the flip moves do not change the parity (evenness or oddness) of the number of quads in a mesh; we have recently proved [22] that for a given polygonal domain, any quad mesh with the same parity is obtainable from any other by a finite sequence of flips. There is a simple parity-switching move as well: exchange a hexagon split into two quads for a hexagon split into three quads all incident to an interior vertex.

The analogous flips for hex meshes exchange the upper and lower hulls of a 4D cubical polytope; they are shown in Figure 24. Some of these moves have already been discovered and applied in an *ad hoc* way [85]. We believe that the full set may

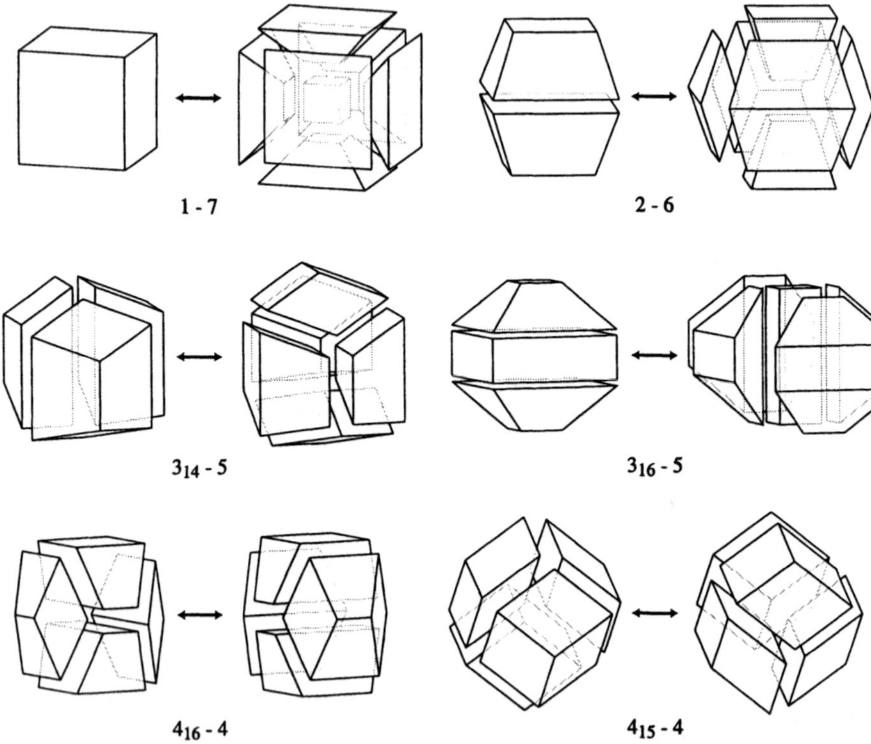

Fig. 24. Flips for hex meshes exchange upper and lower facets of a 4D cube (not shown).

be useful in refining and improving hex meshes. (We also wonder if there is any sort of construction that could be deemed a cubical analogue to Delaunay triangulation.) We have shown that a quad surface mesh can support both even- and odd-parity hex volume meshes, but we do not know if there is a simple parity reversing operation for hex meshes.

5.6 Sliver Removal

Our final topic on mesh improvement is the one that has received the most recent attention from the computational geometry community. Recall that the sliver from Figure 17 is the unique type of bad tetrahedron that appears in a Delaunay triangulation of well-spaced points (say centers of a β-packing for a uniform spacing function). Fortunately slivers turn out to be rather fragile, and there are now several different "provably good" sliver removers.

Chew [37] showed that a randomized form of Delaunay refinement will, with high probability, remove all slivers flatter than a certain constant. The provable constant is very weak, but experiments are encouraging. Chew's sliver removal algorithm inserts a randomly chosen point within a ball around a bad tetrahedron's circumcenter, rather than the exact circumcenter. The randomly chosen point is un-

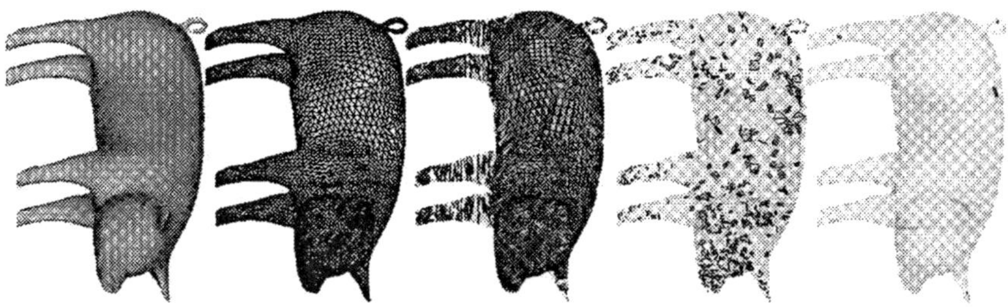

Fig. 25. Sliver exudation modifies the vertex weights in a weighted Delaunay triangulation in order to remove slivers. These figures show (a) the input boundary mesh, (b) a wire-frame of the boundary, (c) initial badly-shaped tetrahedra in a surface-reconstructing tetrahedralization (produced by *Wrap* from Rainbow Geomagic, Inc.), (d) remaining slivers (tetrahedra with a dihedral less than 5°) after sink insertion, and (e) remaining surface slivers after exudation. (D. Guoy)

likely to form a new sliver, because the locus of sliver-forming positions is very flat. As in other forms of Delaunay refinement, if the randomly chosen point falls within the diameter sphere of a boundary edge or equatorial sphere of a boundary facet, then the boundary face is subdivided instead.

Edelsbrunner et al. [35,45] gave a provably good sliver removal algorithm based on weighted Delaunay triangulation. This intriguing algorithm, called *sliver exudation*, removes interior slivers without even adding new Steiner points. We can think of sliver exudation as pushing slivers ("exuding" them) to the boundary of the domain, where they can be fixed by subdividing the boundary. Sliver exudation treats the Delaunay triangulation as a weighted Delaunay triangulation with uniform weights and iteratively adjusts vertex weights in order to remove slivers. The key to the success of this algorithm is that a mild change in the vertex weights of a sliver will radically inflate the orthogonal sphere of the sliver. Thus as weights change, the orthogonal spheres of slivers will penetrate too deeply into vertex spheres and slivers will drop out of the tetrahedralization before the vertices themselves drop out. We can also think of sliver exudation as displacing lifted points in the vertical direction; a mild displacement will remove a sliver facet from the lower convex hull. As in Chew's algorithm, the provable constants are very weak, but experiments [47,58] look fairly persuasive. See Figure 25.

Edelsbrunner et al. [48] also proved that perturbing a well-spaced point set in 3D can remove slivers. As in Chew's algorithm [37] the key observation is that the locus of sliver-forming positions for the fourth vertex of a tetrahedron with three fixed vertices has relatively small volume. Most recently, Li and Teng [72] and Cheng and Dey [34] combined Delaunay refinement (as in Chew's algorithm) with perturbation (as in the two Edelsbrunner et al. algorithms) to incorporate sliver removal into the main meshing loop.

6 Solution-Adaptive Mesh Refinement

Solution-adaptive mesh refinement reduces the sizes of elements in areas where the PDE solution error is large. Local error estimates based on an initial solution are known as *a posteriori* error estimates [10] and can be used to determine which elements should be refined. For elliptic problems these estimators asymptotically bound the true error and can be computed locally using only the information on an element [119].

One way to achieve solution-adaptive refinement is rather anticlimactic: use a spacing function (such as $f(p)$ in Section 5.2 above) to control mesh density, change the spacing function based on the solution, and then completely remesh the domain based on the new spacing function. The drawback of this approach is inefficiency: it is faster to remesh only locally and retain as much of the old mesh as possible. In multigrid methods, local remeshing may enable complete reuse of the coarser levels of the initial mesh.

6.1 Iterative Vertex Insertion

One approach to solution-adaptive mesh refinement inserts extra vertices into the triangulation and then locally repairs the mesh. Reasonable choices of locations to insert new vertices include at edge bisectors as in Bossen and Heckbert's mesher (Section 5.2) and at triangle circumcenters as in Delaunay refinement (Section 5.1). New vertices along the boundaries of curved domains should be computed using the curved boundary rather than the current straight edge, thereby giving a truer approximation of the domain as the mesh refines [31].

Li, Teng, and Üngör [70] show how to adaptively refine and coarsen β-packings, described in Section 5.2 above. (We shall say more about coarsening in Section 7 below.) They change the spacing function $f(p)$ based on the initial solution and then shrink or grow the radii of balls in regions where the spacing function has changed. They then fill gaps that have opened up with new non-overlapping balls in order to restore the β-packing property, and then recompute the Delaunay triangulation. Figure 26 shows two snapshots from a sequence of meshes for a growing circular "shock front".

If balls only shrink, meaning refinement but no coarsening, this method will produce a *node conforming* refinement: each vertex of the initial mesh will appear in the refined mesh. It does not produce an *edge conforming* refinement, one in which each edge of the initial mesh appears (possibly subdivided) in the refined mesh.

6.2 Nested Refinement

In this section, we describe techniques used to produce edge conforming or *nested* refinements. These techniques subdivide elements in the current mesh rather than remeshing, even locally. Nested refinements simplify the interpolation step in the multigrid method, because each element from the refined mesh lies inside exactly

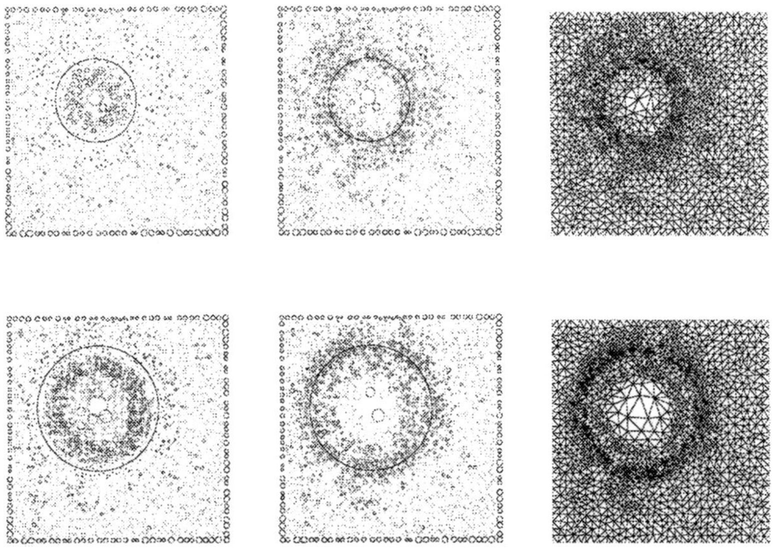

Fig. 26. Ball packings can be used in solution-adaptive remeshing. The top row shows (a) an initial overlapping set of balls, (b) a maximal set of non-overlapping balls (β-packing), and (c) the Delaunay triangulation of ball centers. The bottom row shows a remeshing derived from the initial packing by adjusting radii and filling gaps. (A. Üngör)

one element from the initial mesh. Nested refinements are also close relatives of subdivision meshes which will be described in Section 8. To compute nested refinements, we *split* triangles in need of refinement, by adding the midpoints of sides. The pseudocode below gives the overall approach.

$k = 0$
solve the differential equation on the initial mesh T_0
estimate the error on each triangle
while the maximum error on a triangle is larger than the given tolerance **do**
  based on error estimates, mark a set of triangles S_k to refine
  $\star$ divide the triangles in S_k, along with adjacent invalid triangles to get T_{k+1}
  solve the differential equation on T_{k+1}
  estimate the error on each triangle
  $k = k + 1$
endwhile

There are a number of popular alternatives for step $\star$, in which the current mesh T_k is adaptively refined. In *regular 1:4 refinement* [13,14], the midpoints of the sides of a marked triangle are connected, as in Figure 27(b), to form four similar triangles. Unmarked triangles that received two or three midpoints are split in the same way. Unmarked triangles that received only one midpoint are *bisected* by connecting the midpoint to the opposite vertex as in Figure 27(a). Before the next iteration of $\star$, bisected triangles are glued back together and then marked for refinement; this precaution guarantees that each triangle in T_{k+1} will either be similar to a triangle in T_0 or be the bisection of a triangle similar to a triangle in T_0. Thus, regular refinement—regardless of the number of times through the refinement loop—produces a mesh with minimum angle at least half the minimum angle in T_0. Hence the angles in T_{k+1} are bounded away from 0 and π.

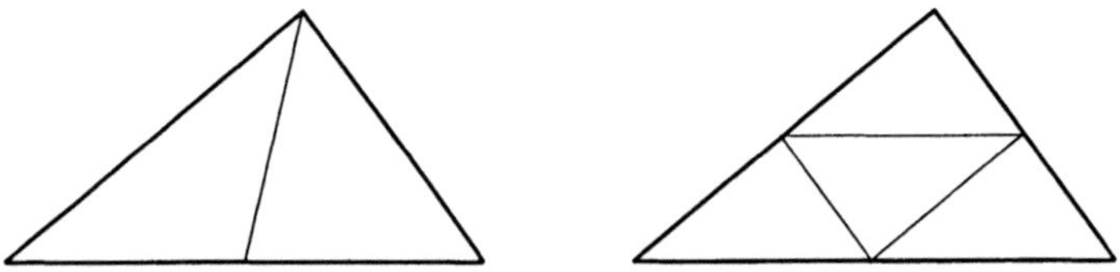

Fig. 27. A triangle divided by (a) bisection, and (b) regular 1:4 refinement.

Rivara [96–98] proposed several alternatives for step $\star$ based on triangle bisection. One method refines each marked triangle by cutting from the opposite vertex to the midpoint of the longest edge. Neighboring triangles are now *invalid*, meaning that one side contains an extra vertex; these triangles are then bisected in the same way. Bisections continue until there are no remaining invalid triangles. Refinement can propagate quite far from marked triangles; however, propagation cannot fall into an infinite loop, because along a propagation path each bisected edge is longer than its predecessor. This approach, like the previous one, produces only a finite number [100] of different triangle shapes—*similarity classes*—and the minimum angle is again at least half the smallest angle in T_0. Quite often longest-edge refinement actually improves angles.

A second Rivara refinement method is given in the pseudocode below and illustrated in Figure 28. This method does not always bisect the longest edge, so bisections tend to propagate less, yet retains the same final angle bound as the first Rivara method.

$i = 0$
$Q_i = S_k$ { *Q denotes "marked" triangles to be refined* }
$R_i = \emptyset$ { *R denotes children of refined triangles* }
while $(Q_i \cup R_i) \neq \emptyset$ **do**
  bisect each triangle in Q_i across its longest edge
  bisect each triangle in R_i across its subdivided edge
  add all invalid children of Q_i triangles to R_{i+1}
  add all other invalid triangles to Q_{i+1}
  $i = i + 1$
endwhile

We now move up to three dimensions and discuss two different generalizations of bisection. To bisect tetrahedron $v_0v_1v_2v_3$ across edge v_0v_1, we add the triangle

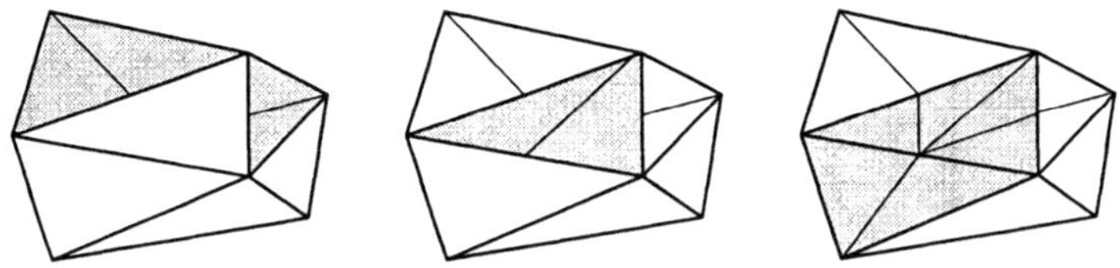

Fig. 28. The bisection algorithm given in the pseudocode splits invalid children of refined triangles to their subdivision points, rather than to their longest edges.

$v_{01}v_2v_3$, where v_{01} is the midpoint of v_0v_1, as shown in Figure 29. This operation creates two child tetrahedra, $v_0v_{01}v_2v_3$ and $v_{01}v_1v_2v_3$, and bisects the faces $v_0v_1v_2$ and $v_0v_1v_3$, which, unless they lie on the domain boundary, are each shared with an adjacent tetrahedron. Two tetrahedra that share a face must agree on how it is to be bisected; otherwise an invalid mesh will be constructed.

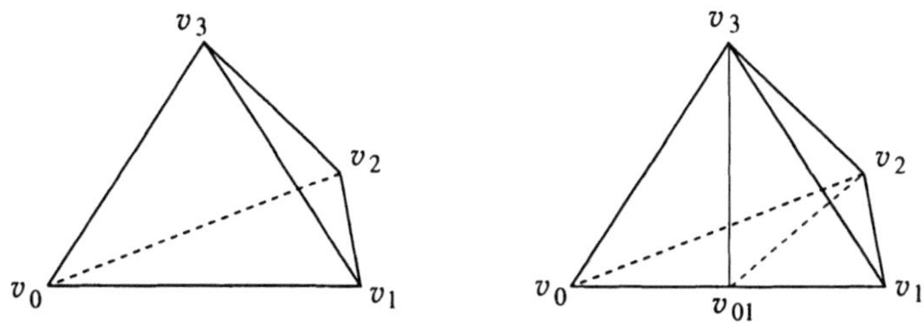

Fig. 29. The tetrahedron on the left is bisected to form two new tetrahedra

A single bisection of a tetrahedron can approximately square the minimum solid angle, unlike in two dimensions where the minimum angle of a triangle is decreased by no more than a factor of two. Consider the wedge tetrahedron with vertex coordinates $(0, \epsilon, 0)$, $(0, -\epsilon, 0)$, $(1, 0, \epsilon)$, and $(1, 0, -\epsilon)$. Bisection of the longest edge of this tetrahedron creates a new tetrahedron with minimum solid angle about ϵ^2.

Rivara and Levin [99] suggested an extension of longest-edge Rivara refinement to three dimensions. Notice that splitting the longest edge in a tetrahedron also splits the longest edge on the two subdivided faces, and thus the bisection of shared faces is uniquely defined. (Ties can be broken by vertex labels.) Neighboring invalid tetrahedra, meaning all those sharing the subdivided longest edge, are refined recursively.

Rivara and Levin provide experimental evidence suggesting that repeated rounds of longest-edge refinement cannot reduce the minimum solid angle below a fixed threshold, but this guarantee has not been proved. The guarantee would follow if it could be shown that the algorithm generates only a finite number of tetrahedron similarity classes.

A bisection algorithm first introduced by Bänsch [16] does indeed generate only a finite number of similarity classes. Before describing the algorithm, we sketch the argument of Liu and Joe [74] which motivates the algorithm. The key observation

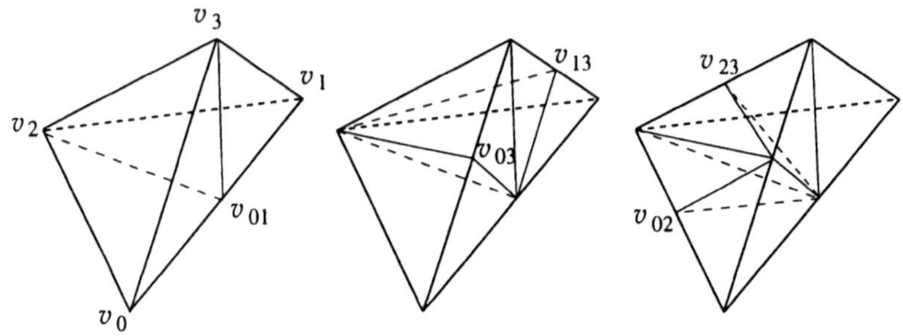

Fig. 30. The first three levels of longest-edge bisection of the canonical tetrahedron. Note that the tetrahedra generated at each level are similar. For the final level of refinement we show only the four tetrahedra obtained from $v_0v_{01}v_2v_3$. Four similar tetrahedra are obtained from $v_{01}v_1v_2v_3$.

is that there exists an affine transformation that maps any tetrahedron to a canonical tetrahedron for which longest-edge bisection generates only a finite number of similarity classes. Consider the canonical tetrahedron t_c with vertices $v_0 = (-1, 0, 0)$, $v_1 = (1, 0, 0)$, $v_2 = (0, 1/\sqrt{2}, 0)$, and $v_3 = (0, 0, 1)$. In Figure 30 we illustrate the first three levels of longest-edge bisection of $t_c = v_0v_1v_2v_3$. It can be shown that all the tetrahedra generated at each level of refinement are similar and that the eight tetrahedra generated after three levels of refinement are similar to t_c. Refinement in the canonical space induces a refinement in the original space with only a finite number of different similarity classes. A subtlety: due to the affine transformation, similarity in the original space requires homothety (identical up to scaling and translation) not just similarity in the canonical space. Hence the number of similarity classes turns out to be 36, corresponding to 3 similarity classes in the canonical space times 12 non-inverting affine transformations [6,80].

Bänsch [16], Liu and Joe [75], and Arnold et al. [6] give essentially equivalent [65] algorithms for generating the bisection order; we follow Bänsch's presentation. Each face in each tetrahedron elects one of its edges to be its *refinement edge*, so that two conditions hold: the choice for a face is consistent between the two tetrahedra that share it, and exactly one edge in each tetrahedon—the *global refinement edge*—is chosen by an adjacent pair of faces of the tetrahedron. These conditions hold initially if each face picks its longest edge and ties are broken in any consistent manner, for example, by vertex or edge label order. and the other refinement edges to either v_0v_3. In the pseudocode below, a *child face* is a triangle like $v_{01}v_1v_3$ in Figure 29, and a *new face* is one like $v_{01}v_2v_3$. The key point is that the refinement edge of a child face is its *old edge*, that is, the edge like v_1v_3 that was in the mesh before bisection; thus the next bisection that splits child face $v_{01}v_1v_3$ will agree with the corresponding split on the canonical tetrahedron.

mark the refinement edge of every face in the current mesh
let T_0 be the set of marked tetrahedra; $i = 0$
while $(T_i \neq \emptyset)$ **do**

bisect each tetrahedron in T_i across its global refinement edge
pick the old edge in each child face as its refinement edge
pick the longest edge in each new face as its refinement edge
$i = i + 1$
let T_i be the set of invalid tetrahedra
enddo

7 Mesh Simplification

We now discuss the reverse of mesh refinement: simplifying or coarsening a mesh. Local *a posteriori* simplification helps reduce the total number of elements when tracking solutions to time-varying differential equations. Global *a priori* simplification can be used to turn a single highly refined mesh into a sequence of meshes with varying levels of detail for multigrid methods or other uses.

In Figure 26 we already saw an example of a straightforward algorithm for *a posteriori* coarsening. This algorithm marks a set of vertices in areas that are more refined than necessary, eliminates all the marked vertices, and then retriangulates to give a node conforming coarse mesh. One potential difficulty is that the shapes of the triangles may degrade as the mesh is coarsened due to increasing disparity between the interior and boundary point densities. Hence the boundary may need to be coarsened at the same time as the interior.

Another method of *a posteriori* simplification, very popular for the surface meshes used in computer graphics [63], is *edge contraction*. An edge contraction coalesces a pair of adjacent vertices, producing a new vertex adjacent to all the neighbors of the pair. The new vertex is usually placed at the midpoint of the contracted edge, so that the resulting mesh is neither node nor edge conforming. (Notice that if the new vertex is coincident with one of the old vertices, then an edge contraction is the same as a vertex removal followed by a particular retriangulation.) It is possible to determine locally whether the edge contraction will change the topological type of the mesh [45]; if it would then perhaps some other edge should be constracted instead.

Garland and Heckbert [54] showed that there is a smarter way to place the vertex than to simply use the edge midpoint. They place the vertex at the position that minimizes *quadric error measure*, the sum of the squares of the distances to the planes supporting the triangles incident to the original pair. These planes are "carried along" with the coalesced vertex, so that if that vertex is then coalesced with another vertex, the next position is the least-squares position taking into account all the relevant planes from the initial triangulation. A priority queue is used to determine which edge to contract next, the edge whose contraction would introduce the least quadric error. Figure 31 shows an example.

Quadric-error simplification may sound expensive, but it is actually quite efficient: each coalesced vertex carries along only a single matrix, the sum of all the matrices of the vertices that went into it. To be more specific, a plane in $\mathbb{R}^3$ can be given by an equation $n_i \cdot u = d_i$, where n_i is a normal vector and d_i is the plane's distance from the origin. The square of the distance from a vertex $v \in \mathbb{R}^3$, given as a column vector, to this plane is then

$$(v,1)^T \cdot (n_i, d_i) \cdot (n_i, d_i)^T \cdot (v,1).$$

The outer product $(n_i, d_i) \cdot (n_i, d_i)^T$ is a 4×4 matrix M_i called the *quadric* matrix. The sum of the squares of the distances from v to a number of different planes, $(n_1, d_2), (n_2, d_2), \ldots$, can be computed as $(v,1)^T \cdot (\sum_i M_i) \cdot (v,1)$. Quadric-error simplification generalizes easily to higher dimensions and even to higher-order basis functions; we think it should be useful in solution-adaptive meshing. Quadric error may also be useful in refinement and smoothing. Refinement could, for example, split a given triangle by adding the point that minimizes the sum of distances to the planes at the three neighboring triangles, and split triangles throughout the mesh according to priority order. In addition to repositioning vertices, smoothing could balance quadric matrices of adjacent vertices, subtracting an outer product off one and adding it to the other, in order to reduce total quadric error.

Teng, Talmor, and Miller [82] gave an algorithm for *a priori* simplification. Their algorithm uses Delaunay triangulations of centers of ball packings, and produces a sequence of bounded-aspect-ratio, node-conforming meshes of approximately minimum depth (number of levels of refinement). They pointed out that the minimum depth of the sequence of meshes may be dictated by the initial mesh and be much larger than $O(\log n)$. Their algorithm recovers a spacing function and then changes the spacing function in order to decide which balls centers to remove. An alternative algorithm [88], which gives a sequence of depth $O(\log n)$ but cannot guarantee bounded aspect ratios, simply removes a maximal independent set of vertices at each stage. (A *maximal independent set* is a set of vertices with no pair adjacent such that no other vertex can be added to the set without forming an adjacency.)

8 Cartesian Meshes and Subdivision Surfaces

In this section we mention two popular types of meshes based on recursive subdivision. Neither of these constructions comes out of the computational geometry community, and they are somewhat out of the scope of this survey (and beyond the range of expertise of the author!), so our remarks will be brief.

A *Cartesian mesh* is a mesh produced by repeatedly splitting a square into four half-size squares or a cube into eight half-size cubes, as in a quadtree or an octree. Squares or cubes are repeatedly subdivided until they are small to enough to capture the desired level of detail in the domain geometry and the expected solution. Typically a grading constraint is imposed, such as no square is next to one more than twice its own size.

Cartesian meshes do not attempt to conform to domain geometry, thereby transferring some burdens from the mesh generator to the solver. Boxes cut by the boundary are classified into a number of patterns by determining which of their vertices

Fig. 31. A 47,904-face brontosaurus model (a) along with a 500-face quadric-based simplification. Notice how the triangles conform to the geometry. (M. Garland and P. Heckbert)

lie interior and exterior to Ω. Each pattern corresponds to a different type of non-hexahedral element. Boxes adjacent to ones half their own size can similarly be classified as non-hexahedral elements, or alternatively the solution value at their subdivision vertices can be treated as implicit variables using Lagrange multipliers [1].

Recent fluid dynamics simulations have used Cartesian meshes quite successfully in both finite element and finite volume formulations of Euler equations [40,41, 123]. The approach can be adapted even to very difficult meshing problems. For example, Berger and Oliger [19] and Berger and Colella [18] have developed adaptive Cartesian-based methods for rotational flows and flows with strong shocks. Cartesian meshes are not suitable for viscous flows with boundary layers, which require precisely conforming meshes.

A *subdivision surface* is a surface produced by starting from an initial very simple surface mesh and then repeatedly splitting facets and displacing vertices in space. One can generate a smooth curved surface by displacing each vertex to a weighted average of neighboring vertices; this process is the 3D analogue of converging to a cubic spline curve by repeatedly cutting off corners of a control polygon. There are a number of choices of subdivision patterns and averaging stencils; subdivision surfaces popular in computer graphics include Catmull-Clark surfaces [32] and Loop surfaces [73]. Subdivision surfaces have numerous advantages

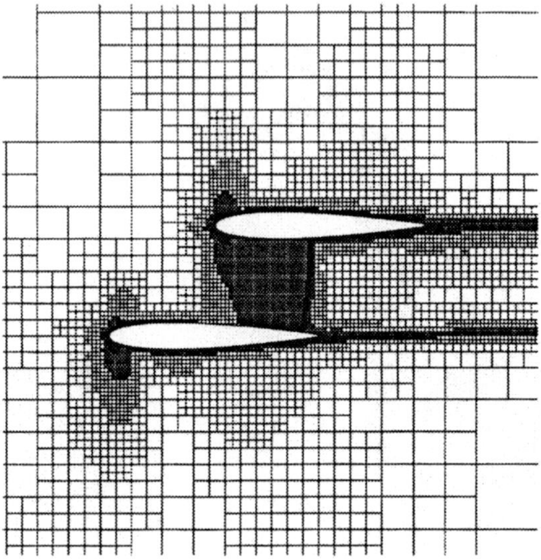

Fig. 32. A Cartesian mesh is a graded quadtree or octree that need not conform to the boundary of the domain. (W. Coirier)

over other sorts of surfaces such as NURBS patches, including multiresolution editing [124] and signal-processing-style filtering and smoothing [60].

At a given stage of refinement, a subdivision surface is a *semi-regular* surface mesh: the initial unstructured mesh, uniformly refined by a subdivision pattern such as the regular 1:4 refinement shown in Figure 27(b). Because a surface mesh does not usually need to conform to a complicated one-dimensional boundary, subdivision surfaces can be more pure than 2D finite element meshes produced by similar techniques (for example, Figure 22).

Lee et al. [69] gave a nice algorithm called MAPS (for "multiresolution adaptive parameterization of surfaces") for remeshing an arbitrary unstructured triangular surface mesh into such a semi-regular surface mesh. The algorithm starts by simplifying the unstructured mesh by iteratively removing maximal independent sets of vertices. When a vertex v is removed, the *star* of faces around v is flattened into the plane by a piecewise-linear approximation of conformal mapping, and the resulting flat polygon is triangulated and folded to fit into the hole produced by the removal of the star of v. The conformal mapping takes the removed vertex to a point in the new triangulation; we can think of the removed point and its incident edges as drawn on the new triangulation. "Drawings" are carried along from triangle to triangle as the mesh simplifies. At the end of the simplification process, we have reduced the initial surface mesh to a very simple base domain such as an octahedron or a polyhedral torus.

We now switch from simplification to refinement. We subdivide the mesh on the base domain using regular 1:4 refinement; newly added midpoints are displaced back into space according to the drawings on the adjacent triangles. The parameterization of the subdivision surface is not smooth across triangles of the base domain. This could be an advantage, for example if the original surface includes creases [67], or it could be an undesirable artifact, in which case the parameterization can be smoothed by weighted averaging [69]. MAPS does not have a provable guarantee; if we are unlucky drawings could fold over on themselves and spoil the bijective parameterization. A provably good version of MAPS would be of interest; a related problem for which there are theoretically justified algorithms [59] asks for a piecewise-linear homeomorphism between two polygonal domains.

9 Geometry Compression

There has been a flurry of recent work in computer graphics on *geometry compression*, meaning data compression of meshes [2,42,57,61,67,101,112,115]. Since meshes can be quite large, geometry compression has a number of potential applications, for example, interactive gaming over the Web. One can think of the problem as having two parts: encoding the topology (connectivity) of the mesh, and encoding the coordinates of the vertices. A completely uncompressed encoding might specify the topology by adjacency lists and coordinates simply by 16-bit integers. (For computer graphics models and all but the most detailed finite element simulations, 16 bits are sufficient.) In such an encoding, an average vertex would require a total of about $9 \times 16 = 144$ bits, because it would use about six 16-bit integers for its neighbors and three more for its coordinates.

One idea for compression [115] uses a spanning tree on the vertices of the mesh, along with the dual tree on the faces that would result from cutting along all the edges of the vertex-spanning tree. The vertex tree can be used to predict (and hence compress) the coordinates of vertices from previous vertex coordinates along an in-order traversal of the tree. Thus this method prefers a long skinny vertex tree with many degree-2 nodes. The face tree can be used to encode the topology of the tree; a few bits per face suffice to encode the adjacency (across which edge is the next face adjacent) of the child faces in the tree. Using this method, the topology encoding drops to only about 2 bits per vertex and the coordinate encoding to about 10–16 bits per vertex, rather than 16 bits per coordinate. Another factor of almost two in the compression can be achieved by predicting vertex coordinates using a more two-dimensional predictor, such as incident triangles in the face tree, or such as layers of triangles determined by a breadth-first traversal from a starting vertex [57,101]. Rossignac has released software called EDGEBREAKER for lossless compression of surface meshes [101].

A recent celebrated paper [67] demonstrated still another four-fold or more improvement in compression, down to about 2–4 bits per vertex for both topology and coordinates. This method, by Khodakovsky, Schröder and Sweldens, remeshes the arbitrary unstructured surface mesh into a semi-regular subdivision surface mesh using the MAPS algorithm of Lee et al. [69] described above. The remeshing is "almost lossless" in the sense that the new surface is visually indistinguishable from the old, even though triangle shapes and orientations may have changed. Due to the change in triangle shapes, this method is probably more appropriate for graphics applications than for physical simulation. Subdivision meshes compress extremely well: the topology after the initial unstructured mesh is automatic in the subdivision process, and the vertex coordinates are small deviations from predictions given by stencil-weighted averages of vertices one level up. (In fact, it is possible to give a general solid while constraining the deviations to be normal to the surface [61] and this observation gives a further improvement in compression [66].) Subdivision meshes are closely related to wavelet representations of functions, and well-developed wavelet compression methods such as zero-trees can be used to give very efficient encodings of coordinates. This compression scheme is also naturally progressive, meaning that initial sequences of the compressed data stream give better and better approximations to the surface.

10 Future Directions

We have mentioned future directions at various places in this survey. Here we gather together some of the previously mentioned directions along with some other possibilities.

1. It should be possible to prove efficiency bounds for solution-adaptive meshes for a given error tolerance, analogous to the "provably good" efficiency bounds for conforming to domain geometry, mentioned in Section 5.1.
2. Is there a tetrahedral meshing algorithm that will guarantee an M-matrix for elliptical PDEs? See Section 2 for a statement of sufficient geometric conditions for both finite element and finite volume formulations.
3. The geometry of cubical meshes (quads and hexes) is not very well understood compared to the geometry of simplicial meshes (triangles and tetrahedra). There are numerous open questions about cubical meshes. For examples: Can the flip moves for quad and hex meshes (Section 5.5) help to generate and improve cubical meshes? Is there a simple parity-switching "flip" for hex meshes? Is there an algorithm for no-small-angle quad meshing with a polynomial bound on the number of quadrilaterals? For no-large-angle quad meshing, there is such an algorithm, in fact one using only $O(n)$ quadrilaterals for an n-vertex polygonal domain (Section 5.2).
4. Related to unstructured quad and hex meshes is the problem of automatic multiblock meshing. Blocks in a multiblock mesh are topological cubes, typically with curved sides, each of which are meshed with a warped rectangular grid of cubes. One difference between multiblock meshing and unstructured quad or hex meshing is that it is often advantageous to curve blocks into horseshoe shapes, for example at leading and trailing edges of a wing, in order to confine small or anisotropic elements in the region where they are needed [77].
5. The stretching approach to generating anisotropic meshes (Section 5.3) suffers from the drawback of producing flat angles as easily as sharp angles. Are there better ways to produce *a priori* or *a posteriori* anisotropy? Section 5.3 mentions two possible approaches: square or rectangle packing and stiffly connected strings of bubbles.

6. So far there has been very little work on hybrid meshes or boundary layers in the computational geometry community. We would be expecially interested in hybrid meshes in which each boundary element has all of its faces either nearly parallel or nearly perpendicular to the boundary. For this to be possible, the domain boundary would have to satisfy certain angle bounds. It may be more convenient, at least in the initial theoretical studies, to assume a smooth curved domain boundary.
7. Adaptive mesh refinement is currently quite easy with unstructured triangular meshes, somewhat more difficult with tetrahedral meshes, and very difficult with quad or hex meshes. Are there better ways to refine tetrahedra, quads, and hexes?
8. So far there has been only a little work on subdivision volume meshes [11]. It would be nice to produce subdivision volume meshes whose boundaries are subdivision surface meshes, but this problem appears to be quite hard for general geometry. We are not aware of any work on geometry compression of 3D volume meshes.

References

1. M. Aftosmis, J. Melton, and M. Berger. Adaptation and surface modeling for Cartesian mesh methods. AIAA Paper 95-1725, 12th AIAA CFD. Conf., San Diego, CA. June, 1995, 1995.
2. P. Alliez and M. Desbrun. Progressive compression for lossless transmission of triangle meshes. In *Proc. SIGGRAPH*, 2001.
3. N. Amenta, M. W. Bern, and D. Eppstein. Optimal point placement for mesh smoothing. In *Proc. 8th ACM-SIAM Symp. Disc. Algorithms*, pages 528–537, 1997.
4. N. Amenta, M. W. Bern, and M. Kamvysselis. A new Voronoi-based surface reconstruction algorithm. In *Proc. SIGGRAPH*, pages 415–421, 1998.
5. E. Amezua, M. V. Hormaza, A. Hernandez, and M. B. G. Ajuria. A method of the improvement of 3d solid finite-element meshes. *Advances in Engineering Software*, 22:45–53, 1995.
6. D. N. Arnold, A. Mukherjee, and L. Pouly. Locally adapted tetrahedral meshes using bisection. *SIAM J. Sci. Computing*, 22:431–448, 2000.
7. D. N. Arnold and A. Mukherjee. Tetrahedral bisection and adaptive finite elements. In M. Bern et al., editor, *Grid Generation and Adaptive Algorithms*, pages 29–42. IMA Volumes in Mathematics and its Applications, Springer, 1999.
8. F. Aurenhammer. Voronoi diagrams—a survey of a fundamental geometric data structure. *ACM Computing Surveys*, 23:345–405, 1991.
9. I. Babuška and A. Aziz. On the angle condition in the finite element method. *SIAM J. Numer. Analysis*, 13:214–227, 1976.
10. I. Babuška and W. C. Rheinboldt. Error estimates for adaptive finite element computations. *SIAM J. Numer. Analysis*, 15:736–754, 1978.
11. C. Bajaj, J. Warren, and G. Xu. A smooth subdivision scheme for hexahedral meshes. http://www.cs.rice.edu/~jwarren/papers/hexahedral.pdf
12. T. J. Baker. Resolution of boundary layers on triangular meshes. In M. Bern et al., editor, *Grid Generation and Adaptive Algorithms*, pages 43–51. IMA Volumes in Mathematics and its Applications, Springer, 1999.

13. R. E. Bank. *PLTMG: A Software Package for Solving Elliptic Partial Differential Equations. Users' Guide 6.0.* SIAM Publications, Philadelphia, Penn., 1990.
14. R. E. Bank, A. H. Sherman, and A. Weiser. Refinement algorithms and data structures for regular local mesh refinement. In R. Stepleman et al., editor, *Scientific Computing*, pages 3–17. IMACS/North-Holland Publishing Company, Amsterdam, 1983.
15. R. E. Bank and R. K. Smith. Mesh smoothing using a posteriori error estimates. *SIAM J. Num. Anal.*, to appear. ftp://math.ucsd.edu/pub/scicomp/reb/ftpfiles/a67.ps.Z.
16. E. Bänsch. Local mesh refinement in 2 and 3 dimensions. *Impact of Computing in Science and Engineering*, 3:181–191, 1991.
17. T. J. Barth. Aspects of unstructured grids and finite-volume solvers for the Euler and Navier-Stokes equations. Technical report, von Karman Institute for Fluid Dynamics, Lecture Series 1994-05, 1994.
18. M. J. Berger and P. Colella. Local adaptive mesh refinement for shock hydrodynamics. *Journal of Computational Physics*, 82:64–84, 1989.
19. M. J. Berger and J. Oliger. Adaptive mesh refinement for hyperbolic partial differential equations. *Journal of Computational Physics*, 53:484–512, 1984.
20. M. Bern, L.P. Chew, D. Eppstein, and J. Ruppert. Dihedral bounds for mesh generation in high dimensions. In *Proc. 6th ACM Symp. Disc. Algorithms*, pages 189–196, 1995.
21. M. Bern and D. Eppstein. Mesh generation and optimal triangulation. In D.-Z. Du and F.K. Hwang, editors, *Computing in Euclidean Geometry, 2nd Edition*, pages 47–123. World Scientific, Singapore, 1995.
22. M. Bern and D. Eppstein. Flipping cubical meshes. In *Proc. 10th International Meshing Roundtable*, 2001.
23. M. Bern and D. Eppstein. Optimal möbius transformations for information visualization and meshing. In *Proc. Workshop on Algorithms and Data Structures*, 2001.
24. M. Bern, D. Eppstein, and J.R. Gilbert. Provably good mesh generation. *J. Comp. System Sciences*, 48:384–409, 1994.
25. M. Bern, D. Eppstein, and J.R. Gilbert. Quadrilateral meshing by circle packing. *Int. J. Comp. Geometry and Applications*, 10:347–360, 2000.
26. M. Bern, S. Mitchell, and J. Ruppert. Linear-size nonobtuse triangulation of polygons. *Disc. and Comp. Geometry*, 14:411–428, 1995.
27. M. Bern and P. Plassmann. Mesh generation. In J.-R. Sack and J. Urrutia, editors, *Handbook of Computational Geometry*, pages 291–332. North-Holland, 2000.
28. M. Berzins. A solution based H^1 norm triangular mesh quality indicator. In M. Bern et al., editor, *Grid Generation and Adaptive Algorithms*, pages 77–89. IMA Volumes in Mathematics and its Applications, Springer, 1999.
29. F. J. Bossen and P. S. Heckbert. A pliant method for anisotropic mesh generation. In *Proc. 5th International Meshing Roundtable*, 1996.
30. S. Canann, M. Stephenson, and T. Blacker. Optismoothing: an optimization-driven approach to mesh smoothing. *Finite Elements in Analysis and Design*, 13:185–190, 1993.
31. M. J. Castro-Diaz, F. Hecht, and B. Mohammadi. New progress in anisotropic grid adaptation for inviscid and viscid flows simulations. In *Proc. 4th International Meshing Roundtable*. Sandia National Laboratories, 1995.
32. E. Catmull and J. Clark. Recursively generated B-spline surfaces on arbitrary topological surfaces. *Computer-Aided Design* 10(6):350–355, 1978.
33. Centaur Grid Generator. http://www.centaursoft.com/
34. S. Cheng and T. Dey. To appear in *Proc. 13th ACM-SIAM Symp. Disc. Algorithms*, 2002.
35. S.W. Cheng, T.K. Dey, H. Edelsbrunner, M.A. Facello, and S. Teng. Sliver exudation. In *Proc. 15th ACM Symp. Comp. Geometry*, pages 1–13, 1999. Also *J. Assoc. Comput. Mach.* 47:883–904, 2000.

36. L.P. Chew. Guaranteed-quality triangular meshes. Technical Report TR-89-983, Comp. Science Dept., Cornell University, 1989.
37. L.P. Chew. Guaranteed-quality Delaunay meshing in 3D. In *Proc. 13th ACM Symp. Comp. Geometry*, pages 391–393, 1997.
38. P.G. Ciarlet. *The Finite Element Method for Elliptic Problems*. North-Holland, 1978.
39. P.G. Ciarlet and P.A. Raviart. Maximum principle and uniform convergence for the finite element method. *Computer Methods in Appl. Mechanics and Engineering*, 2:17–31, 1973.
40. W. J. Coirier. An adaptively-refined, Cartesian cell-based scheme for the Euler and Navier-Stokes equations. NASA Technical Memorandum 106754, NASA, October 1994.
41. W. J. Coirier and K. G. Powell. An accuracy assessment of Cartesian-mesh approaches for the Euler equations. *Journal of Computational Physics*, 117:121–131, 1995.
42. M. Deering. Geometry compression. *Computer Graphics*, pages 13–20, 1995.
43. L. De Floriani and E. Puppo. An on-line algorithm for constrained Delaunay triangulation. *Graphical Models and Image Processing*. 54:290–300, 1992.
44. B. Delaunay. Sur la sphère vide. *Izvestia Akademia Nauk SSSR, VII Seria, Otdelenie Matematicheskii i Estestvennyka Nauk*, 7:793–800, 1934.
45. H. Edelsbrunner. *Geometry and Topology for Mesh Generation*. Cambridge University Press, 2001.
46. H. Edelsbrunner and D. Guoy. Sink-insertion for mesh improvement. In *Proc. 17th ACM Symp. Comp. Geometry*, pages 115–123, 2001.
47. H. Edelsbrunner and D. Guoy. An experimental study of sliver exudation. In *Proc. 10th International Meshing Roundtable*, 2001.
48. H. Edelsbrunner, X.-L. Li, G. L. Miller, A. Stathopoulos, D. Talmor, S.-H. Teng, A. Üngör, and N. Walkington. Smoothing and cleaning up slivers. In *Proc. ACM Symp. Theory of Computing*, 2000.
49. H. Edelsbrunner and T.-S. Tan. An upper bound for conforming Delaunay triangulations. *Disc. and Comp. Geometry*, 10:197–213, 1993.
50. D. A. Field. Laplacian smoothing and Delaunay triangulations. *Communications and Applied Numerical Methods*, 4:709–712, 1988.
51. P. Fleischmann. *Mesh Generation for Technology CAD in Three Dimensions* PhD Thesis, Technical University of Vienna, 1999.
52. On the road with the Opel CFD group. http://www.fluent.com/about/news/newsletters/01v10i1/a1.htm
53. L. A. Freitag, M. T. Jones, and P. E. Plassmann. An efficient parallel algorithm for mesh smoothing. In *Proc. 4th International Meshing Roundtable*, pages 47–58. Sandia National Laboratories, 1995.
54. M. Garland and P. S. Heckbert. Surface simplification using quadric error metrics. In *Proc. SIGGRAPH*, pages 209–216, 1997. http://graphics.cs.uiuc.edu/~garland/papers.html
55. A. S. Glassner. The perils of problematic parameterization. *IEEE Computer Graphics and Applications*, 2:78–83, September/October 1997.
56. L. J. Guibas, D. E. Knuth, and M. Sharir. Randomized incremental construction of Delaunay and Voronoi diagrams. *Algorithmica*, 7:381–413, 1992. Also in *Proc. 17th Int. Colloq. Automata, Languages and Programming*, Springer-Verlag, LNCS 443, 1990, pp. 414–431.
57. S. Gumhold and W. Straßer. Real time compression of triangle mesh connectivity. In *Proc. SIGGRAPH*, pages 133–140, 1998.

58. D. Guoy. *Tetrahedral mesh improvement, algorithms and experiments* PhD Thesis and Tech. Report UIUCDCS-R-2001-2225, University of Illinois, Computer Science Department, 2001.
59. H. Gupta and R. Wenger. Constructing piecewise linear homeomorphisms of simple polygons. *J. Algorithms*, 22(1):142–157, 1997.
60. I. Guskov, W. Sweldens, and P. Schröder. Multiresolution signal processing for meshes. In *Proc. SIGGRAPH*, pages 325–334, 1999. http://multires.caltech.edu/pubs/
61. I. Guskov, K. Vidimce, W. Sweldens, and P. Schröder. Normal meshes. In *Proc. SIGGRAPH*, 2000. http://multires.caltech.edu/pubs/
62. P. Heckbert. Collection of mesh generation links. http://www-2.cs.cmu.edu/afs/cs/user/ph/www/mesh.html
63. H. Hoppe. Progressive meshes. In *Proc. SIGGRAPH*, pages 99–108, 1996.
64. B. Joe. Three-dimensional triangulations from local transformations. *SIAM J. Sci. Stat. Comput.*, 10:718–741, 1989.
65. M. T. Jones and P. E. Plassmann. Adaptive refinement of unstructured finite-element meshes. *Finite Elements in Analysis and Design*, 25(1-2):41–60, March 1997.
66. A. Khodakovsky and I. Guskov. Normal mesh compression. http://multires.caltech.edu/pubs/
67. A. Khodakovsky, P. Schröder, and W. Sweldens. Progressive geometry compression. In *Proc. SIGGRAPH*, 2000. http://multires.caltech.edu/pubs/
68. C. L. Lawson. Software for C^1 surface interpolation. In J. Rice, editor, *Mathematical Software III*, pages 161–194. Academic Press, 1977.
69. A. W. F. Lee, W. Sweldens, Peter Schröder, Lawrence Cowsar, and David Dobkin. MAPS: Multiresolution adaptive parameterization of surfaces. In *Proc. SIGGRAPH*, pages 95–104, 1998.
70. X. Li, S. Teng, and A. Üngör. Simultaneous refinement and coarsening: adaptive meshing with moving boundaries. In *Proc. 7th International Meshing Roundtable*, pages 201–210, 1998.
71. X. Li, S. Teng, and A. Üngör. Biting: advancing front meets sphere packing. In *Proc. 2nd Symp. Trends in Unstructured Mesh Generation (Boulder, Colorado)*, 1999.
72. X.-Y. Li and S. Teng. Generating well-shaped Delaunay meshes in 3D. In *Proc. 12th ACM-SIAM Symp. Disc. Algorithms*, pages 28–37, 2001.
73. C. Loop. *Smooth Subdivision based on Triangles* Master's Thesis, University of Utah, Dept. of Mathematics, 1987.
74. A. Liu and B. Joe. On the shape of tetrahedra from bisection. *Mathematics of Computation*, 63(207):141–154, July 1994.
75. A. Liu and B. Joe. Quality local refinement of tetrahedral meshes based on bisection. *SIAM J. on Scientific Computing*, 16(6):1269–1291, November 1995.
76. S. H. Lo. A new mesh generation scheme for arbitrary planar domains. *Int. J. Numer. Meth. Eng.*, 21:1403–1426, 1985.
77. N. Lu and P. R. Eisman. Interactive high quality grid generation. In *Proc. NASA Workshop on Surface Modeling, Grid Generation, and Related Issues in CFD Solutions*, NASA-CP-3291, pages 833–843, May 1995.
78. I. MacPhedran and R. Young. Internet Finite Element Resources. http://www.engr.usask.ca/~macphed/finite/fe_resources/fe_resources.html
79. D. L. Marcum and N. P. Weatherill. Aerospace applications of solution adaptive finite element method. *Computer Aided Geometric Design*, 12:709–731, 1995.
80. J. M. Maubach. The number of similarity classes created by local n-simplicial bisection refinement. Manuscript, 1996.

81. D. J. Mavriplis. Adaptive mesh generation for viscous flows using Delaunay triangulation. *J. of Computational Physics*, 90(2):271–291, Oct. 1990.
82. G.L. Miller, D. Talmor, and S.-H. Teng. Optimal coarsening of unstructured meshes. In *Proc. 8th ACM-SIAM Symp. Disc. Algorithms*, pages 538–547, 1997.
83. G.L. Miller, D. Talmor, and S.-H. Teng. Data generation for geometric algorithms on nonuniform distributions. *Int. J. Computational Geometry and Applications*, 1998.
84. G.L. Miller, D. Talmor, S.-H. Teng, and N. Walkington. A Delaunay based numerical method for three dimensions: generation, formulation, and partition. In *Proc. 36th IEEE Symp. on Foundations of Comp. Science*, pages 683–692, 1995.
85. S. A. Mitchell and T. J. Tautges. Pillowing doublets: refining a mesh to ensure that faces share at most one edge. In *Proc. 4th International Meshing Roundtable*, pages 231–240. Sandia National Laboratories, 1995.
86. M. Murphy, D. Mount, and C. W. Gable. A point-placement strategy for conforming Delaunay tetrahedralization. In *Symp. on Discrete Algorithms*, pages 67–74, 2000.
87. G.M. Nielson. Tools for triangulation and tetrahedration and constructing functions defined over them. In *Scientific Visualization: Overviews, Methodologies, Techniques*, pages 429–525. *IEEE Computer Society Press*, 1998.
88. C. F. Ollivier-Gooch. Multigrid acceleration of an upwind Euler solver on unstructured meshes. *AIAA Journal*, 33(10):1822–1827, 1995.
89. S. J. Owen. Meshing Research Corner. http://www.andrew.cmu.edu/user/sowen/mesh.html
90. V. N. Parthasarathy and Srinivas Kodiyalam. A constrained optimization approach to finite element mesh smoothing. *Finite Elements in Analysis and Design*, 9:309–320, 1991.
91. J. Peraire, J. Peiró, and K. Morgan. Adaptive remeshing for 3-D compressible flow computations. *J. Comput. Physics*, 103:269–285, 1992.
92. M. Putti and C. Cordes. Finite element approximation of the diffusion operator on tetrahedra. *SIAM J. Sci. Comput.*, 19:1154–1168, 1998.
93. J.F. Randolph. *Calculus and Analytic Geometry*. Wadsworth, 1961.
94. S. Rippa. Minimal roughness property of the Delaunay triangulation. *Computer Aided Geometric Design*, 7:489–497, 1990.
95. S. Rippa. Long and thin triangles can be good for linear interpolation. *SIAM J. Numer. Analysis*, 29:257–270, 1992.
96. M.-C. Rivara. Algorithms for refining triangular grids suitable for adaptive and multigrid techniques. *Int. J. Numer. Meth. Eng.*, 20:745–756, 1984.
97. M.-C. Rivara. Design and data structure of fully adaptive, multigrid, finite-element software. *ACM Transactions on Mathematical Software*, 10(3):242–264, September 1984.
98. M.-C. Rivara. Mesh refinement processes based on the generalized bisection of simplices. *SIAM J. Numer. Analysis*, 21(3):604–613, June 1984.
99. M.-C. Rivara and C. Levin. A 3-d refinement algorithm suitable for adaptive and multigrid techniques. *Communications in Applied Numerical Methods*, 8:281–290, 1992.
100. I. G. Rosenberg and F. Stenger. A lower bound on the angles of triangles constructed by bisecting the longest side. *Math. Comp.*, 29:390–395, 1975.
101. J. Rossignac. Edgebreaker: connectivity compression for triangle meshes. Technical Report GIT-GVU-98-35, Georgia Tech, 1998.
102. J. Ruppert. A Delaunay refinement algorithm for quality 2-dimensional mesh generation. *J. Algorithms*, 18(3):548–585, May 1995.
103. A. Saalfeld. Delaunay edge refinements. In *Proc. 3rd Candadian Conf. Comp. Geometry*, pages 33–36, 1991.

104. R. Schneiders. Mesh Generation and Grid Generation on the Web. http://www-users.informatik.rwth-aachen.de/~roberts/meshgeneration.html
105. E. Schönhardt. Über die Zerlegung von Dreieckspolyedern in Tetraeder. *Math. Annalen*, 98:309–312, 1928.
106. J. R. Shewchuk. A condition guaranteeing the existence of higher-dimensional constrained Delaunay triangulations. In *Proc. 14th ACM Symp. Comp. Geometry*, pages 76–85, 1998.
107. J. R. Shewchuk. Tetrahedral mesh generation by Delaunay refinement. In *Proc. 14th ACM Symp. Comp. Geometry*, pages 86–95, 1998.
108. J. R. Shewchuk. Mesh generation for domains with small angles. In *Proc. 16th ACM Symp. Comp. Geometry*, pages 350–359, 2000.
109. J. R. Shewchuk. Sweep algorithms for constructing higher-dimensional constrained Delaunay triangulations. In *Proc. 16th ACM Symp. Comp. Geometry*, pages 350–359, 2000.
110. K. Shimada. *Physically-Based Mesh Generation: Automated Triangulation of Surfaces and Volumes via Bubble Packing*. PhD thesis, ME Dept., MIT, 1993.
111. K. Shimada and D. C. Gossard. Computational methods for physically-based FE mesh generation. In *Proc. IFIP TC5/WG5.3 8th Int. Conference on PROLAMAT, Tokyo*, 1992.
112. J. Snoeyink and M. van Kreveld. Linear-time reconstruction of Delaunay triangulations with applications. In *Proc. European Symp. on Algorithms*, 1997.
113. K. Stephenson. http://www.math.utk.edu/~kens/
114. D. Talmor. *Well-Spaced Points for Numerical Methods*. PhD thesis, School of Computer Science, Carnegie-Mellon University, 1997.
115. G. Taubin and J. Rossignac. Geometric compression through topological surgery. *ACM Trans. on Graphics*, 17:84–115, 1998.
116. Anisotropic adaptive unstructured mesh generation. Von Karman Institute for Fluid Dynamics. http://www.vki.ac.be/research/themes/cfd99/1.html
117. N. P. Weatherill and O. Hassan. Efficient three-dimensional Delaunay triangulation with automatic point creation and imposed boundary constraints. *Int. J. Numer. Meth. Eng.*, 37:2005–2039, 1994.
118. N. P. Weatherill, O. Hassan, D. L. Marcum, and M. J. Marchant. Grid generation by the delaunay triangulation. *von Karman Institute for Fluid Dynamics, Lecture Series*, 1994.
119. A. Weiser. Local-mesh, local-order, adaptive finite element methods with a posteriori error estimates for elliptic partial differential equations. Technical Report 213, Yale University, New Haven, Connecticut, 1981.
120. W. Welch. *Serious Putty: Topological Design for Variational Curves and Surfaces*. PhD thesis, CS Dept, Carnegie Mellon University, Dec. 1995. CMU-CS-95-217, ftp://reports.adm.cs.cmu.edu/usr/anon/1995/CMU-CS-95-217A.ps, 217B.ps, 217C.ps.
121. J. Xu and L. Zikatanov. A monotone finite element scheme for convection diffiusion equations. *Math. of Computation*, to appear.
122. S. Yamakawa and K. Shimada. High quality anisotropic tetrahedral mesh generation via ellipsoidal bubble packing. In *Proc. 9th International Meshing Roundtable*, pages 263–273, 2000.
123. D. P. Young, R. G. Melvin, M. B. Bieterman, and J. E. Bussoletti. A locally refined rectangular grid finite element method: Application to computational fluid dynamics and computational physics. *Journal of Computational Physics*, 92:1–66, 1991.
124. D. Zorin, P. Schröder, and Wim Sweldens. Interactive multiresolution mesh editing. In *Proc. SIGGRAPH*, pages 259–268, 1997.

Adjoint Error Correction for Integral Outputs

Michael B. Giles[1] and Niles A. Pierce[2]

[1] Oxford University Computing Laboratory,
Oxford OX1 3QD, United Kingdom
[2] Applied & Computational Mathematics,
California Institute of Technology, Pasadena, CA 91125

Abstract. These lecture notes begin by observing that in many cases the most important engineering outputs of CFD calculations are one or two integral quantities, such as the lift and drag. It is then explained that the solution to an appropriate adjoint problem gives the effect of numerical approximations on the output functional of interest, facilitating the calculation of more accurate functional estimates. The theory is presented for both linear and nonlinear differential equations, incorporating a range of numerical examples illustrating the ability to obtain answers with twice the order of accuracy of the underlying numerical solution.

Contents

1 Introduction

1.1 Output functionals

Why do engineers perform CFD calculations? In the case of a transport aircraft at cruise conditions, a calculation might be performed to investigate whether there is an adverse pressure gradient near the leading edge of the wing, causing boundary layer separation and premature transition. Alternatively, one might be concerned about wing/pylon/nacelle integration, in which case one might be looking to see if there are any shocks on the pylon, leading to unacceptable integration losses. In both of these examples, qualitative information is being obtained from the computed flow field to understand and interpret the impact of the phenomena on the quantitative outputs of most concern to the aeronautical engineer, the lift and drag on the aircraft. The quality of the CFD calculation is judged, first and foremost, by the accuracy of the lift and drag predictions. The details of the flow field are much less important, and are used in a more qualitative manner to suggest ways in which the design may be modified to improve the lift or drag. This focus on a few output quantities is even clearer in design optimisation, when one is trying to optimise a single objective function, possibly subject to a number of constraints.

This interest in integral outputs, also referred to as *functionals*, arises in many applications of CFD. Occasionally, volume integrals are of importance. For example, the infra-red signature of a military aircraft will depend in part on a volume integral of some function of the temperature in the thermal wake behind the aircraft. However, usually it is surface integrals that are of most concern, as with lift and drag. Other aeronautical examples include: the roll moment produced by aileron deflection; the mass flow through a compressor blade row; the outflow flux of nitrous oxides from a combustor; the total heat flux into a high pressure turbine blade from the surrounding flow; average noise levels on the ground due to an aircraft landing or taking off.

The idea of output functionals is central to these lecture notes; they are concerned specifically with the analysis of the numerical error in these functionals, and a particular method of correction that very greatly reduces the error, typically doubling the order of accuracy for the functional relative to the underlying flow solution. This distinguishes this kind of error analysis from other approaches that focus on the maximum, root-mean-square or some other measure of the error in the whole flow field. The problem with such measures is that they can have little relation to the errors in the integral outputs of primary concern to the engineer.

As an example, consider the wake behind a wing. To adequately resolve the wake requires a fine grid locally, but it is often the case that the computed wake a chord or two downstream of the wing passes into a region in which the grid resolution is rather coarse. Grid adaptation based on error estimates that look at the whole solution, possibly by looking at the local truncation error, would cause the grid to be further refined in this region. However, the influence of errors in this region on the computed lift and drag would be very small, and a much greater reduction in the lift and drag errors could probably be achieved by adding the grid refinement closer to the wing, possibly near the leading and trailing edge where very small errors can have an enormous impact on the lift and drag.

1.2 *A priori* and *a posteriori* error analysis

The adjoint error correction technique to be described later is applied as a post-processing step, and so it fits into the framework of *a posteriori* error analysis. This is error analysis based on the computed flow solution, as opposed to *a priori* error analysis that is based on some (usually limited) knowledge of the analytic solution without the benefit of any numerical solution.

A priori error analysis leads to an error bound of the form

$$\text{Error} < c\, h^p$$

where h is the representative grid spacing, and c, p are positive constants that do not depend on h.

The main point of *a priori* error analysis is finding the value of p, which determines how rapidly the error reduces as the computational grid is refined, uniformly. For most finite difference and finite volume methods, the error in output functionals is of the same order as the error in the flow solution, so it does not matter which error is considered. On structured grids with smooth flow solutions, the solution error is proportional to the truncation error and its order can be relatively easily determined.

The value of the other constant c depends on the details of the analytic problem being solved, the geometry of the computational domain, the boundary conditions, etc. It is extremely difficult to get a good value for c for anything but the simplest problems. Ridiculously large values such as 10^{10} are not uncommon in the literature. This makes the error bound useless in any practical sense, and as a consequence there is often no attempt made to evaluate c.

One area in which *a priori* error analysis is very helpful is applications with singularities in the solution. For such problems, careful analysis can reveal the degree of local grid refinement that is required to recover the order of accuracy (expressed as a function of the total number of grid points) that would be obtained for non-singular solutions.

With *a posteriori* error analysis, ideally one would like a *guaranteed* error bound of the form

$$\text{Error} < e(u_h)$$

where $e(u_h)$ is a computable function of the numerical solution u_h. If the error being considered is the error in the lift from a CFD calculation, this would enable an engineer to perform a calculation and know, with complete certainty, that the true value for the lift lay within certain limits.

For such a bound to be of use, it needs to be *tight*. The efficiency or tightness of the bound is measured by the ratio $e(u_h)/\text{Error}$. A value of 1 is perfect. In the range 2–10, it is useful, but if it were more than 1000 then it would be fairly useless for practical purposes.

Although *guaranteed* error bounds are the ideal, in practice they are extremely difficult to obtain for anything but the simplest of problems. Nonlinearity causes particular difficulties. Therefore, most *a posteriori* bounds are *asymptotic*, so that

$$\text{Error} < e(u_h) \qquad \text{for all} \quad h < h_0$$

The problem is that the value of h_0 is not known. All that is known is that a positive h_0 does exist, below which the asymptotic error bound will be valid. However, above this value the error may exceed the error bound.

As an example, if

$$\text{Error} = 1.36\,h^2 + 0.77\,h^4,$$

then $1.37\,h^2$ is an asymptotic bound that is valid for $h^2 < 0.01/0.77$, but exceeded when $h = 1$.

The distinction between *guaranteed* and *asymptotic* bounds is important. With asymptotic bounds, a user must exercise their judgement to decide whether the grid is sufficiently fine that the bound is likely to be valid. With complex geometries and complex flow fields, this is not easy, particularly for a novice user. On the other hand, with a guaranteed bound one could start with an extremely coarse grid, and then use the guaranteed error bounds to drive grid adaptation until it produces a numerical solution within a user-specified tolerance. This would require no user judgement other than the choice of the error tolerance.

We conclude this discussion of error bounds with a comment on the issue of error bounds versus error correction. Error bounds based on adjoint solutions require a similar level of computational effort to the adjoint error correction to be discussed in these notes. If one has a precise estimate of the error, this could be used to form a near-perfect asymptotic error bound, or it could be used to correct the leading order terms in the error and thereby obtain a solution with a higher order of accuracy. The latter approach is the one that we follow.

1.3 An introduction to adjoints

The use of adjoints lies at the heart of error analysis for output functionals. The main theory will use adjoint differential equations, but here we introduce the ideas at an algebraic level.

Suppose we want to calculate the value of a vector scalar product

$$g^T u,$$

where the vector u is the solution of the system of linear equations

$$Au = f.$$

An equivalent *dual* treatment is to evaluate the product

$$v^T f,$$

where the v is the solution of the *adjoint* (or *dual*) equations

$$A^T v = g,$$

The equivalence of the two calculations comes from the simple identity

$$v^T (A u) = (A^T v)^T u, \tag{1}$$

from which it follows that

$$v^T f = g^T u.$$

So, to obtain a linear output functional from the solution of a linear system of equations, we can *either* solve the original equations (sometimes referred to as the *primal* equations) *or* solve the adjoint (dual) equations.

This simple result is the basis for all that follows later. With differential equations, the vector product becomes an integral inner product, the transposed matrix A^T becomes the adjoint differential operator, and the adjoint identity includes certain boundary integral terms, but in essence the equivalence is the same.

When the output is desired for a single f and g, there is no benefit in using the adjoint approach. Either method requires the solution of a linear system of equations of the same dimension, with the same computational cost. The benefit arises when the value of the output is wanted for a single g but several different vectors f. The direct approach would require the solution of the primal equations for each value of f, greatly increasing the computational cost, whereas the dual approach would still require just one adjoint calculation, to be followed by an inexpensive vector product $v^T f$, for each f.

We are now going to look at how this result can be used in two different contexts: design optimisation and error analysis. The motivation for beginning with design optimisation is that this is the primary reason why many research groups within academia and industry are developing adjoint Euler and Navier-Stokes codes [Jam95,AV99]. Design optimisation has a clear industrial "pay-off", whereas the benefits of good error analysis are yet to be appreciated. For the same reason, design optimisation is also the most widely known application for the use of adjoints.

Design optimisation Consider design optimisation using a 'discrete' algebraic approach [EP97,NA99,AB99,MP99,GDM01], rather than the 'continuous' differential approach [Jam88,KIH91,TKS92,BE92,Jam95,JPM98,DG00], (see [Gil97,GP00] for a discussion of their relative merits and [NTB+99] for an excellent review of research on adjoint design methods). The starting point of this formulation is that U, the flow variables at a discrete set of points with coordinates X, is the solution of a system of nonlinear equations

$$N(U,X) = 0,$$

that come from the discretisation of the Euler or Navier-Stokes equations, together with appropriate boundary conditions.

Through the grid generation process, the grid coordinates depend on α which represents one or more geometric design variables. In wing design, for example, perturbations to α might change the thickness distribution and the camber of the wing. If there is only one design variable α, we can linearise about a flow solution for the baseline geometry to obtain

$$Au = f$$

where u is the sensitivity of the flow field to changes in α,

$$u \equiv \frac{\mathrm{d}U}{\mathrm{d}\alpha},$$

and

$$A = \frac{\partial N}{\partial U}, \quad f = -\frac{\partial N}{\partial X}\frac{\mathrm{d}X}{\mathrm{d}\alpha}.$$

The aim of design optimisation is to minimise some objective function $J(U,X)$ which, for example, might be a discrete approximation to the drag. Linearising this function gives

$$\frac{\mathrm{d}J}{\mathrm{d}\alpha} = g^T u + \frac{\partial J}{\partial X}\frac{\mathrm{d}X}{\mathrm{d}\alpha},$$

where

$$g^T = \frac{\partial J}{\partial U}.$$

In the adjoint approach, this sensitivity of the objective function to changes in α is obtained from

$$\frac{\mathrm{d}J}{\mathrm{d}\alpha} = v^T f + \frac{\partial J}{\partial X}\frac{\mathrm{d}X}{\mathrm{d}\alpha},$$

where v satisfies the adjoint equations

$$A^T v = g.$$

If there are several design variables, each has a *different* f, but the *same* g, so the adjoint approach is much cheaper, requiring the solution of just one adjoint set of equations.

Error analysis We now return to the original problem of evaluating $g^T u$ with u being the solution of the linear equations

$$Au = f.$$

The corresponding dual problem which is to evaluate $v^T f$ where v is the solution of the adjoint equations

$$A^T v = g.$$

Suppose we have approximate solutions $\tilde{u}, \tilde{v}$ to each of these equations. We can then obtain the following result.

$$\begin{aligned} g^T u &= g^T \tilde{u} + g^T (u - \tilde{u}) \\ &= g^T \tilde{u} + v^T A (u - \tilde{u}) \\ &= g^T \tilde{u} + \tilde{v}^T A (u - \tilde{u}) + (v - \tilde{v})^T A (u - \tilde{u}) \\ &= g^T \tilde{u} - \tilde{v}^T (A\tilde{u} - f) + (v - \tilde{v})^T A (u - \tilde{u}). \end{aligned} \tag{2}$$

The first of the three terms in the final result is the value of the functional using the approximate solution $\tilde{u}$. The second term is also computable since it involves the known approximate solutions $\tilde{u}$ and $\tilde{v}$.

The third term is not computable if the exact solutions u and v are not known. However, if $\tilde{u}$ and $\tilde{v}$ are close approximations to u and v, respectively, then the third term will be very small. Thus, the sum of the first two terms gives a very good approximation to the true value of $g^T u$ – a much better approximation in general than $g^T \tilde{u}$.

Note the form of the second term, which we refer to as the *adjoint error correction term*. $A\tilde{u} - f$ is the residual error in solving the equations $Au = f$. The approximate adjoint solution $\tilde{v}$ provides the appropriate weighting for the residual error, giving the effect of the residual error on the output functional of interest. This inner product of a residual error and an adjoint weighting will be repeated throughout these notes.

To take it a step further, suppose now that we want to evaluate a nonlinear function $J(U)$, where U is the solution of the nonlinear equations

$$N(U) = 0.$$

Given an approximate solution $\widetilde{U}$, we define u to be the solution error,

$$u = \widetilde{U} - U,$$

and then linearise both the nonlinear equations and the functional to obtain

$$N(\widetilde{U}) = N(U+u) \approx \frac{\partial N}{\partial U}\, u,$$

and

$$J(\widetilde{U}) = J(U+u) \approx J(U) + \frac{\partial J}{\partial U}\, u.$$

These can be re-written as

$$Au \approx f,$$

where

$$A = \frac{\partial N}{\partial U}, \quad f = N(\widetilde{U}).$$

and

$$J(U) \approx J(\widetilde{U}) - g^T u,$$

where

$$g^T = \frac{\partial J}{\partial U}.$$

If v is defined to satisfy the adjoint equation

$$A^T v = g,$$

then we obtain

$$J(U) \;\approx\; J(\widetilde{U}) - v^T f \;\approx\; J(\widetilde{U}) - v^T N(\widetilde{U}).$$

Hence, the quantity

$$J(\widetilde{U}) - v^T N(\widetilde{U})$$

is a more accurate estimate for $J(U)$ than $J(\widetilde{U})$ alone. Again note that the adjoint error correction term is a product of an approximate adjoint solution and the residual error from the original nonlinear equations.

Automatic differentiation An introduction to adjoints would not be complete without a mention of Automatic Differentiation (AD). This is a technique, implemented in a number of software packages, that starts with a code to compute a nonlinear vector function $F(U)$, and automatically generates codes to compute either

$$\frac{\partial F}{\partial U}\,\tilde{u}$$

for any $\tilde{u}$ (forward mode), or

$$\left(\frac{\partial F}{\partial U}\right)^T \bar{v}$$

for any $\bar{v}$ (reverse mode).

The forward mode is relatively easy to understand. A computer code can be decomposed into a sequence of binary operations

$$c = \mathrm{op}(a, b),$$

where the operation is addition, subtraction, multiplication or division, plus a few unitary operations

$$c = \mathrm{fn}(a),$$

where the function may be, for example, an exponential or a logarithm. If we treat unitary operations as a special case of binary operations, then linearising a binary operation gives

$$\tilde{c} = \frac{\partial c}{\partial a}\,\tilde{a} + \frac{\partial c}{\partial b}\,\tilde{b} = \begin{pmatrix} \frac{\partial c}{\partial a} & \frac{\partial c}{\partial b} \end{pmatrix} \begin{pmatrix} \tilde{a} \\ \tilde{b} \end{pmatrix}.$$

The forward mode AD software inserts the instructions to compute $\frac{\partial c}{\partial a}$ and $\frac{\partial c}{\partial b}$ and evaluate the output sensitivity $\tilde{c}$ given the sensitivities of the two inputs. Carrying this out throughout the code gives the linear sensitivity of the output of the whole code to a specified combination of linear perturbations to the inputs.

The reverse mode AD software performs a task that seems much harder than the forward mode, but in fact it is only slightly harder. Looking again at the single binary instruction, suppose for simplicity that the variables a, b, c are used only once during the whole code. Let $\bar{a}, \bar{b}, \bar{c}$ denote the sensitivity of the output of the whole code to perturbations in a, b, c. These are then related by the equation

$$\begin{pmatrix} \bar{a} \\ \bar{b} \end{pmatrix} = \begin{pmatrix} \frac{\partial c}{\partial a} & \frac{\partial c}{\partial b} \end{pmatrix}^T \bar{c},$$

which is the transpose of the linear sensitivity equation. The tricky thing with the reverse mode is that the adjoint steps have to be performed in the reverse order to the original nonlinear code. Therefore the AD software has to generate temporary storage for each operation in which to keep the linearisation coefficients such as $\frac{\partial c}{\partial a}$ and $\frac{\partial c}{\partial b}$. Other than this, the computational cost of reverse mode AD is similar to forward mode AD, typically no more than a factor 4 greater than the original nonlinear code.

The main AD packages are ADIFOR [BCH+98], Odyssée [FP98] and TAMC [GK98]. For further information, the reader is referred to the documentation for each of these, and the excellent book by Griewank [Gri00], one of the original developers of ADIFOR.

The significance of these packages is that they can greatly simplify the task of writing an adjoint CFD code. For examples of the use of AD to generate such codes see [MP99,Moh97,CFG98]. However, there are limitations to their ability to carry out automatic differentiation of codes that use iterative solvers [Gil01], so it is best to view them as aides rather than a black-box solution.

1.4 A brief overview of the literature

Here we give a very brief overview of some of the main developments in the literature concerning the use of adjoints for error analysis.

The subject begins in 1967 with the work of Aubin and Nitsche (see [SF73]), who used a suitably defined adjoint problem to derive *a priori* optimal order proofs of L_2 convergence of finite element approximations of elliptic p.d.e.'s. In 1978, Babuška and Rheinboldt [BR78b,BR78a] built on this to develop an *a posteriori* error analysis that they applied to finite element approximations of the Poisson and Cauchy-Riemann equations.

In 1984, Babuška and Miller [BM84a,BM84b] were perhaps the first to focus attention on integral functional outputs. Because their primary interest was in point functionals such as the maximum stress in structural analysis applications, they used "extraction functions" to convert the point quantities into integrals. A key feature of these papers is the *a priori* analysis of the superconvergence of the finite element approximations of the integral functionals. This will be discussed later in these lecture notes, but the essence is that the adjoint error correction term outlined previously is zero because of a particular feature of Galerkin finite element methods known as "Galerkin orthogonality". As a result, the order of accuracy of the values for integral functionals is roughly double that of the underlying finite element solution.

In extending this work to the convection-diffusion equation, Barrett and Elliott [BE87] were the first to analyse a problem that is not self-adjoint, (i.e. one for which the adjoint differential operator is not the same as the original differential operator). This step was vital for CFD applications, none of which are self-adjoint.

The late 1990's saw an explosion of interest and research into *a posteriori* analysis of errors in integral functionals and related methods for optimal grid adaptation. Süli [GLLS97,MS98,Sül98,HRS00], Johnson [Joh95,JRB95] and Rannacher and Becker [BR96,BR98,BR99,Ran00,BKR00,BR01] have used finite element methods that exhibit natural superconvergence and have focussed their attention on using *a posteriori* error bounds to derive good grid adaptation indicators. In outline, their approaches are similar, but with significant differences in the details.

Patera and Peraire [PP97,PPP97,PP99] also focus on finite element methods, but they use a completely different *a posteriori* approach to derive error bounds for the functional computed on a "truth mesh" that is defined to be sufficiently fine that the discretisation errors may be neglected. Yet another approach for bounding the errors in functional outputs from finite element methods is that of Oden and Prudhomme [OP99,OP00].

These lecture notes cover the adjoint error correction ideas developed by Giles and Pierce [GP98,GP99,PG00,Gil00]. One way in which they may be viewed is that they extend to finite volume methods the superconvergence that is natural for many finite element methods. This is achieved through the explicit evaluation of the adjoint correction term which is non-zero because of the lack of "orthogonality". However, as will be shown later, it is also possible to apply the technique with finite element solutions to obtain functional values that are even more accurate than the superconvergent values that arise naturally from the finite element computation.

Venditti & Darmofal [VD00,VD01] have used an algebraic version of the adjoint error correction to correct the functional errors computed on a "truth mesh" using a solution interpolated from the original mesh. They have also used the approach to derive grid adaptation criteria. This will be discussed later in these notes, and sample results will be shown.

2 Linear adjoint error correction

In this section we develop the adjoint correction theory for linear differential equations. We begin with a restricted version without boundary terms because it has the greatest similarity to the algebraic error correction presented in the previous section.

2.1 Theory without boundary terms

Let u be the solution of the linear differential equation

$$Lu = f,$$

on some domain Ω, subject to homogeneous boundary conditions for which the problem is well-posed when $f \in L_2(\Omega)$ (meaning that f is a square-integrable function).

The adjoint differential operator L^* and associated homogeneous boundary conditions are defined by the identity

$$(v, Lu) = (L^* v, u), \tag{3}$$

that must hold for all u, v satisfying the respective boundary conditions. Here the notation $(.,.)$ denotes an integral inner product over the domain Ω, i.e.

$$(v, Lu) \equiv \int_\Omega v^T Lu \,\mathrm{d}V,$$

allowing for the possibility that u, and hence v, may be a vector function rather than just a scalar.

The appropriate definition for L^* can be constructed by integration by parts, starting from (v, Lu), until all of the derivatives are acting on v rather than u. In the process, the adjoint boundary conditions come from the requirement that the boundary terms that arise from the integration by parts must be zero. Examples of this will be given later.

Suppose now that we are concerned with the value of the functional $J = (g, u)$, for a given function $g \in L_2(\Omega)$. An equivalent dual formulation of the problem is to evaluate the functional $J = (v, f)$, where v satisfies the adjoint equation

$$L^* v = g,$$

subject to the homogeneous adjoint boundary conditions. The equivalence of the two forms of the problem follows immediately from the definition of the adjoint operator.

$$(v, f) = (v, Lu) = (L^* v, u) = (g, u).$$

Suppose that u_h and v_h are approximations to u and v, respectively, and satisfy the homogeneous boundary conditions. The subscript h indicates that the approximate solutions are derived from a numerical computation using a grid with average spacing h. When using finite difference or finite volume methods, u_h and v_h might be created by interpolation through computed values at grid nodes. With finite element solutions, one might simply use the finite element solutions themselves, or one could again use interpolation through nodal values and thereby obtain approximate solutions that are smoother than the finite element solutions.

It is assumed that u_h and v_h are sufficiently smooth that Lu_h and $L^* v_h$ lie in $L_2(\Omega)$. If u_h and v_h were equal to u and v, then the *residual errors* $Lu_h - f$ and $L^* v_h - g$ would be zero. Thus, the magnitude of the residual errors is a computable indication of the extent to which u_h and v_h are not the true solutions.

Now, using the definitions and identities given above, we obtain the following expression for the functional:

$$\begin{aligned}(g, u) &= (g, u_h) - (L^* v_h, u_h - u) + (L^* v_h - g, u_h - u) \\ &= (g, u_h) - (v_h, L(u_h - u)) + (L^*(v_h - v), u_h - u) \\ &= (g, u_h) - (v_h, Lu_h - f) + (v_h - v, L(u_h - u)).\end{aligned} \tag{4}$$

The first term in the final expression is the value of the functional obtained from the approximate solution u_h. The second term is an inner product of the residual error $Lu_h - f$ and the approximate adjoint solution v_h. The adjoint solution gives the weighting of the contribution of the local residual error to the overall error in the computed functional. Therefore, by evaluating and subtracting this adjoint error term we obtain a more accurate value for the functional.

The third term is the remaining error after making the adjoint correction. If $Lu_h - f = L(u_h - u)$ is of the same order of magnitude as $u_h - u$ then the remaining error has a bound that is proportional to the product $\|u_h - u\| \, \|v_h - v\|$ (using L_2 norms), and thus the corrected functional value is superconvergent. For example, if the solution errors $u_h - u$ and $v_h - v$ are both $O(h^p)$ then the error in the functional is $O(h^{2p})$.

Furthermore, the remaining error term can be expressed as

$$\begin{aligned}(v_h - v, L(u_h - u)) &= \left(v_h - v, L\, L^{-1}(Lu_h - f)\right) \\ &= \left(L^*(v_h - v), L^{-1}(Lu_h - f)\right) \\ &= \left(L^* v_h - g, L^{-1}(Lu_h - f)\right).\end{aligned}$$

This has the computable *a posteriori* bound $\|L^{-1}\| \, \|Lu_h - f\| \, \|L^* v_h - g\|$. The problem with this bound is obtaining a value for the operator norm $\|L^{-1}\|$. This can be calculated analytically in the simplest cases, but for harder problems it may be necessary to estimate it numerically.

Note the similarity between this analysis and the algebraic version in the first section. The adjoint identities (1) and (3) look almost identical, as do the expressions for the functional, (2) and (4).

2.2 Galerkin finite element methods

If the approximate solutions u_h and v_h are the finite element solutions from a Galerkin finite element discretisation, then the correction term

$$(v_h, Lu_h - f)$$

is automatically zero, due to the requirement that the finite element residual is orthogonal to all members of the finite element space [SF73]. Thus, the Galerkin finite element method gives naturally superconvergent estimates for integral outputs, in the sense that a single order of accuracy improvement in the solution, through increasing the degree of the polynomials in the finite element space, leads to two orders of accuracy improvement in the value of the functional.

However, there is usually a loss of accuracy because of a lack of smoothness in the finite element solution. Typically, if the solution errors are $O(h^p)$, then the residual error $Lu_h - f$ is $O(h^{p-m})$ where m is the degree of the differential operator, the degree of the highest derivative in the operator. Hence, the remaining error in the functional is $O(h^{2p-m})$.

If one takes the finite element solution and reconstructs smoother solutions u_h and v_h, then there is the possibility of recovering $O(h^{2p})$ accuracy for the functional, at the cost of carrying out an adjoint calculation to evaluate the adjoint error correction. This will be demonstrated in the second of the two examples to follow.

2.3 First example: 1D Poisson equation

The first example is the one-dimensional Poisson equation,

$$\frac{\mathrm{d}^2 u}{\mathrm{d}x^2} = f,$$

on the unit interval $[0,1]$ subject to the homogeneous boundary conditions $u(0)=u(1)=0$.

The dual problem is the Poisson equation,

$$\frac{\mathrm{d}^2 v}{\mathrm{d}x^2} = g,$$

subject to the same homogeneous boundary conditions. The adjoint identity is easily verified, taking into account that u and v are zero at each end.

$$\int_0^1 v\, \frac{\mathrm{d}^2 u}{\mathrm{d}x^2}\, \mathrm{d}x = -\int_0^1 \frac{\mathrm{d}v}{\mathrm{d}x}\, \frac{\mathrm{d}u}{\mathrm{d}x}\, \mathrm{d}x = \int_0^1 \frac{\mathrm{d}^2 v}{\mathrm{d}^2 x}\, u\, \mathrm{d}x.$$

The Poisson equation is approximated numerically on a uniform grid, with spacing h, using a second order finite difference discretisation,

$$h^{-2}\delta_x^2 u_j = f(x_j).$$

The approximate solution $u_h(x)$ is then defined by cubic spline interpolation through the nodal values u_j. The adjoint solution v_h is obtained in exactly the same manner.

Numerical results have been obtained for the case

$$f = x^3(1-x)^3, \qquad g = \sin(\pi x).$$

Figure 1 shows the residual error $Lu_h - f$ when $h = \frac{1}{32}$, as well as the three Gaussian quadrature points on each sub-interval that are used in the numerical integration of the inner product $(v_h, Lu_h - f)$. Since u_h is a cubic spline, $f_h \equiv \frac{\mathrm{d}^2 u_h}{\mathrm{d}x^2}$ is continuous and piecewise linear. The best piecewise linear approximation to f has an approximation error whose dominant term is quadratic on each sub-interval; this explains the scalloped shape of the residual error. Figure 2 shows the approximate adjoint solution v_h, which simply illustrates that the residual error in the center of the domain contributes most to the overall error in the functional.

Figure 3 is a log-log plot of three quantities versus the number of cells: the error in the base value of the functional (g, u_h); the remaining error after subtracting the adjoint correction term $(v_h, Lu_h - f)$; the *a posteriori* error bound $\|L^{-1}\| \, \|Lu_h - f\| \, \|L^* v_h - g\|$. The superimposed lines have slopes of -2 and -4, confirming that the base solution is second order accurate while the error in the corrected functional and the error bound are both fourth order. It is also worth noting that on a grid with 16 cells, which might be a reasonable choice for practical computations, the error in the corrected value of the functional is over 200 times smaller than the uncorrected error.

2.4 Second example: 2D Poisson equation

The second example is the two-dimensional Poisson equation,

$$\nabla^2 u = f,$$

on the unit square $[0,1] \times [0,1]$ subject to homogeneous Dirichlet boundary conditions. The dual problem is

$$\nabla^2 v = g,$$

with the same boundary conditions, and the adjoint identity is again easily verified,

$$\int_\Omega v \; \nabla^2 u \; \mathrm{d}A = -\int_\Omega \nabla v \cdot \nabla u \; \mathrm{d}A = \int_\Omega \nabla^2 v \; u \; \mathrm{d}A.$$

For this example, the equations are approximated using a Galerkin finite element method with piecewise bilinear elements on a uniform Cartesian grid. Finite element error analysis reveals that the solution error for the primal problem, and the error in the computed functional using the finite element solution are both $O(h^2)$. However, using bi-cubic spline interpolation through the computed nodal values, one can reconstruct an improved approximate solution $u_h(x,y)$ with an error that is still $O(h^2)$, but much smoother, so that the residual error is also $O(h^2)$. Using a similarly reconstructed approximate adjoint solution $v_h(x,y)$, one can then compute the adjoint error correction term resulting in a corrected functional whose accuracy is $O(h^4)$. All inner product integrals are approximated by 3×3 Gaussian quadrature on each square cell to ensure that the numerical quadrature errors are of a higher order.

Figure 4 shows the numerical results obtained for the functions

$$f(x,y) = x(1-x)y(1-y), \qquad g(x,y) = \sin(\pi x)\,\sin(\pi y).$$

The ordinate is the log of the number of cells in each dimension, and lines of slope -2 and -4 are again superimposed. As predicted by the analysis, the

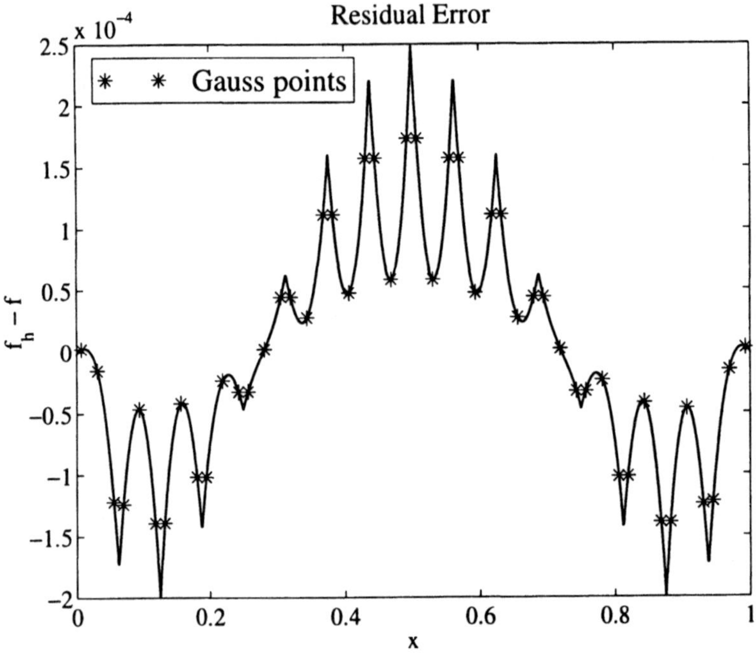

Fig. 1. Residual error for 1D Poisson equation.

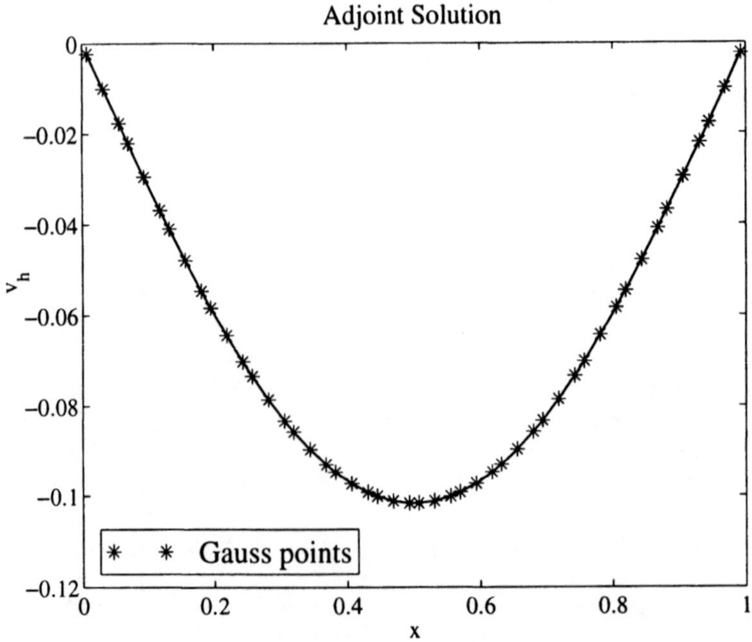

Fig. 2. Adjoint solution for 1D Poisson equation.

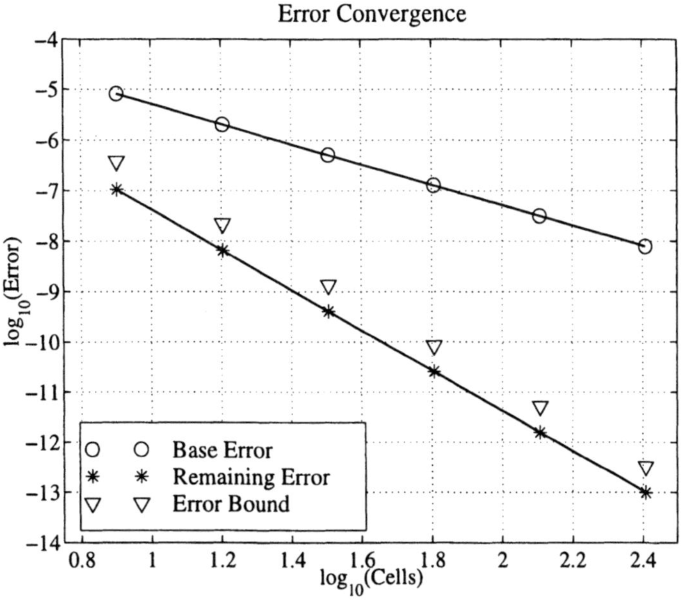

Fig. 3. Functional error convergence for 1D Poisson equation.

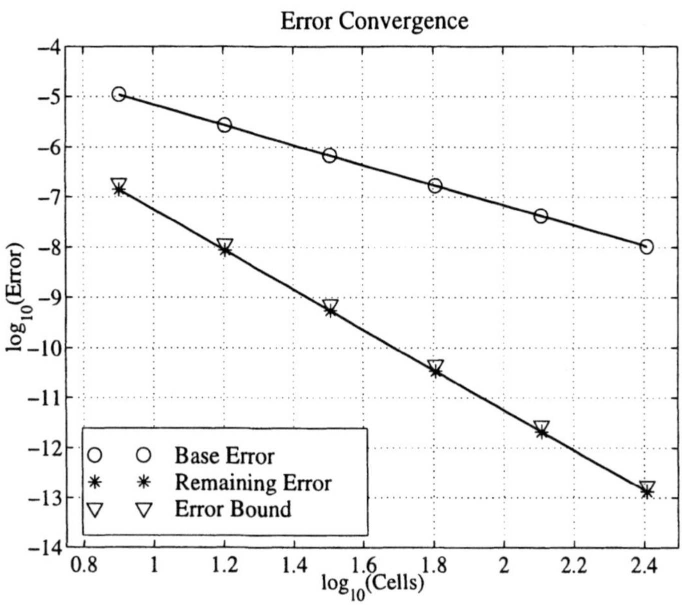

Fig. 4. Functional error convergence for 2D Poisson equation.

base error in the functional is clearly second order whereas the error in the corrected value of the functional as well as the error bound are again fourth order.

2.5 Theory with boundary terms

We now extend the theory to include inhomogeneous boundary conditions for the primal and dual problems, and boundary integrals in their output functionals.

Let u be the solution of the linear differential equation

$$Lu = f,$$

in the domain Ω, subject to the linear boundary conditions

$$Bu = e,$$

on the boundary $\partial\Omega$. In general, the dimension of the operator B may be different on different parts of the boundary (e.g. inflow and outflow sections for the convection p.d.e.).

The output functional of interest is taken to be

$$J = (g, u) + (h, Cu)_{\partial\Omega},$$

where $(.,.)_{\partial\Omega}$ represents an integral inner product over the boundary $\partial\Omega$. The boundary operator C may be algebraic (e.g. $Cu \equiv u$) or differential (e.g. $Cu \equiv \frac{\partial u}{\partial n}$), but must have the same dimension as the adjoint boundary condition operator B^* to be defined shortly. Note that the components of h may be set to zero if the functional does not have a boundary integral contribution.

The corresponding linear adjoint problem is

$$L^* v = g,$$

in Ω, subject to the boundary conditions

$$B^* v = h,$$

on the boundary $\partial\Omega$. The fundamental identity defining L^*, B^* and the boundary operator C^* is

$$(v, Lu) + (C^* v, Bu)_{\partial\Omega} = (L^* v, u) + (B^* v, Cu)_{\partial\Omega}, \qquad (5)$$

for all u, v. This identity is obtained by integration by parts. Examples will be given later, but see also [GP97] for the construction of the appropriate adjoint operators for the linearised Euler and Navier-Stokes equations.

Using the adjoint identity, one immediately obtains the equivalent dual form of the output functional,

$$J = (v, f) + (C^* v, e)_{\partial\Omega}.$$

Given approximate solutions u_h, v_h, we obtain the following result for the functional.

$$\begin{aligned}
(g, u) + (h, Cu)_{\partial\Omega} &= (g, u_h) + (h, Cu_h)_{\partial\Omega} \\
&\quad -(L^* v_h, u_h - u) - (B^* v_h, C(u_h - u))_{\partial\Omega} \\
&\quad +(L^* v_h - g, u_h - u) + (B^* v_h - h, C(u_h - u))_{\partial\Omega} \\[2ex]
&= (g, u_h) + (h, Cu_h)_{\partial\Omega} \\
&\quad -(v_h, L(u_h - u)) - (C^* v_h, B(u_h - u))_{\partial\Omega} \\
&\quad +(L^*(v_h - v), u_h - u) + (B^*(v_h - v), C(u_h - u))_{\partial\Omega} \\[2ex]
&= (g, u_h) + (h, Cu_h)_{\partial\Omega} \\
&\quad -(v_h, Lu_h - f) - (C^* v_h, Bu_h - e)_{\partial\Omega} \\
&\quad +(v_h - v, L(u_h - u)) + (C^*(v_h - v), B(u_h - u))_{\partial\Omega}.
\end{aligned}$$

In the final result, the first line is the functional based on the approximate solution u_h. The second line is the computable adjoint error correction that now includes a term related to the residual error in satisfying the primal boundary conditions. The third line is the remaining error. In principle, an *a posteriori* error bound for this can again be found, but the main point is that each of the terms involves an inner product of two small quantities, so we again have the result that the corrected functional is superconvergent relative to the underlying solutions to the primal and adjoint problems.

2.6 Example: 2D Laplace equation

For the 2D Laplace equation

$$\nabla^2 u = 0,$$

with Dirichlet boundary conditions $u = e$, and functional

$$J = \int_{\partial\Omega} h \frac{\partial u}{\partial n} \, \mathrm{d}s,$$

we have the operators

$$Lu = \nabla^2 u, \qquad Bu = u, \qquad Cu = \frac{\partial u}{\partial n}.$$

Integrating by parts gives

$$\begin{aligned}
\int_\Omega v \; Lu \; \mathrm{d}A &= -\int_\Omega \nabla v \cdot \nabla u \; \mathrm{d}A + \int_{\partial\Omega} v \frac{\partial u}{\partial n} \, \mathrm{d}s \\
&= \int_\Omega \nabla^2 v \; u \; \mathrm{d}A + \int_{\partial\Omega} \left(v \frac{\partial u}{\partial n} - \frac{\partial u}{\partial n} v \right) \, \mathrm{d}s,
\end{aligned}$$

so the adjoint identity is satisfied for all u, v if we define

$$L^* v = \nabla^2 v, \qquad B^* v = v, \qquad C^* v = \frac{\partial v}{\partial n}.$$

To construct an analytic testcase with curved boundaries and a singularity in the solution, we use a conformal mapping. We start by defining the domain in a complex Z-plane to be the region between two circles centered at $(X, Y) = (-0.1, 0)$ with radii of $R_1 = 1.1$ and $R_2 = 3.0$. Application of the Joukowski mapping

$$z = Z + \frac{1}{Z},$$

then produces a computational domain between a cusped airfoil ($\partial\Omega_{z1}$) and a smooth outer boundary ($\partial\Omega_{z2}$). Using cylindrical coordinates R, θ defined by

$$X + 0.1 = R\cos\theta, \quad Y = R\sin\theta,$$

the function

$$U(X, Y) = \frac{R^2 - R_1^2}{R} \sin\theta,$$

is a solution of the Laplace equation in the Z-plane. Furthermore, by a well-known feature of conformal mappings, the function $u(x, y) = U(X, Y)$ is also a solution of the Laplace equation in the z-plane.

Evaluating $u(x, y)$ on the inner and outer boundaries gives the Dirichlet boundary condition for the test problem. As illustrated in Figure 6, the solution corresponds to the stream function for incompressible inviscid flow around the airfoil, with zero circulation.

The functional, expressed in the Z-plane, is chosen to be

$$\int_0^{2\pi} \sin\theta \left.\frac{\partial U}{\partial n}\right|_{R=R_1} d\theta.$$

Its analytic value is -2π. When mapped into the z-plane, the corresponding expression for the functional is

$$\left(h, \frac{\partial u}{\partial n}\right)_{\partial\Omega},$$

where $h = R_1^{-1}\sin\theta$, on the inner boundary $\partial\Omega_{z1}$, and $h = 0$ on the outer boundary. Hence the dual problem is the Laplace equation subject to the Dirichlet boundary conditions $v = h$.

The numerical results for both the primal and dual problems are calculated using the bilinear Galerkin finite element method. Figure 5 illustrates one of the grids used. The grid points in the z-plane are generated by the conformal mapping of a regular polar grid in the Z-plane, but the use of isoparametric elements in the z-plane means the cells in the z-plane are quadrilaterals, and do not have curved edges.

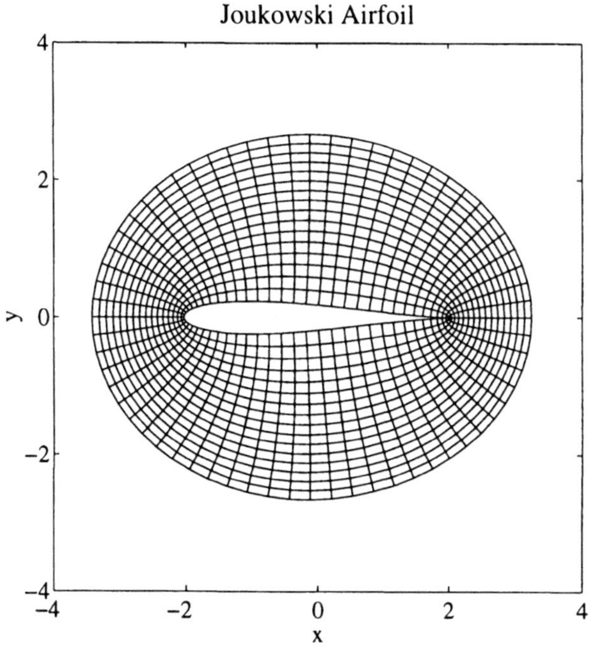

Fig. 5. The computational grid for a 2D Laplace problem around a Joukowski airfoil.

Figure 6 presents the primal and dual solutions. It can be seen that the gradient of the adjoint solution is singular at the cusped trailing edge of the airfoil.

The approximate solutions u_h and v_h are formed by bi-cubic spline interpolation. The coordinate data is also splined, so that the solutions and coordinates, u_h, v_h, x_h, y_h, are all defined parametrically as functions of the two spline coordinates ξ, η. Given these, ∇u_h can be calculated from

$$\begin{pmatrix} \frac{\partial u_h}{\partial \xi} \\ \\ \frac{\partial u_h}{\partial \eta} \end{pmatrix} = \begin{pmatrix} \frac{\partial x_h}{\partial \xi} & \frac{\partial y_h}{\partial \xi} \\ \\ \frac{\partial x_h}{\partial \eta} & \frac{\partial y_h}{\partial \eta} \end{pmatrix} \begin{pmatrix} \frac{\partial u_h}{\partial x} \\ \\ \frac{\partial u_h}{\partial y} \end{pmatrix}$$

and $\nabla^2 u_h$ can be computed similarly.

The error correction integral, which in this case is simply

$$(v_h, \nabla^2 u_h)$$

is evaluated by transforming it into an integral over (ξ, η), and then using 3×3 Gaussian quadrature on each quadrilateral cell.

The errors in the functional are shown in Figure 7. The superimposed lines of slope -2 and -4 show that the base solution is again second order accurate

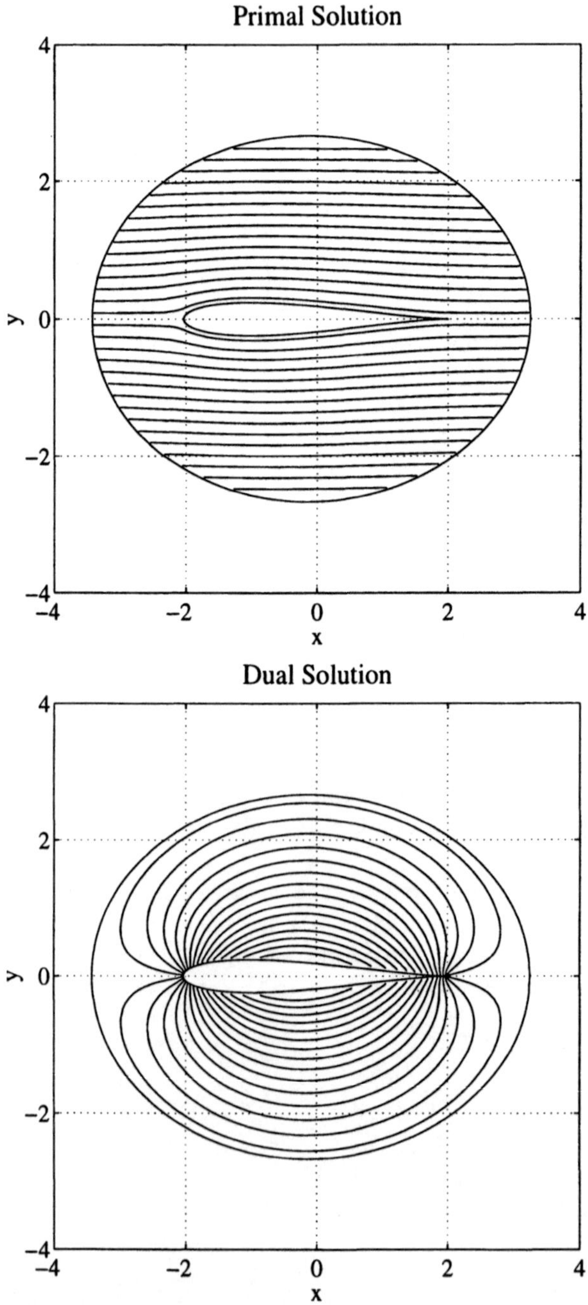

Fig. 6. The reconstructed primal and dual solutions for a 2D Laplace problem around a Joukowski airfoil.

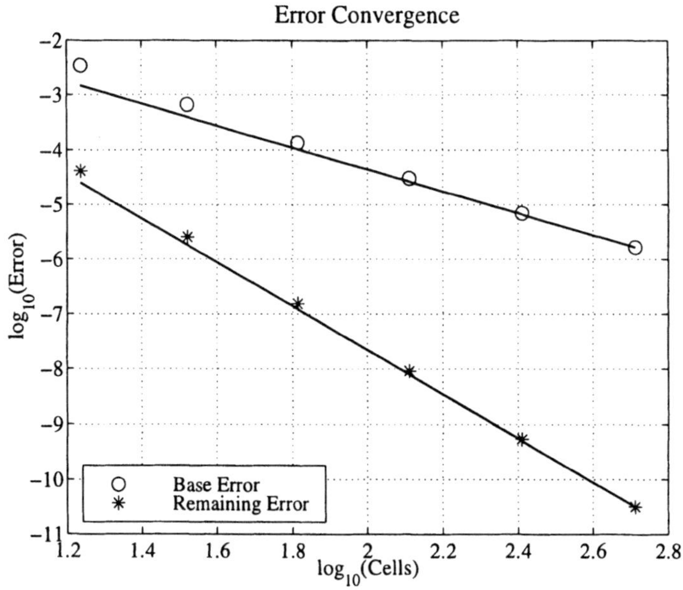

Fig. 7. Error convergence of a boundary functional for a 2D Laplace problem around a Joukowski airfoil.

whereas the corrected value for the functional is fourth order accurate. This improvement is achieved despite the presence of the singularity at the trailing edge, and the added complications of the curved boundaries and the boundary functional.

3 Linear defect error correction

Adjoint error correction is not the only means of improving the accuracy of numerical calculations. In this section, based on Reference [Gil00], we look at the use of defect correction [BMM88,Kor88,Ske81,Ste78], and show that it can be extremely effective in reducing the errors in a model 1D Helmholtz problem; the combination of defect and adjoint error correction is even better.

The primary motivation for this investigation is the need for high order accuracy for aeroacoustic and electromagnetics calculations. In steady CFD calculations, grid adaptation can be used to provide high grid resolution in the limited areas that require it. However, using standard second order accurate methods, the wave-like nature of aeroacoustic and electromagnetic solutions would lead to grid refinement throughout the computational domain in order to reduce the wave dispersion and dissipation to acceptable levels. The preferable alternative is to use higher order methods, allowing one to use fewer points per wavelength, which can lead to a very substantial reduction in the total number of grid points for 3D calculations. The difficulty with this is that one often wants to use unstructured grids because of their geometric flexibility, and the construction of higher order approximations on unstructured grids is complicated and computationally expensive.

3.1 Problem description and Galerkin method

The model problem to be solved is the 1D Helmholtz equation

$$u'' + \pi^2 u = 0, \quad 0 < x < 10,$$

subject to the Dirichlet boundary condition $u=1$ at $x=0$ and the radiation boundary condition $u' - \mathrm{i}\pi u = 0$ at $x=10$. The analytic solution is $u = \exp(\mathrm{i}\pi x)$ and the domain contains precisely five wavelengths. The output functional of interest is the value $u(10)$ at the right hand boundary. This can be viewed as a model of a far-field boundary integral giving the radiated acoustic energy in aeroacoustics, or the radar cross-section in electromagnetics [MS98].

Integrating by parts, the weak form of the inhomogeneous equation

$$u'' + \pi^2 u = f, \quad 0 < x < 10,$$

subject to the same boundary conditions is

$$-(w', u') + \pi^2 (w, u) + \mathrm{i}\pi w^*(10)\, u(10) = (w, f),$$

for any differentiable $w(x)$ with $w(0) = 0$. One important feature of this Helmholtz problem is that the solution is complex. Therefore the inner product (w, u) is defined as

$$(w, u) \equiv \int_0^{10} w^* u \; \mathrm{d}x,$$

with w^* denoting the complex conjugate of w.

The Galerkin solution on the irregular grid $x_j, j = 0, 1, 2, \ldots, N$, is defined as

$$U(x) = \sum_{j=0}^{N} U_j \phi_j(x)$$

where the $\phi_j(x)$ are the usual piecewise linear 'hat' functions for which $\phi_j(x_i) = \delta_{ij}$. The value U_0 is given by the Dirichlet boundary condition. The values of the other coefficients U_j for $j > 0$ are obtained from the equations

$$-(\phi_i', U') + \pi^2 (\phi_i, U) + \mathrm{i}\pi\phi_i(10)\, U(10) = 0, \quad i = 1, 2, \ldots, N.$$

It is well established that this discretisation is second order accurate, producing dispersion but no dissipation on a uniform grid.

3.2 Defect correction

The first step in the defect correction is to define a new approximate solution $u_h(x)$ by cubic spline interpolation of the nodal values U_j. The choice of end conditions for the cubic spline is very important. A natural cubic spline would have $u_h''=0$ at both ends, but this would introduce small errors at each end since $u''\neq 0$ for the analytic solution. Instead, at $x=10$ we require the splined solution to satisfy the analytic boundary condition by imposing $u_h' - i\pi u_h = 0$. At $x=0$, the analytic boundary condition is already imposed through having the correct value for the end point $U(0)$. Therefore, here we require that $u_h'' + \pi^2 u_h = 0$ so the splined solution satisfies the o.d.e. at the boundary.

The solution error, $e = u(x) - u_h(x)$ satisfies the inhomogeneous Helmholtz equation

$$e'' + \pi^2 e = -(u_h'' + \pi^2 u_h), \quad 0 < x < 10,$$

the right-hand-side of which is the residual error of the approximation $u_h(x)$. Given the homogeneous Dirichlet boundary condition at $x=0$, and the same radiation boundary condition at $x=10$, the Galerkin approximation to the error is given by the equations

$$-(\phi_i', E') + \pi^2(\phi_i, E) + \mathrm{i}\pi\phi_i(10)\,E(10) = -(\phi_i, u_h'' + \pi^2 u_h), \quad i = 1, 2, \ldots, N.$$

Adding the nodal corrections E_j to the original nodal values U_j gives a corrected solution. The whole procedure can then be repeated to improve the accuracy. This follows the procedure described by Barrett *et al* who also showed that it converges to a solution of an appropriately defined Petrov-Galerkin discretisation, with the trial space being the space of cubic splines, while the test space is the space of piecewise linear functions [BMM88].

3.3 Adjoint error correction

To apply the linear theory to the Helmholtz problem, the first step is to construct the appropriate adjoint problem. Integration by parts reveals that the Helmholtz equation is self-adjoint, so

$$L^* v \equiv v'' + \pi^2 v,$$

and

$$(v, Lu) - (L^* v, u) = \left[\boldsymbol{v}^H \boldsymbol{A} \boldsymbol{u} \right]_0^{10},$$

where

$$\boldsymbol{u} = \begin{pmatrix} u \\ \frac{\mathrm{d}u}{\mathrm{d}x} \end{pmatrix}, \quad \boldsymbol{v} = \begin{pmatrix} v \\ \frac{\mathrm{d}v}{\mathrm{d}x} \end{pmatrix},$$

and

$$\boldsymbol{A} = \begin{pmatrix} 0 & 1 \\ -1 & 0 \end{pmatrix}.$$

At $x = 10$ we have

$$Bu \equiv u' - \mathrm{i}\pi u \equiv \boldsymbol{B}\boldsymbol{u}, \quad \boldsymbol{B} = (-\mathrm{i}\pi \;\; 1),$$

and

$$Cu \equiv u \equiv \boldsymbol{C}\boldsymbol{u}, \quad \boldsymbol{C} = (1 \;\; 0).$$

To satisfy the adjoint identity (5) we require $\boldsymbol{B}^*$ and $\boldsymbol{C}^*$ such that

$$\boldsymbol{A} = \left(\frac{-\boldsymbol{C}^*}{\boldsymbol{B}^*}\right)^H \left(\frac{\boldsymbol{B}}{\boldsymbol{C}}\right).$$

Solving this gives

$$\left(\frac{-\boldsymbol{C}^*}{\boldsymbol{B}^*}\right) = \left(\frac{\boldsymbol{B}}{\boldsymbol{C}}\right)^{-H} \boldsymbol{A}^H = \begin{pmatrix} 1 & 0 \\ -\mathrm{i}\pi & -1 \end{pmatrix}$$

and hence $B^*v \equiv -v' - \mathrm{i}\pi v$ and $C^*v \equiv -v$. Similarly, at $x = 0$, we obtain $B^*v = v$ and $C^*v = v'$.

Now, noting that in our application $f = g = 0$, and h has value 0 at $x = 0$ and 1 at $x = 10$, then the full specification of the adjoint problem is

$$v'' + \pi^2 v = 0, \quad 0 < x < 10,$$

with $v = 0$ at $x = 0$ and $-v' - \mathrm{i}\pi v = 1$ at $x = 10$.

Let v_h be an approximate solution of this problem, obtained by the same Galerkin and cubic spline reconstruction approach as u_h, with or without defect correction. Noting that the cubic spline reconstruction ensures that the boundary conditions are satisfied exactly, the corrected approximation to the value $u(10)$ is

$$u_h(10) - (v_h, u_h'' + \pi^2 u_h).$$

The theory gives the error in this corrected functional as being

$$(v_h - v, u_h'' + \pi^2 u_h).$$

In the absence of defect correction, both terms in this inner product are second order in the average grid spacing and so the error is fourth order. With defect correction, the first term is fourth order while the second term remains second order. Therefore, the error remaining after the adjoint error correction is sixth order.

3.4 Numerical results

Numerical results have been obtained for grids with 4, 8, 16, 32, 64 and 128 points per wavelength. To test the ability to cope with irregular grids, the coordinates for the grid with N intervals are defined as

$$x_0 = 0, \quad x_N = 10, \quad x_j = \frac{10}{N}(j + \sigma_j), \; 0 < j < N,$$

where σ_j is a uniformly distributed random variable in the range $[-0.3, 0.3]$.

Figure 1 shows the L_2 norm of the error in the reconstructed cubic spline solution before and after defect correction. Without defect correction, the error is second order, while with defect correction it is fourth order. Note that a second application of defect correction makes a significant reduction in the error even though it remains fourth order. This is because one application of the defect correction procedure gives a correction that is second order in magnitude, with a corresponding error that is second order in relative magnitude and therefore fourth order in absolute magnitude. It is this error that is corrected by a second application of the defect correction procedure.

Figure 2 shows the error in the numerical value for the output functional $u(10)$. Without any correction, the error is second order. Using either defect correction or adjoint error correction on their own increases the order of accuracy to fourth order, but using them both increases the accuracy to sixth order. Note that the calculation with 8 points per wavelength plus both defect and adjoint error correction gives an error which is approximately 2×10^{-3}. This is more accurate than the calculation with 128 points per wavelength and no corrections, and comparable to the results using 14 points and defect correction, or 30 points with adjoint error correction.

In 3D, the computational cost is proportional to the cube of the number of points per wavelength, so this indicates the potentially huge savings offered by the combination of defect and adjoint error correction. The cost of computing the corrections is five times the cost of the original calculation, due to the additional two calculations for the defect correction, and the one adjoint calculation plus its two defect corrections. In practice, the second defect correction for the primal and adjoint calculations make negligible difference to the value obtained after the adjoint error correction, so these can be omitted, reducing the cost of the corrections to just three times the cost of the original calculation.

4 Nonlinear adjoint error correction

This section looks at the extension from the linear theory to handle nonlinear problems. It begins with some preliminaries that address the key issues in linearising nonlinear functions and operators.

4.1 Preliminaries

If u is a scalar variable and $f(u)$ is a nonlinear scalar function then a standard Taylor series expansion gives

$$f(u_2) = f(u_1) + f'(u_1)\,(u_2 - u_1) + O((u_2 - u_1)^2).$$

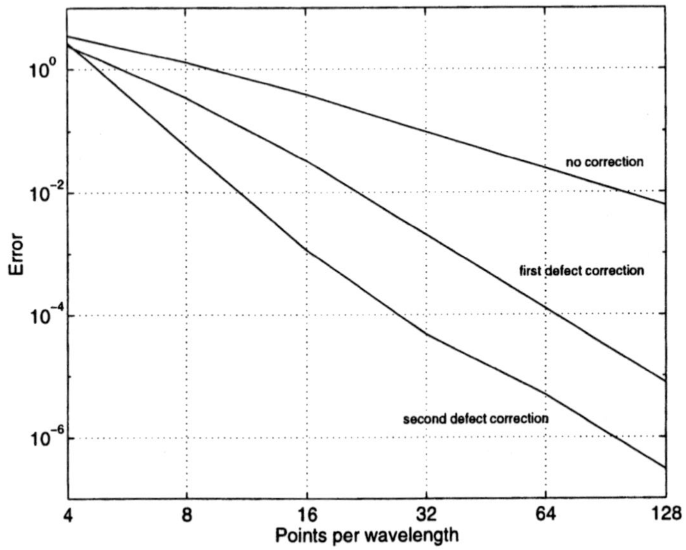

Fig. 8. L_2 error in the numerical approximation to $u(x)$

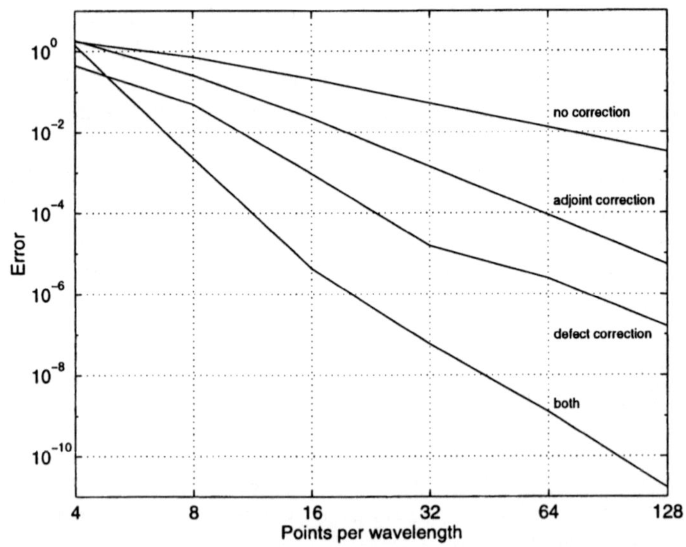

Fig. 9. Error in the numerical approximation to $u(10)$

However, one can obtain an exact expression without any remainder terms by starting from

$$\frac{d}{d\theta} f\left(u_1+\theta(u_2-u_1)\right) = f'\left(u_1+\theta(u_2-u_1)\right)\ (u_2 - u_1),$$

and then integrating this from $\theta=0$ to $\theta=1$ to obtain

$$f(u_2) - f(u_1) = \overline{f'}_{(u_1,u_2)}\ (u_2 - u_1),$$

where

$$\overline{f'}_{(u_1,u_2)} \equiv \int_0^1 f'\left(u_1+\theta(u_2-u_1)\right)\ d\theta.$$

If u and f are vectors, we need to define the Jacobian matrix

$$A_u = \left.\frac{\partial f}{\partial u}\right|_u,$$

with the subscript u denoting the fact that the value of the Jacobian matrix depends on the value of u around which $f(u)$ is linearised. We then obtain

$$\frac{d}{d\theta} f\left(u_1+\theta(u_2-u_1)\right) = A_{u_1+\theta(u_2-u_1)}\ (u_2-u_1)$$

so integrating over θ gives

$$f(u_2) - f(u_1) = \overline{A}_{(u_1,u_2)}\ (u_2-u_1),$$

where

$$\overline{A}_{(u_1,u_2)} = \int_0^1 \left.\frac{\partial f}{\partial u}\right|_{u_1+\theta(u_2-u_1)} d\theta.$$

The next step is to consider a nonlinear operator $N(u)$. The linearised operator L_u is called a Fréchet derivative, and it is formally defined by

$$L_u\tilde{u} \equiv \lim_{\varepsilon\to 0} \frac{N(u+\varepsilon\tilde{u}) - N(u)}{\varepsilon}$$

Again the subscript u denotes the fact that the linear operator matrix depends on the value of u around which $N(u)$ is linearised. For example, if

$$N(u) = \frac{\partial}{\partial x}\left(\frac{1}{2}u^2\right) - \nu\,\frac{\partial^2 u}{\partial x^2}$$

then

$$L_u\tilde{u} = \frac{\partial}{\partial x}\left(u\,\tilde{u}\right) - \nu\,\frac{\partial^2 \tilde{u}}{\partial x^2}$$

The final step in these preliminaries is to start from

$$\frac{d}{d\theta} N\left(u_1+\theta(u_2-u_1)\right) = L_{u_1+\theta(u_2-u_1)}\ (u_2-u_1)$$

and then integrate over θ to obtain

$$N(u_2) - N(u_1) = \overline{L}_{(u_1,u_2)}\ (u_2 - u_1),$$

where

$$\overline{L}_{(u_1,u_2)} = \int_0^1 L|_{u_1+\theta(u_2-u_1)}\ \mathrm{d}\theta.$$

Thus $\overline{L}_{(u_1,u_2)}$ is the average value of the linear operator L_u over the "path" from u_1 to u_2.

4.2 Nonlinear theory

Let u be the solution of the nonlinear differential equation

$$N(u) = 0,$$

in the domain Ω, subject to the nonlinear boundary conditions

$$D(u) = 0,$$

on the boundary $\partial\Omega$.

The linear differential operators L_u and B_u are defined to be the Fréchet derivatives of N and D, respectively,

$$L_u\,\tilde{u} \equiv \lim_{\epsilon\to 0} \frac{N(u+\epsilon\tilde{u}) - N(u)}{\epsilon},$$

$$B_u\,\tilde{u} \equiv \lim_{\epsilon\to 0} \frac{D(u+\epsilon\tilde{u}) - D(u)}{\epsilon}.$$

It is assumed that the nonlinear functional of interest, $J(u)$, has a Fréchet derivative of the following form,

$$\lim_{\epsilon\to 0} \frac{J(u+\epsilon\tilde{u}) - J(u)}{\epsilon} = (g(u),\tilde{u}) + (h, C_u\tilde{u})_{\partial\Omega}.$$

Here the dimension of the operator C_u (which may be differential) is required to equal the dimension of the adjoint boundary operator B_u^*, to be defined shortly.

The corresponding linear adjoint problem is

$$L_u^* v = g(u)$$

in Ω, subject to the boundary conditions

$$B_u^* v = h$$

on the boundary $\partial\Omega$. The adjoint identity defining L_u^*, B_u^* and the boundary operator C_u^* is

$$(v, L_u\tilde{u}) + (C_u^* v, B_u\tilde{u})_{\partial\Omega} = (L_u^* v, \tilde{u}) + (B_u^* v, C_u\tilde{u})_{\partial\Omega}, \tag{6}$$

for all $\tilde{u}, v$.

We now consider approximate solutions u_h, v_h of the primal and dual problems, respectively. The analysis will use the quantities

$$L_{u_h}^* v_h, \qquad B_{u_h}^* v_h, \qquad C_{u_h}^* v_h.$$

Note that these can be evaluated since u_h and v_h are both known, whereas we would not be able to evaluate the Fréchet derivatives based on the unknown analytic solution u.

The analysis also requires averaged Fréchet derivatives defined by

$$\begin{aligned}
\overline{L}_{(u,u_h)} &= \int_0^1 L|_{u+\theta(u_h-u)}\, d\theta, \\
\overline{B}_{(u,u_h)} &= \int_0^1 B|_{u+\theta(u_h-u)}\, d\theta, \\
\overline{C}_{(u,u_h)} &= \int_0^1 C|_{u+\theta(u_h-u)}\, d\theta, \\
\overline{g}(u,u_h) &= \int_0^1 g(u+\theta(u_h-u))\, d\theta,
\end{aligned}$$

so that, as explained in the preliminaries,

$$\begin{aligned}
N(u_h) - N(u) &= \overline{L}_{(u,u_h)}\,(u_h-u), \\
D(u_h) - D(u) &= \overline{B}_{(u,u_h)}\,(u_h-u), \\
J(u_h) - J(u) &= (\overline{g}(u,u_h), u_h-u) + (h, \overline{C}_{(u,u_h)}(u_h-u))_{\partial\Omega}.
\end{aligned}$$

We now obtain the following:

$$
\begin{aligned}
J(u_h) - J(u) &= (\overline{g}(u,u_h), u_h - u) + (h, \overline{C}_{(u,u_h)}(u_h - u))_{\partial\Omega} \\
&= (L^*_{u_h} v_h, u_h - u) + (B^*_{u_h} v_h, C_{u_h}(u_h - u))_{\partial\Omega} \\
&\quad -(L^*_{u_h} v_h - \overline{g}(u,u_h), u_h - u) \\
&\quad -(h, (C_{u_h} - \overline{C}_{(u,u_h)})(u_h - u))_{\partial\Omega} \\
&\quad -(B^*_{u_h} v_h - h, C_{u_h}(u_h - u))_{\partial\Omega} \\
&= (v_h, L_{u_h}(u_h - u)) + (C^*_{u_h} v_h, B_{u_h}(u_h - u))_{\partial\Omega} \\
&\quad -(L^*_{u_h} v_h - \overline{g}(u,u_h), u_h - u) \\
&\quad -(h, (C_{u_h} - \overline{C}_{(u,u_h)})(u_h - u))_{\partial\Omega} \\
&\quad -(B^*_{u_h} v_h - h, C_{u_h}(u_h - u))_{\partial\Omega} \\
&= (v_h, \overline{L}_{(u,u_h)}(u_h - u)) + (C^*_{u_h} v_h, \overline{B}_{(u,u_h)}(u_h - u))_{\partial\Omega} \\
&\quad -(L^*_{u_h} v_h - \overline{g}(u,u_h), u_h - u) \\
&\quad -(h, (C_{u_h} - \overline{C}_{(u,u_h)})(u_h - u))_{\partial\Omega} \\
&\quad -(B^*_{u_h} v_h - h, C_{u_h}(u_h - u))_{\partial\Omega} \\
&\quad +(v_h, (L_{u_h} - \overline{L}_{(u,u_h)})(u_h - u)) \\
&\quad +(C^*_{u_h} v_h, (B_{u_h} - \overline{B}_{(u,u_h)})(u_h - u))_{\partial\Omega} \\
&= (v_h, N(u_h)) + (C^*_{u_h} v_h, D(u_h))_{\partial\Omega} \\
&\quad -(L^*_{u_h} v_h - \overline{g}(u,u_h), u_h - u) \\
&\quad -(h, (C_{u_h} - \overline{C}_{(u,u_h)})(u_h - u))_{\partial\Omega} \\
&\quad -(B^*_{u_h} v_h - h, C_{u_h}(u_h - u))_{\partial\Omega} \\
&\quad +(v_h, (L_{u_h} - \overline{L}_{(u,u_h)})(u_h - u)) \\
&\quad +(C^*_{u_h} v_h, (B_{u_h} - \overline{B}_{(u,u_h)})(u_h - u))_{\partial\Omega}.
\end{aligned}
$$

In the final result, the first line is the adjoint correction term taking into account the residual errors in satisfying both the p.d.e. and the boundary conditions. The other lines are the remaining errors, which include the consequences of nonlinearity in L, B, C and g as well as residual errors in approximating the adjoint problem.

If the solution errors for the nonlinear primal problem and the linear adjoint problem are of the same order, and they are both sufficiently smooth that the corresponding residual errors are also of the same order, then the order of accuracy of the functional approximation after making the adjoint correction is twice the order of the primal and adjoint solutions. However, rigorous *a priori* and *a posteriori* analysis of the remaining errors is much harder than in the linear case [PG98] and practical *a posteriori* error bounds have yet to be obtained for the quasi-1D and 2D Euler equations.

4.3 Quasi-1D Euler equations

The steady quasi-1D Euler equations for the flow of an ideal compressible fluid in a variable area duct are

$$\frac{\mathrm{d}}{\mathrm{d}x}(AF) - \frac{\mathrm{d}A}{\mathrm{d}x} P = 0,$$

where $A(x)$ is the cross-sectional area of the duct and U, F and P are defined as

$$U = \begin{pmatrix} \rho \\ \rho q \\ \rho E \end{pmatrix}, \qquad F = \begin{pmatrix} \rho q \\ \rho q^2 + p \\ \rho q H \end{pmatrix}, \qquad P = \begin{pmatrix} 0 \\ p \\ 0 \end{pmatrix}.$$

Here ρ is the density, q is the velocity, p is the pressure, E is the total energy and H is the stagnation enthalpy. The system is closed by the equation of state for an ideal gas.

The functional of interest is the 'lift'

$$J = \int p \, \mathrm{d}x.$$

The Fréchet derivative operator is

$$L_u \tilde{u} \;\equiv\; \frac{\mathrm{d}}{\mathrm{d}x}\left(A \frac{\partial F}{\partial u} \tilde{u} \right) - \frac{\mathrm{d}A}{\mathrm{d}x} \frac{\partial P}{\partial u} \tilde{u},$$

and therefore the corresponding adjoint equations are

$$L_u^* v \;\equiv\; -A \left(\frac{\partial F}{\partial u} \right)^T \frac{\mathrm{d}v}{\mathrm{d}x} - \frac{\mathrm{d}A}{\mathrm{d}x} \left(\frac{\partial P}{\partial u} \right)^T v \;=\; \left(\frac{\partial p}{\partial u} \right)^T.$$

The equations are approximated using a standard second order finite volume method with characteristic smoothing on a uniform computational grid. The linear adjoint problem is approximated by the so-called 'continuous' method, in which one discretises the analytic adjoint equations on the same uniform grid as the flow solution [AV99,Jam95]. In the alternative 'discrete' approach, one starts with the discretised nonlinear flow equations, linearises them and then uses the transpose of the linear matrix as the discrete adjoint operator [EP97]. Previous research has shown that both approaches produce approximate solutions which converge to the analytic adjoint solution, which has been determined in closed form for the quasi-1D Euler equations [GP01].

The approximate solution $u_h(x)$ is constructed from the discrete flow solution by cubic spline interpolation of the nodal values of the three components of the state vector U. Similarly, the approximate adjoint solution $v_h(x)$ is obtained by cubic spline interpolation of the nodal values of the three components of the discrete adjoint solution. The integrals that form the base value for the functional and the adjoint correction are approximated by 3-point Gaussian quadrature.

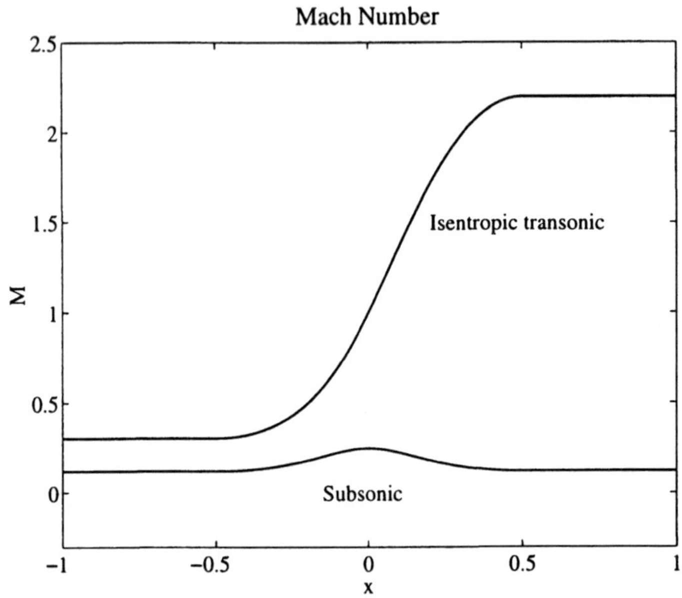

Fig. 10. Mach number distributions for quasi-1D Euler equation test cases.

Subsonic flow The first case is smooth subsonic flow in a converging-diverging duct corresponding to the Mach number distribution depicted in Figure 10. Figure 11 shows the error convergence for the computed functional. The superimposed lines of slope -2 and -4 show that the base error is second order whereas the error in the corrected functional is fourth order. This is in agreement with an *a priori* error analysis [PG98] that proves that $u_h - u$, $v_h - v$ and their first derivatives are all $O(h^2)$ for the particular finite volume scheme that is used, and hence the error in the corrected functional is $O(h^4)$.

Isentropic transonic flow Figure 12 shows the error convergence for a transonic flow in a converging-diverging duct corresponding to the Mach number distribution of Figure 10. The flow is subsonic at the inflow boundary and upstream of the throat (located at $x = 0$), and supersonic downstream of the throat and at the outflow boundary. Again the results show that the base error is second order while the remaining error after the adjoint correction is fourth order, even though there is logarithmic singularity in the adjoint solution at the throat [GP01].

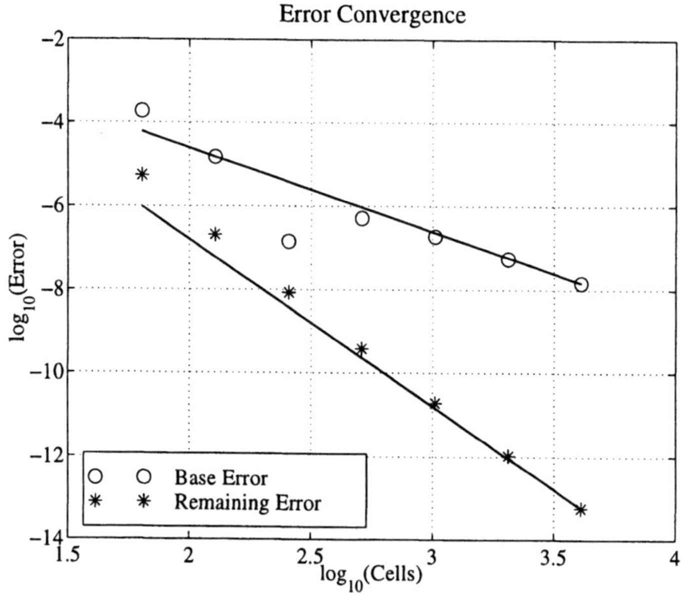

Fig. 11. Error convergence for quasi-1D subsonic flow.

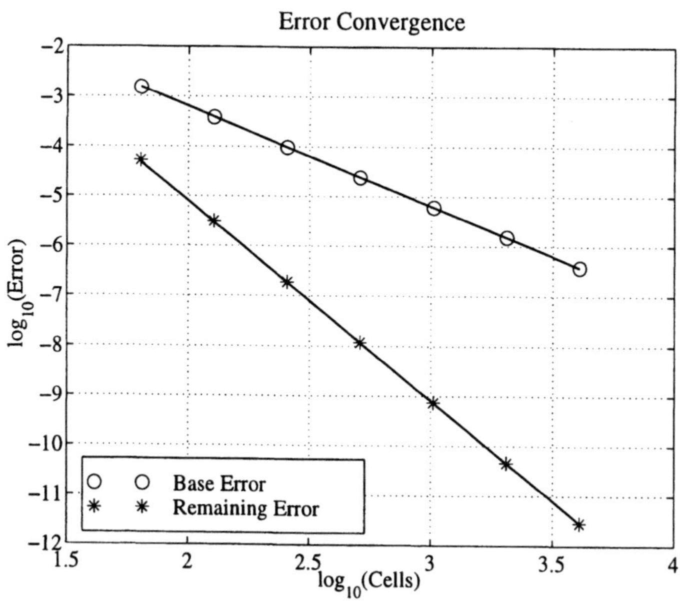

Fig. 12. Error convergence for quasi-1D shock-free transonic flow.

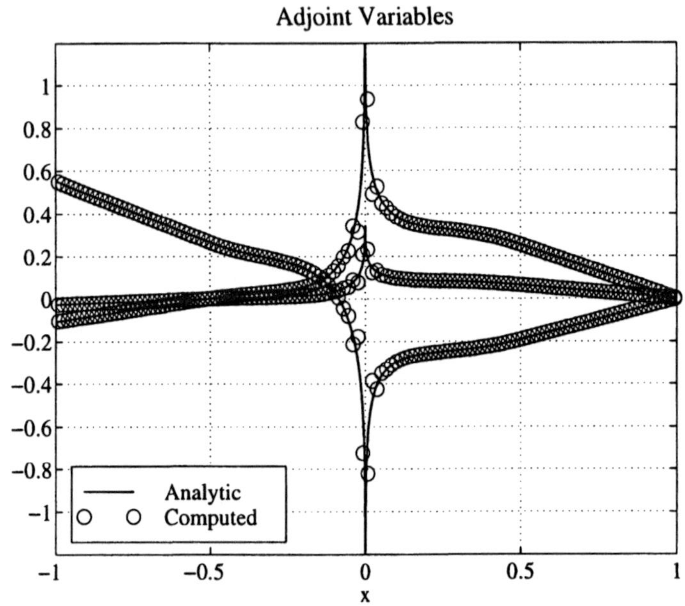

Fig. 13. Numerical and analytic adjoint solution for quasi-1D shock-free transonic flow.

4.4 Nonlinear thermal diffusion

The computational domain for this problem is the circular annulus $1 \leq r \leq 3$, and the p.d.e. is the nonlinear diffusion equation

$$\nabla \cdot (u \, \nabla u) = 0,$$

subject to the requirement that u is positive. Dirichlet boundary conditions are specified at the inner and outer boundaries so as to agree with the analytic solution

$$u(r,\theta) = \left(1 + \left(\frac{r}{4} - \frac{1}{r} \right) \cos\theta \right)^{1/2}.$$

The functional of interest is

$$J(u) = \int_0^{2\pi} \left. \frac{\partial u}{\partial n} \right|_{r=1} d\theta,$$

and the corresponding dual problem is

$$L_u^* v \equiv u \nabla^2 v = 0,$$

with Dirichlet boundary conditions of $1/u$ and 0 on the inner and outer boundaries, respectively.

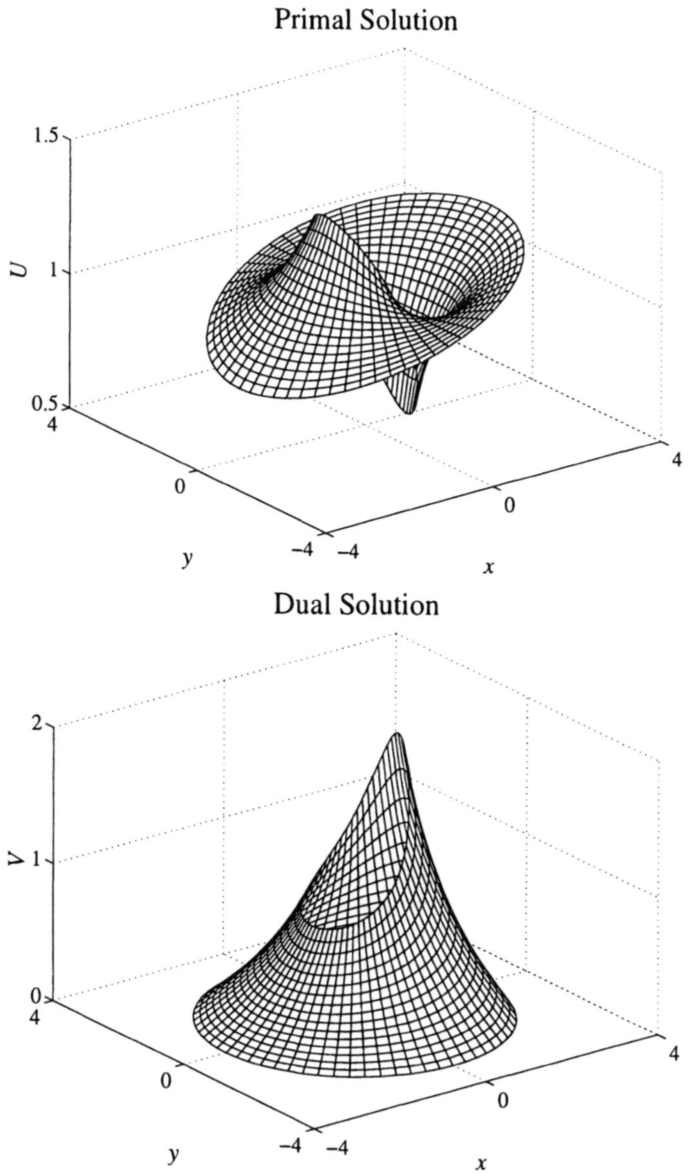

Fig. 14. The primal and solutions for a 2D nonlinear thermal diffusion problem.

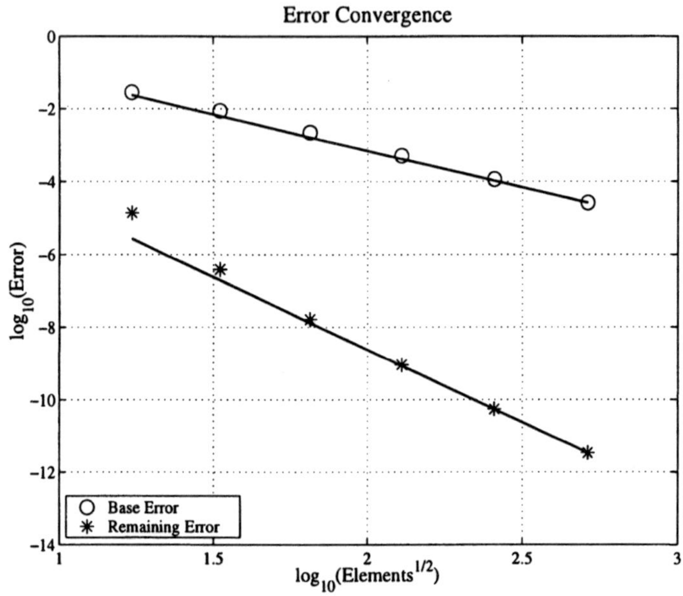

Fig. 15. Error convergence of the boundary functional for a 2D nonlinear thermal diffusion problem.

The primal and dual solutions shown in Figure 14 are obtained by a bilinear Galerkin finite element formulation using 3×3 Gaussian quadrature to evaluate the mass and stiffness matrices. The nonlinear equations are solved using a full approximation scheme multigrid method. Bi-cubic spline interpolation and 3x3 Gaussian quadrature are then used to calculate the functional with and without the adjoint correction.

The error in the functional is shown in Figure 15 on a log-log plot versus the square root of the total number of cells; this is a measure of h^{-1}, the inverse of the average mesh spacing. The superimposed lines of slopes -2 and -4 show second order accuracy for the basic finite element solution and fourth order accuracy after the inclusion of the adjoint error correction. For a 128×32 mesh, the error decreases by a factor of more than 10^5.

5 Grid adaptation

In this section we very briefly outline a number of possible strategies for grid adaptation based on *a posteriori* error estimates for output functionals. In each case we express reservations about the justification of using the strategy, while recognising that in the end an acceptable justification may be that it provides a robust refinement criterion in practice.

5.1 Option 1: magnitude of correction term

For nonlinear adjoint error correction, the dominant correction term is the interior integral inner product $(v_h, N(u_h))$, which can be expressed as a sum of contributions from each cell in the domain

$$(v_h, N(u_h)) \equiv \sum_{\alpha} (v_h, N(u_h))_{\alpha}.$$

One adaptive strategy is to subdivide those cells for which

$$(v_h, N(u_h))_{\alpha}$$

is bigger than some tolerance.

Müller and Giles have tried this approach for subsonic and transonic airfoil calculations [MG01], but the results are little better than using an *ad hoc* method based on pressure differences across each cell.

One questionable aspect of this strategy is that the purpose of the adjoint error correction is to evaluate and correct for this term, so what is the justification for trying to make it small? Is it not better to try to make the remaining error small?

5.2 Option 2: estimated remaining error term

After making the adjoint error correction, the main remaining error term can be expressed as

$$(v - v_h, N(u_h)) .$$

The difficulty with this expression is that the analytic adjoint solution v is not known. One option therefore is to estimate it and then adapt those cells in which

$$(v - v_h, N(u_h))_{\alpha}$$

is greater than some tolerance.

The problem is how to estimate v. Rannacher *et al* [BR96,BR01] use a quadratic reconstruction to estimate v, having used a piecewise linear finite element solution for v_h. However, if the quadratic reconstruction is a better approximation to v than the piecewise linear one, might it not be better to use the quadratic reconstruction as the approximate solution v_h and thereby get a more accurate adjoint error correction?

It might appear that another possible criticism of this approach is that it assumes that the solution error $v - v_h$ is primarily a local interpolation error, whereas it may be due to truncation errors in an entirely different part of the grid. However, for a Galerkin finite element method, because of orthogonality the quantity

$$(v - v_h, N(u_h))$$

has the same value for *any* v_h in the appropriate finite element space. Therefore, it is permissible to consider a different v_h which is an interpolant of v, so $v-v_h$ is then an interpolation error which can be estimated using the computed adjoint solution.

The approach used by Venditti & Darmofal [VD01] is an extension of that used by Rannacher *et al.* An alternative, approximately equivalent form for the dominant part of the remaining error is

$$(R_h, u-u_h),$$

where $R_h \equiv L^*_{u_h} v_h - g(u_h)$ is the residual error in satisfying the adjoint p.d.e. Therefore, they adapt any cell in which the sum

$$\left|(v-v_h, N(u_h))\right|_\alpha + \left|(R_h, u-u_h)\right|_\alpha$$

is greater than some threshold. The analytic solutions u and v are again approximated by a higher order reconstruction. Because they use a finite volume method to calculate u_h and v_h, the replacement of $v-v_h$ by a local interpolation error does not have the same theoretical justification as with the Galerkin method. On the other hand, it does seem an excellent idea to take into account the residual errors of the adjoint problem, and they do use the more accurate reconstruction of the approximate solution to obtain the correction to the value of the functional. The numerical results they obtain are very good; Figure 16, taken from [VD01], illustrates the results they obtain for a three-element airfoil calculation. It shows the adapted grid obtained using their refinement criterion, as well as the improved accuracy of the lift prediction with and without adjoint correction, compared to a sequence of uniformly-refined grids.

5.3 Option 3: coarse grid error estimates

Using the residual errors from both the original and adjoint problems, the dominant remaining error can be expressed as

$$\left(L^{*\,-1}_{u_h} R_h,\ N(u_h)\right) \approx \left(R_h,\ L^{-1}_{u_h} N(u_h)\right)$$

The problem with using this in an adaptive approach is that $L^{-1}_{u_h}$ is a global operator. However, it might be possible to use a coarse grid to approximately evaluate $L^{-1}_{u_h} N(u_h)$ and $L^{*\,-1}_{u_h} R_h$, and then adapt in any cell α for which

$$\left|\left(L^{*\,-1}_{u_h} R_h,\ N(u_h)\right)_\alpha\right| + \left|\left(R_h,\ L^{-1}_{u_h} N(u_h)\right)_\alpha\right|$$

is bigger than some tolerance. This approach may also give a useful *a posteriori* bound on the total remaining error.

The criticism that can be levelled at this idea is that it requires the calculation of the quantity $L^{-1}_{u_h} N(u_h)$ which is essentially a defect correction.

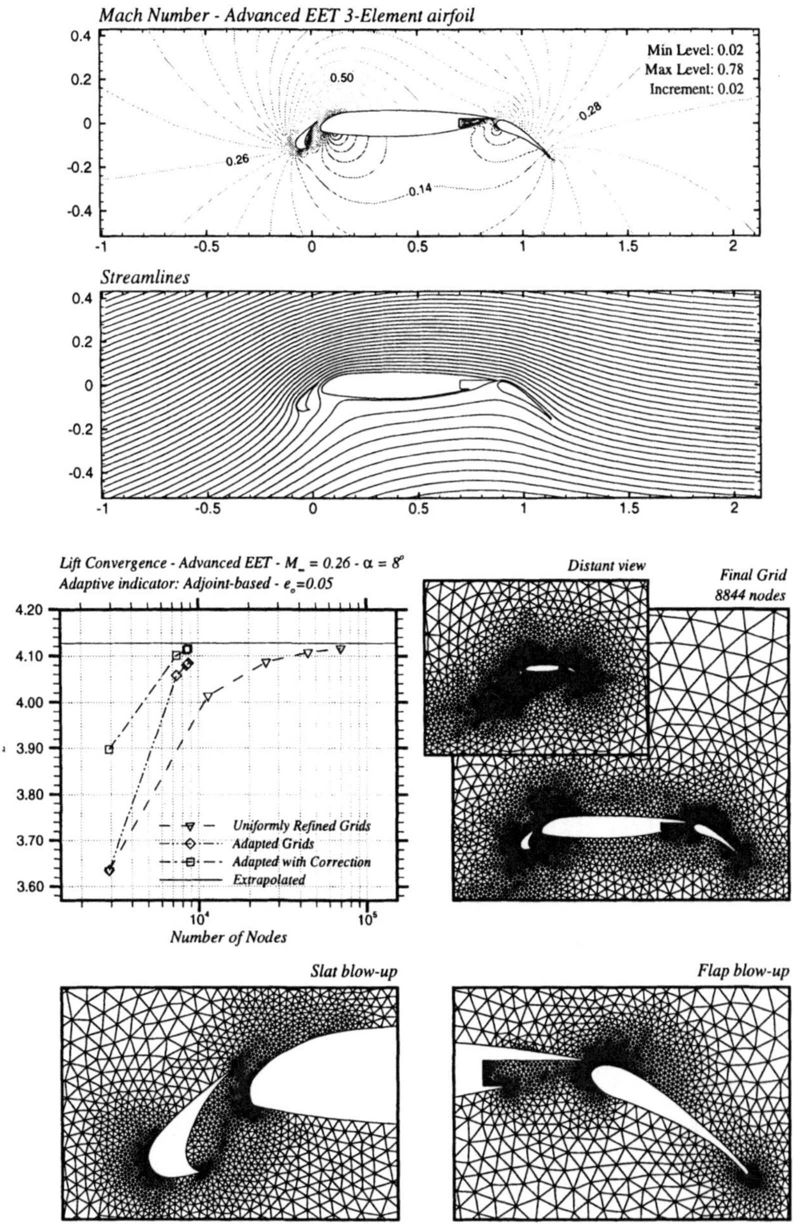

Fig. 16. Venditti / Darmofal test case (Figures 2 and 3 from [VD01]).

In that case, is it not better to use this to further improve the accuracy of the solution and hence the functional, rather than bound the remaining error?

Yet again, it appears there is a choice to be made between improved accuracy or a tight bound. If we choose the improved accuracy then we can still use the suggested measure as a refinement criterion, but we are refining based on the estimated magnitude of one part of the corrected error, rather than on the estimated remaining error. It may still prove to be a useful adaptation criterion, but its rational basis has been compromised.

5.4 Multiple functionals

A last comment concerns the situation in which there are several different functionals of concern, such as lift, drag and pitching moment. How should all of these be incorporated into the adaptation criterion?

One could perform a separate adjoint calculation for each. This would be necessary if one wished to perform an adjoint error correction to improve the accuracy of each of the output functionals. Alternatively, following options 1 and 2 above, one could adapt whenever

$$V_\alpha \int_\alpha |N(u_h)| \; \mathrm{d}A$$

exceeds some threshold, with V_α representing the magnitude of the typical adjoint solution in cell α (big near the airfoil, tending to zero far away) if following option 1, or a corresponding interpolation error estimate for option 2.

6 Future research directions

In this concluding section, we offer some thoughts about directions for future research. Some of these are specific to adjoint error correction, but most apply more generally to the subject of *a posteriori* error analysis and grid adaptation for functional outputs.

6.1 Grid adaptation

As indicated in the previous section, grid adaptation remains a topic for future research. Being practical, there is no need for a "perfect" adaptive strategy, even if one could define what that would be. The important criterion for success is that the strategy should be robust and produce consistently good results for a wide variety of problems. It is also important that for problems with singularities, for which it is known that the standard order of accuracy (expressed in terms of the number of nodes/cells used) can be recovered with the appropriate degree of local grid resolution, the adaptive strategy should automatically accomplish this.

This issue of singularities is particularly important because in two dimensions the adjoint solution to the Euler equations has a singularity along the stagnation streamline upstream of the leading edge of an airfoil [GP97], and another singularity at the trailing edge reflecting the sensitivity of the Kutta condition to changes in the trailing edge geometry. Thus, the grid adaptation must consider the needs of both the original nonlinear solution and the adjoint solution, unless one chooses to use different grids for each [GS02].

Two other adaptation issues to be addressed are anisotropic refinement and grid redistribution. Currently, most adaptive strategies use grid refinement, adding additional nodes/cells through an isotropic refinement process that locally refines equally in each direction, giving cells with a bounded aspect ratio. This is good for many applications, but far from ideal for others.

One example is the inviscid flow around a wing. Here the grid resolution normal to the leading edge needs to be much finer than the spanwise resolution. In this case, anisotropic refinement is probably the best solution. This means adding nodes in such a way that the resolution normal to the leading edge is greater than in the spanwise resolution. The question this poses is how to decide which direction requires additional resolution? There are already *ad hoc* refinement methods that address this. The challenge will be to extend the *a posteriori* adjoint-based refinement indicators to give a more quantitative answer to this question.

Another more extreme example of the need for anisotropic resolution is a boundary layer on a wing, where there is clearly a need for much better resolution across the boundary layer than in the other two directions. In this case, the best solution may well be grid redistribution, moving existing grid nodes to provide the resolution where it is needed. Again there are good existing *ad hoc* methods for doing this and the challenge is how to develop new methods using *a posteriori* error estimates.

6.2 Asymptotic error bounds

Ideally, we would like to have our cake and eat it too! We would like to use smooth reconstruction and adjoint error correction to generate extremely accurate functional values, and at the same time still be able to bound the remaining error with bounds that are at least asymptotically valid and fairly tight.

As suggested in the previous section, this may be unachievable. There may be a choice to be made between using smooth reconstructed solutions for defect and/or adjoint error correction, or for tight error bounds. If so, then our preference would be for the increased accuracy. However, this definitely merits further research.

6.3 Smooth reconstruction on unstructured grids

One key issue is going to be the smooth reconstruction of approximate solutions in multiple dimensions on unstructured grids. On a structured grid, cubic spline interpolation can be used in each direction, but on an unstructured grid there is a need for a suitable generalisation of cubic spline interpolation to produce a reconstructed solution of sufficient smoothness.

Venditti & Darmofal have achieved some success with a piecewise quadratic reconstruction using least squares minimization in the H_1 Sobolev norm [VD01]. In unpublished research, this local approximation has been sufficient for them to obtain results doubling the order of accuracy of functional outputs from a Galerkin approximation of a convection-diffusion equation.

However, there is reason to believe that in general the solution at the nodes of an unstructured grid may not be very smooth. In particular, the solution error,

$$e_i = u_i - u(x_i)$$

may not be very smooth, and therefore even if $e_i = O(h^2)$, any interpolation may lead to a reconstruction error

$$e(x) = u_h(x) - u(x)$$

for which $\nabla e = O(h)$. This loss of accuracy because of the loss of smoothness would completely negate the ability of the adjoint error correction to improve upon the accuracy of finite element solutions.

To avoid this, it seems likely that it will be necessary to use some form of smoothed spline reconstruction, in which the spline does not interpolate the nodal values, but instead compromises between the twin objectives of matching the computed data and maintaining smoothness. For example, if U_h is a second order accurate piecewise linear finite element solution, then the reconstruction u_h could be defined by some suitable approximation to the equation

$$h^2\nabla^2(\nabla^2 u_h) + u_h - U_h = 0.$$

The purpose of the bi-harmonic term is to ensure the smoothness of the solution. The h^2 weighting ensures that this is not achieved at the expense of sacrificing the second order accuracy of the underlying solution. Some preliminary numerical analysis for problems with periodic boundary conditions [GS02] confirms that this should have the desired smoothing effect, but it has yet to be implemented and tested.

6.4 Shocks

One last challenge we wish to highlight is the problem of shocks and other discontinuities.

With the quasi-1D Euler equations, it can be proved that with an appropriate conservative formulation, and a numerical discretisation that is second order accurate when the solution is smooth, the accuracy of output functionals such as the integrated pressure is also second order [Gil96]. However, numerical evidence suggests this is not the case in multiple dimensions, and instead there is an error in quantities such as the lift on a transonic airfoil that is proportional to the local grid spacing at the shock. Thus, to get even second order accuracy would require anisotropic grid adaptation so that the grid spacing at the shock is $O(h^2)$, with h here being the average grid spacing in the rest of the grid.

Application of adjoint error correction ideas raises another problem. The reconstructed solution will be continuous, whereas the true solution is discontinuous. Therefore, it is unavoidable that there is an $O(1)$ error in the approximate solution at the shock. This violates the whole basis for the adjoint error correction since it relies on a linearisation of the nonlinear equations that is valid only for small perturbations.

We are currently working on this problem. Our approach is to numerically approximate a "viscous" shock with the level of viscosity being $O(h^2)$. The adjoint error correction then has to correct for the numerical error in approximating the viscous shock, plus the analytic error in using the viscous shock problem to approximate the inviscid shock problem. This latter part requires the use of matched asymptotic expansions to understand that to leading order there is a linear dependence of integral functionals on the level of viscosity. This error can be compensated for by using the viscous adjoint to give the sensitivity of the lift to a change in the level of the viscosity.

Acknowledgements

In the course of the research discussed in these notes, we have benefited from stimulating discussions with many people, including in particular Endre Süli, Paul Houston, David Darmofal and Rolf Rannacher.

Our research has been supported by EPSRC grants GR/K91149 and GR/L95700, and by NASA/Ames Cooperative Agreement No. NCC2-5431.

References

[AB99] W.K. Anderson and D.L. Bonhaus. Airfoil design on unstructured grids for turbulent flows. *AIAA Journal*, 37(2):185–191, 1999.

[AV99] W.K. Anderson and V. Venkatakrishnan. Aerodynamic design optimization on unstructured grids with a continuous adjoint formulation. *Computers & Fluids*, 28(4-5):443–480, 1999.

[BCH+98] C.H. Bischof, A. Carle, P.D. Hovland, P. Khademi, and A. Mauer. ADIFOR 2.0 User's Guide (Revison D). Technical Report 192, Mathematics and Computer Science Division, Argonne National Laboratory, 1998.

[BE87] J.W. Barrett and C.M. Elliott. Total flux estimates for a finite-element approximation of elliptic equations. *IMA Journal of Numerical Analysis*, 7:129–148, 1987.

[BE92] O. Baysal and M. Eleshaky. Aerodynamic design optimization using sensitivity analysis and computational fluid dynamics. *AIAA Journal*, 30(3):718–725, 1992.

[BKR00] R. Becker, H. Kapp, and R. Rannacher. Adaptive finite element methods for optimal control of partial differential equations: basic concepts. *SIAM Journal of Control and Optimization*, 39:113–132, 2000.

[BM84a] I. Babuška and A. Miller. The post-processing approach in the finite element method – Part 1: calculation of displacements, stresses and other higher derivatives of the displacements. *International Journal for Numerical Methods in Engineering*, 20:1085–1109, 1984.

[BM84b] I. Babuška and A. Miller. The post-processing approach in the finite element method – part 2: the calculation of stress intensity factors. *International Journal for Numerical Methods in Engineering*, 20:1111–1129, 1984.

[BMM88] J.W. Barrett, G. Moore, and K.W. Morton. Optimal recovery in the finite element method, part 2: Defect correction for ordinary differential equations. *IMA Journal of Numerical Analysis*, 8:527–540, 1988.

[BR78a] I. Babuška and W.C. Rheinboldt. *A posteriori* error estimates for the finite element method. *International Journal for Numerical Methods in Engineering*, 12:1597–1615, 1978.

[BR78b] I. Babuška and W.C. Rheinboldt. Error estimates for adaptive finite element computations. *SIAM Journal of Numerical Analysis*, 15(4):736–754, 1978.

[BR96] R. Becker and R. Rannacher. A feed-back approach to error control in finite element methods: basic analysis and examples. *East-West Journal of Numerical Mathematics*, 4:237–264, 1996.

[BR98] R. Becker and R. Rannacher. Weighted *a posteriori* error control in finite element methods. In *et al* H.G. Block, editor, *Proceedings of ENUMATH-97*, pages 621–637. World Scientific Publishing, 1998.

[BR99] R. Braack and R. Rannacher. Adaptive finite element methods for low Mach number flows with chemical reactions. VKI Lecture Series 1999-03, 1999.

[BR01] R. Becker and R. Rannacher. An optimal control approach to error control and mesh adaptation. In A. Iserles, editor, *Acta Numerica 2001*. Cambridge University Press, 2001.

[CFG98] A. Carle, M. Fagan, and L.L. Green. Preliminary results from the application of automated code generation to CFL3D. AIAA Paper 98-4807, 1998.

[DG00] A. Dadone and B. Grossman. Progressive optimization of inverse fluid dynamic design problems. *Computers & Fluids*, 29(1), 2000.

[Ell98] J. Elliott. *Aerodynamic optimization based on the Euler and Navier-Stokes equations using unstructured grids.* PhD thesis, MIT Dept. of Aero. and Astro., 1998.

[EP97] J. Elliott and J. Peraire. Practical 3D aerodynamic design and optimization using unstructured meshes. *AIAA Journal*, 35(9):1479–1485, 1997.

[FP98] C. Faure and Y. Papegay. Odyssée User's Guide version 1.7. Technical Report RT-0224, INRIA, Sophia-Antipolis, 1998.

[GDM01] M.B. Giles, M.C. Duta, and J.-D. Müller. Adjoint code developments using the exact discrete approach. AIAA Paper 2001-2596, 2001.

[Gil96] M.B. Giles. Analysis of the accuracy of shock-capturing in the steady quasi-1D Euler equations. *International Journal of Computational Fluid Dynamics*, 5(2):247–258, 1996.

[Gil97] M.B. Giles. Aerospace design: a complex task. Lecture notes for VKI Lecture Course on Inverse Design. Technical Report NA97/07, Oxford University Computing Laboratory, Wolfson Building, Parks Road, Oxford, OX1 3QD., 1997.

[Gil00] M.B. Giles. Defect and adjoint error correction. Invited paper at the ICCFD conference, Kyoto, 2000.

[Gil01] M.B. Giles. On the iterative solution of adjoint equations. In G. Corliss, C. Faure, A. Griewank, L. Hascoët, and U. Naumann, editors, *Automatic Differentiation: From Simulation to Optimization*. Springer, 2001.

[GK98] R. Giering and T. Kaminski. Recipes for adjoint code construction. *ACM Transactions on Mathematical Software*, 24(4):437–474, 1998.

[GLLS97] M.B. Giles, M.G. Larson, M. Levenstam, and E. Süli. Adaptive error control for finite element approximations of the lift and drag in a viscous flow. Technical Report NA 97/06, Oxford University Computing Laboratory, Wolfson Building, Parks Road, Oxford, OX1 3QD., 1997.

[GP97] M.B. Giles and N.A. Pierce. Adjoint equations in CFD: duality, boundary conditions and solution behaviour. AIAA Paper 97-1850, 1997.

[GP98] M.B. Giles and N.A. Pierce. Superconvergent lift estimates using the adjoint Euler equations. In *Proceedings of Third Asian CFD Conference*, 1998.

[GP99] M.B. Giles and N.A. Pierce. Improved lift and drag estimates using adjoint Euler equations. AIAA Paper 99-3293, 1999.

[GP00] M.B. Giles and N.A. Pierce. An introduction to the adjoint approach to design. *Flow, Turbulence and Control*, 65(3/4):393–415, 2000.

[GP01] M.B. Giles and N.A. Pierce. Analytic adjoint solutions for the quasi-one-dimensional Euler equations. *Journal of Fluid Mechanics*, 426:327–345, 2001.

[Gri00] A. Griewank. *Evaluating derivatives : principles and techniques of algorithmic differentiation*. SIAM, 2000.

[GS02] M.B. Giles and E. Süli. Adjoints methods for PDEs: *a posteriori* error analysis and postprocessing by duality. In A. Iserles, editor, *Acta Numerica 2002*. Cambridge University Press, 2002.

[HRS00] P. Houston, R. Rannacher, and E. Süli. A posteriori error analysis for stabilised finite element approximations of transport problems. *Computer Methods in Applied Mechanics and Engineering*, 190(11-12):1483–1508, 2000.

[Jam88] A. Jameson. Aerodynamic design via control theory. *Journal of Scientific Computing*, 3:233–260, 1988.

[Jam95] A. Jameson. Optimum aerodynamic design using control theory. In M. Hafez and K. Oshima, editors, *Computational Fluid Dynamics Review 1995*, pages 495–528. John Wiley & Sons, 1995.

[Joh95] C. Johnson. On computability and error control in CFD. *International Journal for Numerical Methods in Fluids*, 20(8-9):777–788, 1995.

[JPM98] A. Jameson, N. Pierce, and L. Martinelli. Optimum aerodynamic design using the Navier-Stokes equations. *Journal of Theoretical and Computational Fluid Mechanics*, 10:213–237, 1998.

[JRB95] C. Johnson, R. Rannacher, and M. Boman. Numerics and hydrodynamic stability – toward error control in computational fluid dynamics. *SIAM Journal of Numerical Analysis*, 32(4):1058–1079, 1995.

[KIH91] V.M. Korivi, A.C. Taylor III, and G.W. Hou. Sensitivity analysis, approximate analysis and design optimization for internal and external viscous flows. AIAA Paper 91-3083, 1991.

[Kor88] B. Koren. Defect correction and multigrid for an efficient and accurate computation of airfoil flows. *Journal of Computational Physics*, 76, 1988.

[MG01] J.-D. Müller and M.B. Giles. Solution adaptive mesh refinement using adjoint error analysis. AIAA Paper 2001-2550, 2001.

[Moh97] B. Mohammadi. Optimal shape design, reverse mode of automatic differentiation and turbulence. AIAA Paper 97-0099, 1997.

[MP99] B. Mohammadi and O. Pironneau. Mesh adaption and automatic differentiation in a CAD-free framework for optimal shape design. *International Journal for Numerical Methods in Fluids*, 30(2):127–136, 1999.

[MS98] P. Monk and E. Süli. The adaptive computation of far field patterns by a posteriori error estimates of linear functionals. *SIAM Journal of Numerical Analysis*, 36(1):251–274, 1998.

[NA99] E. Nielsen and W.K. Anderson. Aerodynamic design optimization on unstructured meshes using the Navier-Stokes equations. *AIAA Journal*, 37(11):957–964, 1999.

[NTB+99] J.C. Newman, A.C. Taylor, R.W. Barnwell, P.A. Newman, and G. J.-W. Hou. Overview of sensitivity analysis and shape optimization for complex aerodynamic configurations. *Journal of Aircraft*, 36(1):87–96, 1999.

[OP99] J.T. Oden and S. Prudhomme. New approaches to error estimation and adaptivity for the Stokes and Oseen equations. *International Journal for Numerical Methods in Fluids*, 31(1):3–15, 1999.

[OP00] J.T. Oden and S. Prudhomme. Goal-oriented error estimation and adaptivity for the finite element method. *Computers and Mathematical Applications*, 41(5-6):735–756, 2000.

[PG98] N.A. Pierce and M.B. Giles. Adjoint recovery of superconvergent functionals from approximate solutions of partial differential equations. Technical Report NA98/18, Oxford University Computing Laboratory, 1998.

[PG00] N.A. Pierce and M.B. Giles. Adjoint recovery of superconvergent functionals from PDE approximations. *SIAM Review*, 42(2):247–264, 2000.

[PP97] J. Peraire and A.T. Patera. Bounds for linear-functional outputs of coercive partial differential equations: local indicators and adaptive refinement. In P. Ladeveze and J.T. Oden, editors, *New Advances in Adaptive Computational Methods in Mechanics*. Elsevier, 1997.

[PP99] P. Peraire and A. Patera. Asymptotic a posteriori finite element bounds for the outputs of noncoercive problems: the Helmholtz and Burgers equations. *Computer Methods in Applied Mechanics and Engineering*, 171(1-2):77–86, 1999.

[PPP97] M. Paraschivoiu, J. Peraire, and A. Patera. A posteriori finite element bounds for linear-functional outputs of elliptic partial differential equations. *Computer Methods in Applied Mechanics and Engineering*, 150(1-4):289–312, 1997.

[Ran00] R. Rannacher. Adaptive Galerkin finite element methods for partial differential equations. *Journal of Computational and Applied Mathematics*, 1-2:205–233, 2000.

[SF73] G. Strang and G. Fix. *An Analysis of the Finite Element Method.* Prentice-Hall, 1973.

[Ske81] R.D. Skeel. A theoretical framework for proving accuracy results for deferred corrections. *SIAM Journal of Numerical Analysis*, 19:171–196, 1981.

[Ste78] H.J. Stetter. The defect correction principle and discretization methods. *Numerische Mathematik*, 29:425–443, 1978.

[Sül98] E. Süli. *A posteriori* error analysis and adaptivity for finite element approximations of hyperbolic problems. In D. Kröner, M. Ohlberger, and C. Rohde, editors, *An Introduction to Recent Developments in Theory and Numerics for Conservation Laws*, volume 5 of *Lecture Notes in Computational Science and Engineering*, pages 123–194. Springer-Verlag, 1998.

[TKS92] S. Ta'asan, G. Kuruvila, and M.D. Salas. Aerodynamic design and optimization in one shot. AIAA Paper 92-0025, 1992.

[VD00] D. Venditti and D. Darmofal. Adjoint error estimation and grid adaptation for functional outputs: application to quasi-one-dimensional flow. *Journal of Computational Physics*, 164:204–227, 2000.

[VD01] D. Venditti and D. Darmofal. A grid adaptive methodology for functional outputs of compressible flow simulations. AIAA Paper 01-2659, 2001.

Adaptive Finite Element Methods for Incompressible Fluid Flow

Johan Hoffman[1] and Claes Johnson[1]

Mathematics Department and Chalmers Finite Element Center
Chalmers University of Technology, 412 96 Göteborg, Sweden

Abstract. We present recent work on the following issues of CFD: (i) discretization of the non-stationary incompressible Navier-Stokes equations, (ii) solution of the discrete system at each time step, (iii) hydrodynamic stability, (iv) adaptive error control and a posteriori error estimates, (v) transition to turbulence and (vi) turbulence modeling.

Contents

1 Introduction

The Navier-Stokes equations give the basic model for fluid flow and describe a variety of phenomena in hydro and aero-dynamics, processing industry, biology, oceanography, meteorology, geophysics and astrophysics. Fluid flow may contain features of *incompressible* and *compressible* flow, *Newtonian* and *non-Newtonian* flow, and *turbulent* and *laminar* flow, with turbulent flow being irregular with rapid fluctuations in space and time and laminar flow being more organized. *Computational Fluid Dynamics* CFD concerns the simulation of fluid flow by solving the Navier-Stokes equations numerically. The basic issues of CFD is *computability* relating to errors from numerical computation, and *predictability* relating to errors from imprecision in given data. The basic question for a given flow situation may be formulated as follows: what quantity can be computed/predicted to what tolerance to what cost? We emphasize the *quantitative* aspects concerning both the choice of quantity to compute, the error tolerance and the computational cost. We may expect a turbulent flow to be more computationally demanding than a laminar flow, and a pointwise quantity (e.g the viscous stresses at a specific point) more demanding than an average quantity (e.g. the drag or lift).

In these notes we give a survey of our recent work on this question, continuing in particular our AGARD 92 lecture notes [17], with focus on adaptive finite element methods for non-stationary incompressible Newtonian laminar and beginning turbulent flow with medium large Reynolds numbers in the range 100 – 1000. The basic issues entering are the following: (i) discretization of the incompressible Navier-Stokes equations, (ii) solution of the discrete system at each time step, (iii) hydrodynamic stability, that is the growth of errors/perturbations from computation/data, (iv) adaptive error control and a posteriori error estimates, (v) transition to turbulence and (vi) turbulence modeling. We now give an introductory overview of the basic topics (i)-(vi). For an overview of adaptive finite element methods including references, we refer to the survey articles [2] and [1], see also [4] and [5], containing many details on various aspects of finite element adaptivity omitted in these notes. For an overview of finite element methods for the incompressible Navier-Stokes equations including references, we refer to [25]. For a survey of turbulence modeling we refer to [6], [29] and [8] and references therein.

1.1 Discretization

To discretize the Navier-Stokes equations to get a discrete system of equations on each time step, we use the general stabilized Galerkin/least squares space-time finite element method developed over the years together with Hughes, Tezduyar and coworkers. With continuous piecewise polynomials in space of order p and discontinuous or continuous piecewise polynomials in time of order q, we refer to this method as cG(p)dG(q) or cG(p)cG(q). In the computations presented in these notes we use cG(1)cG(1) with continuous piecewise linears in space for both velocity and pressure on tetrahedral meshes and Crank-Nicolson time stepping. Discretization introduces *discretization errors* containing Galerkin piecewise polynomial approximation errors and quadrature errors. For turbulent flow, some form of *turbulence modeling* is necessary, introducing also a *modeling error*. We may view the discretization and modeling error together to form the *computational error*.

1.2 Discrete solvers

The discrete system in velocity-pressure on each time step is solved using a fixed point/defect correction outer iteration, and for the corresponding inner iteration linear system we use GMRES in the velocity with the pressure solved in terms of the velocity using a multigrid Poisson solver. We also perform the inner iteration with defect correction in the velocity with given pressure. This method is efficient and robust for the non-stationary medium Reynolds number flows considered in these notes. The discrete solver introduces additional discretization errors from approximate solution of the discrete equations.

1.3 Hydrodynamic stability

Hydrodynamic stability concerns the stability features of a given solution of the Navier-Stokes equations (base flow), that is the growth of perturbations from data or computation in a given flow. Hydrodynamic stability features may be expressed by different *stability factors* measuring the growth of different perturbations/errors from data, modeling or discretization. The Galerkin discretization error basically has the form of a product of *residuals* connecting to *interpolation errors* depending on the smoothness of the solution, and related stability factors. The stability factors may be computed by numerically solving linearized (forward or dual backward) Navier-Stokes equations with appropriate data, linearized at (an approximation of) the base flow. Stability factors normally increase with increasing *Reynolds number* $Re = \frac{UL}{\nu}$, where U is a reference velocity, L is a reference length, and ν is the *kinematic viscosity*, and with increasing time scale T. The Reynolds number Re may range from small (of size 1) for very viscous flows, to very large (up to 10^9 or more) in aero or hydro-dynamics, with turbulent flow often appearing for Re larger than $10^2 - 10^3$. The stability factors may vary from small (of size 1) to medium (size 10^2 say) to large. Stability factors may grow linearly with KT for laminar flow, and (momentarily) exponentially with KT for pointwise quantities in turbulent flow, where K is a measure of the velocity gradient. Typically, K is small for laminar flow (outside layers) and large for turbulent flow (locally of the order $Re^{3/4}$), and T is typically large (of the order Re) for laminar flow. Fluid dynamics with Re medium/large, which is our main focus here, thus usually has KT medium/large, with stability factors growing linearly with KT for laminar flow and (momentarily) exponentially for pointwise quantities in turbulent flow. Turbulent flow thus has non-smooth solutions and large stability factors for pointwise quantities, and thus is pointwise uncomputable in the sense that the required mesh size would be too small (smaller than say 10^{-3} on the unit cube) and thus require too many operations and too much memory for today's computers. The aim of *turbulence modeling* is to compute solution averages which are smoother and have smaller stability factors using feasible mesh sizes (say 10^{-2} on the unit cube). The process of turbulence modeling introduces a *modeling error*, which together with the discretization error forms the total computational error.

1.4 Adaptivity and a posteriori error estimates

We present here a general approach to adaptive error control based on a posteriori error estimates. We then consider a mathematical model of the form $A(u) = f$, where A is a *differential operator*, f is given *data*, and u is the *solution*. The model is subject to perturbations from *data* represented by $\hat{f}$, *modeling* represented by $\hat{A}$, and *discretization* represented by U viewed as a numerical approximate solution to a perturbed problem $\hat{A}(\hat{u}) = \hat{f}$ with exact solution $\hat{u}$, with U obtained using a Galerkin finite element method. We say that the *data/modeling error* is equal to $u - \hat{u}$ and the *discretization error* is equal to $\hat{u} - U$, and that the total error $u - U = u - \hat{u} + \hat{u} - U$, thus has a contribution from data/modeling and a contribution from discretization. The model perturbation $\hat{A}$ may represent a turbulence model in applications to fluid flow.

An *adaptive method* for solving $A(u) = f$ includes a *feed-back process*, where the quality of computed solutions U of perturbed models $\hat{A}(\hat{u}) = \hat{f}$, are investigated with the objective of decreasing the modeling error $u - \hat{u}$ by improving the model $\hat{A}$, and/or the discretization error $\hat{u} - U$, the latter typically by appropriately modifying the local mesh size. An adaptive method is based on *a posteriori error estimates* estimating the data/modeling and discretization errors in terms of computable residuals such as $f - A(U)$ or $\hat{f} - \hat{A}(U)$, or estimated residuals such as $f - A(\hat{u})$.

Adaptive feed-back in modeling and discretization may be viewed as one aspect of *optimization* with the objective of decreasing the modeling and discretization errors. Adding also aspects of optimization of solutions, which is often the main objective, one gets a full picture of solution optimization including optimization of modeling and discretization. This problem is of the same general form with now the equation $A(u) = f$ representing a Lagrange system of equations characterizing solution optimality.

We now present the key steps in the derivation of an a posteriori error estimate for the discretization error in a Galerkin finite element method for the equation $\hat{A}(\hat{u}) = \hat{f}$ of the form: Find $U \in V_h$ such that $(\hat{A}(U), v) = (\hat{f}, v)$ for $v \in V_h$, where V_h is a finite dimensional subspace, on a mesh with mesh size h, of a Hilbert space V with scalar product $(\cdot,\cdot)$ and norm $\|\cdot\|$, and $\hat{A} : V \to V$ is Frechet differentiable with derivative $A' : V \to V$. Supposing that we want to estimate the quantity (e, ψ) where $e = \hat{u} - U$ and ψ is a given element in V, we write

$$\begin{aligned}\hat{A}(\hat{u}) - \hat{A}(U) &= \int_0^1 \frac{d}{ds}\hat{A}(s\hat{u} + (1-s)U)\,ds \\ &= \int_0^1 A'(s\hat{u} + (1-s)U)\,ds\,e \equiv A'(\hat{u}, U)e,\end{aligned}$$

and let $\varphi \in V$ be the solution to the *dual linearized problem* $(A'(\hat{u}, U)w, \varphi) = (w, \psi)$ for all $w \in V$, and obtain choosing $w = e$ the following *error representation*

$$(e,\psi) = (A'(\hat{u}, U)e, \varphi) = (\hat{A}(\hat{u}) - \hat{A}(U), \varphi) = (\hat{f} - \hat{A}(U), \varphi) = (\hat{R}(U), \varphi),$$

in terms of the residual $\hat{R}(U) = \hat{f} - \hat{A}(U)$. We then use *Galerkin orthogonality* to obtain, with $\Phi \in V_h$ an interpolant of φ satsifying an *interpolation error estimate* of the form $\|h^{-2}(\varphi - \Phi)\| \le C_i\|D^2\varphi\|$,

$$(e,\psi) = (\hat{R}(U), \varphi - \Phi) \le C_i\|h^2\hat{R}(U)\|\,\|D^2\varphi\| \le C_i S\|h^2\hat{R}(U)\|\,\|\psi\|,$$

where D^2 represents a second derivative, $h(x,t)$ is the local mesh size, $S = \frac{\|D^2\varphi\|}{\|\psi\|}$ is a stability factor and C_i an interpolation constant. Normalizing ψ, we obtain an a posteriori error estimate of the form

$$|(e,\psi)| \le C_i S\|h^2\hat{R}(U)\|$$

estimating the error in terms of the residual $\hat{R}(U)$, the mesh size h and the stability and interpolation factors S and C_i. The interpolation factor C_i only depends on the finite elements used, while we have to compute the dual problem, depending on both u and U and ψ, to get the stability factor S.

An *adaptive method* typically involves a *stopping criterion* guaranteeing that the error measure is less than a given tolerance, and a *modification strategy* to be applied if the stopping criterion is not satisfied. Both the stopping criterion and modification strategy may be based on an a posteriori error estimates of the form just given, involving the residuals $f - A(U)$ and $\hat{f} - \hat{A}(U)$ of computed solutions, and/or *a priori* error estimates involving estimates of the exact solution $\hat{u}$ or u, and including appropriate stability factors. A modification strategy may concern quantities related to discretization such as the local mesh size, or quantities related to turbulence modeling such as turbulent viscosities. The modification strategy for the local mesh size is often based on *equidistribution* with the objective of satisfying the stopping criterion with a minimal number of degrees of freedom, or largest possible local mesh size.

The dual problem is linearized at a mean value of the exact solution u and the computed solution U. In practice, when solving the dual problem to compute stability factors, we have to replace u by U and thus linearize at the computed solution U. In order for the corresponding linearization effect to be small, we expect to require U to approximate u pointwise sufficiently well. We return to this issue shortly in the context of turbulence modeling, where U may be expected to be a pointwise approximation to a local average u^h of u, but not u itself. It appears natural to choose the size of the average so that the discretization and modeling errors balance, with the discretization error decreasing and the modeling error increasing with increasing averaging.

1.5 Transition to turbulence

The phenomenon of transition from laminar to turbulent flow, studied intensively by Reynolds more than hundred years ago in the case of pipe flow, has long been an outstanding open problem in science, where today computational methods shed new light. Reynolds observed by injecting dye into water flowing through a transparent pipe, that sometimes the flow changed quite abruptly from organized laminar to turbulent fluctuating flow at some point downstream from the inlet, with the transition being identified by the deviation of the dye from a straight line into a rapidly fluctuating path. Seemingly similar sudden changes may be observed in the raising smoke from a cigarett, or in a stock market crash, or a sudden break up of a long-lasting marriage.

The basic question in all these cases is: why does the transition take place at a specific point in space or time, or not at all? Observing the straight line of the dye in Reynolds experiment before the transition, does not appear to give any signal of emerging instability, and a stock market crash necessarily must be a surprise for the majority of the market actors. The research on transition to turbulence in fluid flow has largely focussed on finding a relation between the Reynolds number and transition, with ideally a so called critical Reynolds number for each type of flow, identified by the fact that transition to turbulence takes place if and only if the actual Reynolds number is larger than the critical Reynolds number. Reynolds himself had little reason to believe in the existence of such critical Reynolds numbers noting that in his own experiments the transition took place in one pipe and not in another at the same Reynolds number. Nevertheless, most text books in fluid mechanics today present critical Reynolds numbers for various flows, such as 5772 for Pouiseuille flow between two parallel fixed plates (with parabolic velocity profile), and ∞ for Couette flow between two moving parallel plates (with linear velocity profile), both however at severe variance with experiments. For example, Couette flow may go turbulent in experiments for a wide range of Reynolds numbers starting at around 300, depending on the experimental set-up, and similarly Pouiseuille flow starting around 1000. The stated critical Reynolds numbers come out of a so called normal mode stability analysis of 2d linearized equations, referred to as the Orr-Sommerfeld equations, based on identifying exponentially growing eigenmodes. The striking difference in the theoretical predictions and the practical experiments for transition in parallel flow, has driven the classical study of hydrodynamic stability into a severe crisis.

However, during the last decades new insight on the importance of 3d perturbations and non-modal growth has been developing, see [28], giving a better understanding of the process of transition from laminar to turbulent flow in almost parallel flow, such as pipe flow and boundary layer flow. We will present pieces of this new picture as an application of a quantitative mathematical/computational analysis of hydrodynamic stability. In particular, the analysis will show that even if the dye injected into the pipe indicates that fluid particles follow (almost) straight lines prior to the transition, the flow actually gets considerably reorganized as a necessary preparation before the transition. We will refer to the reorganization as resulting from the *Taylor-Görtler mechanism* through which small transversal velocity perturbations after some time result in big perturbations in the streamwise velocity, which is the initial and crucial phase of the transition process, and which results from the non-normality of the linearized Navier-Stokes equations when linearized at parallel flow. Of course we may expect to find similar reorganizations preparing transitions into stock market and marriage crashes.

A possible reason for the survival of the classical misleading normal mode stability analysis for parallel flow, despite its lack of experimental support, is probably the fact that there are some other cases, where the same type of analysis in fact is correct and conforms with experiments, namely the *bifurcating* Benard and Taylor-Couette flows, changing from one configuration to another at a certain well defined Reynolds number. The bifurcation of Bernard flow involves the development of organized patterns of convective rolls of fluid in motion. A bifurcation involves a change from one configuration loosing stability, to a new stable configuration, which is different from the process of transition to turbulence, with the new configuration being increasingly unstable. Now, a bifurcation in general may be detected through a normal mode analysis based on finding for the linearized equations an eigenvalue with zero real part. In particular, the critical Reynolds number for the first bifurcations in Taylor-Couette and Bernard flow, may be found analytically this way. As indicated this approach does however not work for parallel Couette or Poiseuille flow, which do not bifurcate to find new stable configurations, but instead go into turbulent unstable motion. It appears that the success of the mathematical theory in the bifurcating cases, has overshadowed the failure in the non-bifurcating cases.

1.6 Turbulence modeling

The design of *turbulence models* has been a major open problem in fluid mechanics since the initial efforts in this direction by Reynolds at end of the 19th century. Turbulent flow appears to be pointwise uncomputable on todays computers because of the presence of small-scale features and large stability factors. The basic problem of *turbulence modeling* is to represent the effects of the unresolved scales on resolvable scales in computable models. Today, computational methods open new possibilities of approaching this fundamental problem of mathematical modeling, in the form of *Dynamic Large Eddy Simulation* DLES, where the the turbulence model is constructed as a part of the computational process with dynamic feed-back from the computation.

The problem of turbulence modeling may be described as follows: Let u^h be an approximation of the exact solution u corresponding to a local running average of size h representing the finest computationally resolvable scale. We seek to compute a pointwise accurate approximation U of u^h using a mesh of size h, in a situation where a pointwise accurate approximate solution of the exact solution u itself would be impossible because u contains significant scales finer than h (unresolvable subgrid scales with effects on the resolvable scales). We start seeking an equation satisfied by u^h by making an ansatz of the form $A(u^h) + A(u)^h - A(u^h) = f^h$, where we need to approximate $F_h(u) \equiv A(u)^h - A(u^h)$ in terms of u^h in a *subgrid model* $\hat{F}_h(u^h)$, where $\hat{F}_h(u^h)$ is an approximation of $F_h(u)$, to get a model of the form $\hat{A}(\hat{u}) = A(\hat{u}) + \hat{F}_h(\hat{u}) = f^h = \hat{f}$. The quantity $F_h(u)$ has the form of a (generalized) *covariance* and we thus face a problem of modeling covariances. In the setting of the Navier-Stokes equations, the term $F_h(u)$ corresponds to (derivatives of) the exact *Reynolds stresses*, and the turbulence modeling problem is to find an expression $\hat{F}_h(u^h)$ approximating $F_h(u)$ in terms of u^h. Having determined a turbulence model in the form of a function $\hat{F}_h$, we solve the Galerkin equation: Find $U \in V_h$ such that $(A(U) + \hat{F}_h(U), v) = (\hat{f}, v)$ for all $v \in V_h$. The total error will then connect to a modeling residual $F_h(u) - \hat{F}_h(U)$ and a discretization residual $A(U)+\hat{F}_h(U)-\hat{f}$ in a posteriori error estimates, where the modeling residual has to be approximated by extrapolation, while the discretization residual is directly computable from U.

The problem of turbulence modeling thus may be viewed to be to construct the function $\hat{F}_h$. The basic approaches use ***similarity models*** based on extrapolation from coarser scales assuming a scale regularity, or ***turbulent viscosity models*** with the Reynolds stresses approximated as viscous resolvable stresses given by a ***turbulent viscosity***, or ***mixed models*** with a combination of similarity and turbulent viscosity models.

We now return to the question of linearization in connection to a posteriori error estimates, assuming the approximate solution U to be a pointwise accurate approximation of u^h, but not the turbulent solution u itself, because of unresolvable subgrid scales. Aiming then at an a posteriori error estimate for $u^h - U$, we will have a modeling error contribution from the quantity $F_h(u)$, and we will need to estimate this quantity even if we do not use a turbulence model. Using a turbulence model with $\hat{F}_h(U)$ an approximation of $F_h(u)$, we instead need to estimate $F_h(u) - \hat{F}(U)$. At any rate, we need to estimate or model $F_h(u)$. With the terminology used above, we view the turbulence modeling error as a part of the total computational error with ideally a balance of discretization and modeling errors.

2 The incompressible Navier-Stokes equations

The incompressible Navier-Stokes equations expressing conservation of momentum and incompressibility of a unit density constant temperature Newtonian fluid with constant kinematic viscosity $\nu > 0$ enclosed in a volume Ω in $\mathbb{R}^3$, take the form: find (u, p) such that

$$\begin{aligned} D_{u,t}u - \nu\Delta u + \nabla p &= f \qquad && \text{in } \Omega \times I, \\ \operatorname{div} u &= 0 && \text{in } \Omega \times I, \\ u &= w && \text{on } \partial\Omega \times I, \\ u(\cdot, 0) &= u^0 && \text{in } \Omega, \end{aligned} \tag{1}$$

where $u(x,t) = (u_i(x,t))$ is the *velocity* vector and $p(x,t)$ the *pressure* of the fluid at (x,t), and f, w, u^0, $I = (0,T)$, is a given driving force, Dirichlet boundary data, initial data and time interval, respectively. Further,

$$D_{u,t}v \equiv \dot{v} + (u \cdot \nabla)v \tag{2}$$

is the *particle derivative* of $v(x,t)$ measuring the rate of change $\frac{d}{dt}v(x(t),t)$ of $v(x(t),t)$ along the trajectory $x(t)$ of a fluid particle with velocity u, satisfying $\dot{x}(t) = u(x(t),t)$, where as usual $\dot{v} = \partial v/\partial t$. The quantity $\nu\Delta u - \nabla p$ represents the total fluid force, and may alternatively be expressed as

$$\nu\Delta u - \nabla p = \operatorname{div}\sigma(u,p), \tag{3}$$

where $\sigma(u,p) = (\sigma_{ij}(u,p))$ is the *stress tensor*, with components $\sigma_{ij}(u,p) = 2\nu\epsilon_{ij}(u) - p\delta_{ij}$, composed of the *stress deviatoric* $2\nu\epsilon_{ij}(u)$ with zero trace and an isotropic pressure: Here $\epsilon_{ij}(u) = (u_{i,j} + u_{j,i})/2$ is the *strain tensor*, with $u_{i,j} = \partial u_i/\partial x_j$, and δ_{ij} is the usual Kronecker delta, the indices i and j ranging from 1 to 3. A Neumann type boundary condition, corresponding to the boundary stress being prescribed, takes the form $\sigma \cdot n = g$, where $(\sigma \cdot n)_i = \sum_j \sigma_{ij} n_j$ and $g = (g_i)$ is a given boundary stress with g_i the force component in the x_i-direction.

In the model (1) we assume that the *temperature* T is constant. In the general case with variable density ρ and temperature T, (1) is modified by replacing $D_{u,t}$ by $\rho D_{u,t}$, and adding the following equations expressing conservation of mass and energy:

$$\begin{aligned} D_{u,t}\rho &= 0 && \text{in } \Omega \times I, \\ D_{u,t}T - \nabla \cdot (\mu \nabla T) &= F && \text{in } \Omega \times I, \end{aligned} \tag{4}$$

together with boundary and initial conditions, where μ is a heat conduction coefficient and F a heat source, assuming the heat capacity is equal to one. We note that since $\nabla \cdot u = 0$, we have $D_{u,t}\rho = \dot{\rho} + \nabla \cdot (\rho u) = 0$, which is the usual equation expressing mass conservation.

We assume that (1) is normalized so that the reference velocity and typical length scale are both equal to one. The Reynolds number Re is then equal to ν^{-1}. Of course, the specification of the length scale may not be very obvious and thus the Reynolds number may not have a very precise quantitative meaning.

2.1 Existence and uniqueness of solutions

Existence and uniqueness of solutions of the 3d incompressible Navier-Stokes equations (1) is an outstanding open problem in the mathematical literature, now on the Clay Institute's list of $\$10^6$ prize problems. One may argue, that little progress has been made on this problem since the work by Leray [22] from 1934 proving existence, but not uniqueness, of a certain type of weak solution. In contrast, it is known that for various so called regularized Navier-Stokes equations, existence and uniqueness of classical solutions may be proved mathematically using standard techniques from the theory of partial differential equations and functional analysis, see [9]. A common regularization consists in replacing the constant viscosity ν by the modified viscosity

$$\hat{\nu} = \nu + h^2 |\epsilon(u)|^2, \tag{5}$$

depending on the strain $\epsilon(u)$ and the positive regularization parameter h, representing a smallest possible spatial scale. The regularization introduces a modification of the viscosity for highly strained flows, away from the Newtonian constant viscosity. In this model the basic energy estimate (assuming

$f = w = 0$) obtained by multiplying the momentum equation by u and integrating:

$$\int_\Omega |u(x,T)|^2\,dx + 4\int_0^T\int_\Omega (\nu|\epsilon(u)|^2 + h^2|\epsilon(u)|^4)\,dx\,dt \le \int_\Omega |u^0|^2\,dx,$$

is strong enough to prove uniqueness using a standard Grönwall type argument based on a Sobolev inequality of the form

$$(w\cdot\nabla u, w) \le C\|\nabla w\|^{3/2}\|w\|^{1/2}\|\epsilon(u)\| \le \frac{\nu}{2}\|\nabla w\|^2 + \frac{C}{\nu^3}\|\epsilon(u)\|^4\|w\|^2,$$

with $(\cdot,\cdot)$ and $\|\cdot\|$ appropriate $L_2(\Omega)$-scalar products/norms, $w = u - v$ the difference of two solutions u and v, and C a constant of moderate size. The resulting estimate of the difference (perturbation) $w(t)$ has the form

$$\|w(t)\|^2 \le C\exp\Big(\frac{Ct}{h^2\nu^3}\Big)\|w(0)\|^2.$$

We note that the growth factor is extremely large in the case of main interest with ν and h small, which puts doubts to the physical meaningfulness of the uniqueness proof. If a similar mathematical uniqueness proof was possible for the limiting case of the Navier-Stokes equations with $h = 0$, which would seem to give the Clay prize, it would most likely suffer from the same presence of enormous constants, and the value of the proof could be seriously questioned.

The uniqueness question is of course intimately connected to hydrodynamic stability, which concerns the growth of perturbations (such as $w(t)$ in the above discussion) in the Navier-Stokes equations. Only solutions which are stable in some sense may be expected to exist as observable flows. Hydrodynamic stability has a long tradition going back to Reynolds, Rayleigh and Lord Kelvin, but offers surprisingly few concrete results of significance. In general, both the perturbation growth and the complexity of the flow increases with increasing Reynolds number (decreasing viscosity), with eventually the flow becoming turbulent with very strong perturbation growth and very fine scales developing. We shall show below that for even for basic types of parallel laminar flow, the growth of perturbations may be linear in the Reynolds number, indicating that the 3d Euler equations, corresponding to vanishing viscosity, cannot have any meaning.

3 Hydrodynamic stability

Hydrodynamic stability concerns the quantitative stability properties of the incompressible Navier-Stokes equations, which are of basic importance for both error control in computation and modeling, and for the understanding of phenomena of fluid flow such as bifurcation or transition to turbulence. The basic study of hydrodynamic stability concerns the linearized Navier-Stokes equations for perturbations (φ, q) of a given solution (u, p) of (1) corresponding to the initial data u_0 and right hand side f, obtained by subtracting (u, p) from the solution $(u + \varphi, p + q)$ corresponding to the perturbed initial data $u^0 + \varphi^0$ and right hand side $f + g$, and omitting the quadratic perturbation term $(\varphi \cdot \nabla)\varphi$:

$$\begin{aligned} D_{u,t}\varphi + (\varphi \cdot \nabla)u - \nu\Delta\varphi + \nabla q &= g && \text{in } \Omega \times I, \\ \nabla \cdot \varphi &= 0 && \text{in } \Omega \times I, \\ \varphi &= 0 && \text{on } \partial\Omega \times I, \\ \varphi(\cdot, 0) &= \varphi^0 && \text{in } \Omega, \end{aligned} \tag{6}$$

where $(\varphi \cdot \nabla)u = (\sum_{j=1}^{3} \varphi_j u_{i,j})_{i=1}^{3}$ with $v_{,j} = \partial v / \partial x_j$.

The basic question in hydrodynamic stability is to estimate the solution (φ, q) of (6) in various norms in terms of appropriate corresponding norms of the data (g, φ_0), for example in terms of certain *stability factors*. A basic example is given by the *weak stability factor* $S_0(u, T, \varphi^0)$ depending on the base flow u, final time T and the perturbation φ^0, defined by

$$S_0(u, T, \varphi^0) = \frac{\|\varphi\|_I}{\|\varphi^0\|}, \tag{7}$$

or the factor $S_0(u, T)$ depending on the base flow u and the final time T with a maximization over all perturbations, defined by

$$S_0(u, T) = \sup_{\varphi^0 \in L_2} \frac{\|\varphi\|_I}{\|\varphi^0\|}, \tag{8}$$

where φ is the solution of (6) with $g = 0$ and initial data $\varphi^0 \neq 0$, and $\|v\| = \|v\|_{L_2(\Omega)}$, $\|v\|_I = \sup_{0<t<T} \|v(\cdot, t)\|$. The factor $S_0(u, T, \varphi^0)$ measures the growth over the time interval $(0, T)$ of the perturbation φ^0 of initial data, and the factor $S_0(u, T)$ measures the maximal growth over the time interval $(0, T)$ of a perturbation of initial data. We refer to these stability factors as *weak* because we measure the solution itself and not derivatives thereof. *Strong* stability factors measure derivatives of the solution of dual linearized equations, connect to Galerkin discretization errors and occur as multiplicative factors in a posteriori error estimates.

We now give estimates of the stability factor $S_0(u, T)$ in two extreme cases: a *worst case* with exponential dependence in KT related to non-smooth turbulent flow, and a *best case* with linear dependence in KT related to smooth laminar flow. The strong stability factors coupled to Galerkin discretization show the same span of variation. Assuming that $K = 1$ and $T = \nu^{-1} = Re$, the dependence can be expressed as an exponential or linear dependence in the Reynolds number Re, with the exponential dependence indicating uncomputability even for moderately large Reynolds numbers, while the linear dependence make smooth laminar flows computable.

3.1 Worst case exponential perturbation growth

Multiplying the first equation of (6) by φ and integrating over $\Omega \times (0,t)$, using the incompressibility of both u and φ, one gets for $t > 0$:

$$\|\varphi(\cdot,t)\|^2 \leq -2\int_0^t \int_\Omega (\varphi\cdot\nabla)u\cdot\varphi\, dxds + \|\varphi^0\|^2,$$

from which follows by the Grönwall inequality that $S_0(u,T) \leq \exp(CKT)$, with $C \approx 1$, which is a worst case exponential estimate. We note the exponential growth is generated by the presence of the zero order term $(\varphi\cdot\nabla)u$, as in the simple scalar ODE $\dot{\psi} = K\psi$ with solution $\psi(t) = \psi(0)\exp(Kt)$. A flow with this very strong perturbation growth cannot exist as a stable flow. Since there are some more or less stable flows observable in nature, it must be possible in special cases to obtain reduced growth rates by using particular features of the zero order coupling term $(\varphi\cdot\nabla)u$. A basic such case arises in nearly parallel flow, with a particular coupling of the perturbations of the velocities in streamwise and transversal directions, which we now turn to.

3.2 Linear perturbation growth for nearly parallel flow

Nearly parallel flow is a basic type of flow, also referred to as *shear flow*, occuring in pipe flow and boundary layer flow, where the streamlines are almost parallel straight lines and the transversal variation of the streamline flow velocity is balanced by a shear force. We now show that for such flows the weak stability factor $S_0(u,T)$ defined by (8) satisfies $S_0(u,T) \approx KT$. This estimate underlies the first crucial step in the scenario of transition to turbulence in nearly parallel flow to be presented, showing that a perturbation growth $\approx \nu^{-1}$ over time intervals of length $T = \nu^{-1}$ is possible even if for smooth flows with $K = 1$, indicating that a small initial perturbation (of size ν say) in fact may cause the base flow to change significantly if we only wait long enough (over a time interval $\approx \nu^{-1}$).

We consider a smooth parallel stationary base flow (u,p) in an infinitely long straight pipe $\Omega = \mathbb{R}\times\omega$, where ω in the (x_2,x_3)-plane is the cross-section (with smooth boundary) of the pipe of diameter of size 1, and the axis of the pipe is oriented along the x_1-axis, and u vanishes on the boundary of the pipe. We assume that the base flow (u,p) is independent of x_1 and satisfies the following assumptions

$$\|u_1\| \approx 1, \qquad \|\bar{\nabla}u_1\|_\infty = C, \qquad \|\bar{u}\|_\infty + \|\bar{\nabla}\bar{u}\|_\infty \leq c\nu, \tag{9}$$

where $\|\cdot\|_\infty$ denotes the maximum norm, $\bar{u} = (u_2,u_3)$, and $\bar{\nabla} = (\partial/\partial x_2, \partial/\partial x_3)$ is the gradient with respect to (x_2,x_3). Here and below, c and C denote positive constants of moderate size, which are independent of ν. The assumption (9) including a smooth streamwise velocity $u_1 \approx 1$ in the x_1 direction being independent of x_1, and smooth small transversal velocities $\bar{u}$ of size $\approx \nu$, may be viewed as a basic characteristic of nearly parallel flow. A further characteristic may be that the derivatives in the streamwise direction x_1 are one order smaller in ν, so that $u_{1,1} \sim \nu$ and $u_{2,1}, u_{3,1} \sim \nu^2$. We will return to this feature below in the presentation of the scenario of transition to turbulence. We further assume as already indicated that $T \sim 1/\nu = Re$.

Assuming that also the perturbations (φ, q) are independent of x_1, the linearized equations (6), take the following form:

$$\begin{aligned} D_{u,t}\varphi_1 + (\bar\varphi \cdot \bar\nabla)u_1 - \nu\Delta\varphi_1 &= 0 && \text{in } \omega \times I, \\ D_{u,t}\bar\varphi + (\bar\varphi \cdot \bar\nabla)\bar u + \bar\nabla q - \nu\Delta\bar\varphi &= 0 && \text{in } \omega \times I, \\ \bar\nabla \cdot \bar\varphi \equiv \varphi_{2,2} + \varphi_{3,3} &= 0 && \text{in } \omega \times I, \\ \varphi &= 0 && \text{on } \partial\omega \times I, \\ \varphi(\cdot, 0) &= \varphi_0 && \text{on } \omega. \end{aligned} \tag{10}$$

These equations have a very particular structure. First, the equations for the transversal velocity $\bar\varphi$ are fully decoupled from the equation for the streamwise velocity φ_1, and have zero order terms with small coefficients because $|\bar\nabla\bar u| \leq c\nu$. Secondly, the zero order term $(\bar\varphi \cdot \bar\nabla)u_1$ in the equation for φ_1 does not contain φ_1, because $u_{1,1} = 0$. This means that the zero order terms in (10) have a special form, which makes it possible to reduce the general worst case exponential growth of $S_0(T)$, to a linear growth. The basic structure of the equations (10) is present in the system of ordinary differential equations $\dot\varphi_1 - \varphi_2 = 0$, $\dot\varphi_2 = 0$, for $t > 0$, $\varphi^0 = (0, \varphi_2^0)$ with solution $\varphi_1(t) = t\varphi_2^0$, $\varphi_2(t) = \varphi_2^0$, showing a linear growth of φ_1. The growth in this system is very different from the exponential growth obtained changing the first equation to $\dot\varphi_1 - \varphi_1 = 0$, with the exponentially growing solution $\varphi_1(t) = \exp(t)\varphi_1^0$, assuming now $\varphi_1^0 \neq 0$. Clearly, the change from linear to exponential growth is related to the nature of the coupling, with the direct coupling $\dot\varphi_1 = \varphi_1$ being much stronger than the indirect coupling $\dot\varphi_1 = \varphi_2$, where $\dot\varphi_2 = 0$.

We now prove a basic estimate giving a linear growth bound in time of the streamwise velocity perturbation φ_1 generated by a small transversal perturbation $\bar\varphi^0$. We refer to the physical phenomena causing this perturbation growth as the *Taylor-Görtler mechanism*, which has a crucial role in transition to turbulence. The bound is based on an energy estimate using the decoupling of φ_1 and $\bar\varphi$, resulting from the fact that $q_{,1} = 0$ and $\varphi_{1,1} = 0$. Below we present computations showing that the bound is sharp and that linear perturbation growth actually occurs.

Theorem 1. *The stability constant $S_0(u,T)$, defined by (8) in the context of x_1-independent nearly parallel pipe flow (u,p) satisfying (9), satisfies the following bound for $T = \nu^{-1}$:*

$$S_0(u,T) \leq C\nu^{-1}, \tag{11}$$

where C depends on the constant c in (9). If the constant c is small enough, then the estimate (11) holds for $T \geq \nu^{-1}$ with ν^{-1} replaced by T.

Proof. First, multiplying the equation for $\bar{\varphi}$ by $\bar{\varphi}$, and integrating over ω using the fact that $\bar{\nabla}\cdot\bar{u}=\bar{\nabla}\cdot\bar{\varphi}=0$, shows that

$$\frac{1}{2}\frac{d}{dt}\|\bar{\varphi}\|^2+\nu\|\bar{\nabla}\bar{\varphi}\|^2\leq c\nu\|\bar{\varphi}\|^2.$$

Using Grönwall's inequality, we then find that

$$\|\bar{\varphi}(\cdot,t)\|^2\leq\exp(C\nu t)\|\bar{\varphi}^0\|^2,\qquad 0<t\leq T.$$

Next, multiplying the equation for φ_1 by φ_1 and using again the fact that $\bar{\nabla}\cdot\bar{u}=\bar{\nabla}\cdot\bar{\varphi}=0$, we get

$$\frac{1}{2}\frac{d}{dt}\|\varphi_1\|^2+\nu\|\bar{\nabla}\varphi_1\|^2\leq C(\frac{1}{2}\nu\|\varphi_1\|^2+\frac{1}{2}\nu^{-1}\|\bar{\varphi}\|^2),$$

from which the desired estimate follows by integration. The modification with c sufficiently small is left to the reader.

A challenge is to extend the above result to different base flows (u,p) with slight x_1-dependence. As a small contribution to this problem we present the following example: we assume in addition to (9) that

$$\|u_{1,1}\|_\infty\leq c\nu,\qquad \|\bar{u}_{,1}\|_\infty\leq c\nu^2, \tag{12}$$

where c is a positive constant, and we allow the perturbation velocity φ to depend on x_1, but we assume for the pressure part q that $q_{,1}=0$ and that correspondingly the incompressibility condition reduces to $\varphi_{2,2}+\varphi_{3,3}=0$, which corresponds to a slight compressibility of the original fluid with a pressure perturbation q, which is constant in the x_1-direction. In this case the linearized perturbation equations take the form:

$$\begin{aligned}
D_{u,t}\varphi_1+(\varphi\cdot\nabla)u_1-\nu\Delta\varphi_1&=0 &&\text{in } \Omega\times I,\\
D_{u,t}\bar{\varphi}+(\varphi\cdot\nabla)\bar{u}+\bar{\nabla}q-\nu\Delta\bar{\varphi}&=0 &&\text{in } \Omega\times I,\\
\varphi_{2,2}+\varphi_{3,3}&=0 &&\text{in } \Omega\times I,\\
\varphi&=0 &&\text{on } \partial\Omega\times I,\\
\varphi(\cdot,0)&=\varphi_0 &&\text{on } \Omega,
\end{aligned}$$

which again decouples and thus is amenable to analysis as above.

Remark 1. The Orr-Sommerfeld equations are the linearized Navier-Stokes equations linearized at x_1-directed parallel flow $u=(u_1(x_2),0,0)$ between two parallel plates with normal in the x_2 direction, assuming the perturbations are independent of the transversal direction x_3 parallel to the plates and also that $\varphi_3=0$: find $(\varphi(x_1,x_2,t),p(x_1,x_2,t))$ such that for $|x_2|<d$, $x_1\in\mathbb{R}$, $t>0$

$$\begin{aligned}
\dot{\varphi}_1+u_{1,2}\varphi_2-\nu\Delta\varphi_1+p_{,1}&=0,\\
\dot{\varphi}_2-\nu\Delta\varphi_2+p_{,2}&=0,\\
\varphi_{1,1}+\varphi_{2,2}&=0,
\end{aligned}\tag{13}$$

with $\varphi(x_1, \pm d) = 0$, and the initial condition $\varphi(x_1, x_2, 0) = \varphi^0(x_1, x_2)$, and where $2d$ is the distance between the plates. In the case of Couette flow $u_1(x_2) \propto x_2$ and for Pouiseuille flow $u_1(x_2) \propto (1 - (x_2/d)^2)$. The stability factor $S_0(u, T)$ turns out to be much smaller than the corresponding factor for the linearized problem (10) with x_1 independent perturbations. We conclude that x_2-independent perturbations seem to be less significant than x_1-independent perturbations, and thus conclude that the Orr-Sommerfeld equations do not seem to be so relevant in initial transition to turbulence in nearly parallel flow.

4 Computability of the Lorenz system

The meteorologist E. Lorenz presented in his famous 1972 talk *Does the flap of a butterfly's wings in Brazil set off a tornado in Texas?*, the following 3×3-ODE model of the Navier-Stokes equations with the purpose of explaining well-known uncertainties in weather prediction:

$$\begin{cases} \dot{x} = \sigma(y - x), \\ \dot{y} = rx - y - xz, \\ \dot{z} = xy - bz, \end{cases} \tag{14}$$

with initial data $(x(0), y(0), z(0)) = (1, 0, 0)$, and $\sigma = 10$, $b = 8/3$ and $r = 28$. The solution $u(t) = (x(t), y(t), z(t))$ turns out to be very sensitive to perturbations and in our terminology thus has stability factors with very rapid (exponential) growth in time. The computational challenge is to solve the Lorenz system accurately on a time interval $[0, T]$ with T as large as possible. We present results on this question from [23], which we expect to be of some relevance also for the Navier-Stokes equations.

In Figure 1 we plot the x-component of the Lorenz system on $[0, 40]$ computed using cG(q) and dG(q) with $q = 1, 2, 3, 4, 5, 10$ using a constant time step $k = 0.001$ and double precision arithmetic. We indicate a correct solution with a solid line, and an incorrect solution with a dashed line. We note that increasing q does not help to produce a correct solution beyond $T = 40$, nor does it help to decrease the time step. This is because at every time-step we make a relative round off error of at least 10^{-16} using double precision, and the only possibility to get beyond $T = 40$ would be to take fewer time-steps using a very high order method.

In Figure 2 we present solutions with high order cG(q) with the larger time step $k = 0.1$, which are accurate to almost $T = 50$. Increasing the time step further does not help to get beyond $T = 50$.

In Figure 3 we plot the stability factor $S(T)$ measuring the accumulation of round-off or quadrature errors over the time interval $[0, T]$, which is computed by solving an appropriate dual linearized problem. The stability factor grows with an overall exponential rate, but the growth is different in different parts of the trajectory. We see that $S(T)$ reaches the value 10^{16} for

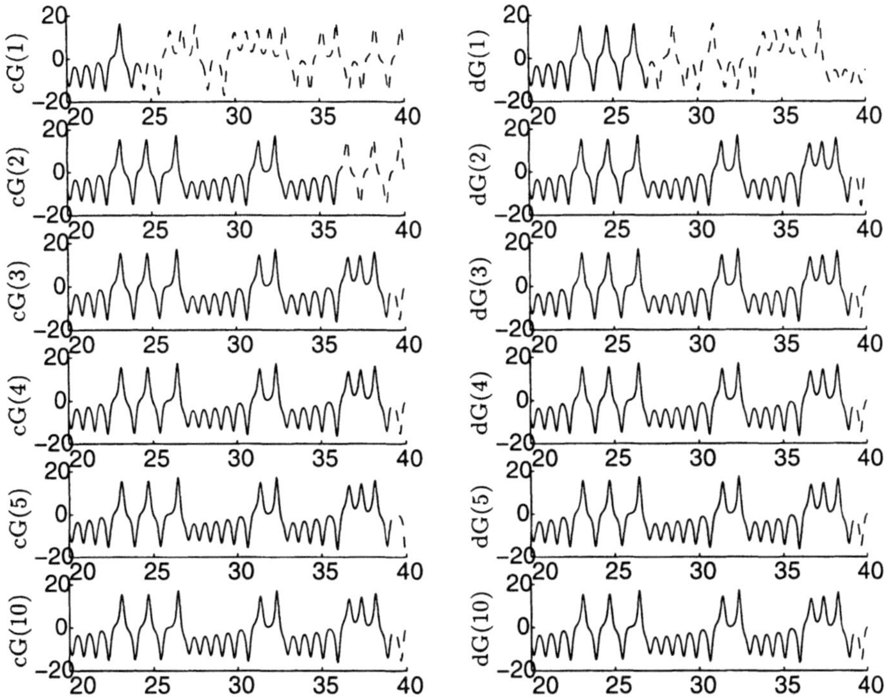

Fig. 1. The x-component of Lorenz solutions on $[20, 40]$ computed with different methods using the constant time-step $k = 0.001$

$T = 50$, which agrees with the double precision limit found above. Computing in quadruple precision would allow accurate computation to $T = 100$.

We note that a worst case estimate using a relevant Lipschitz constant (which is of size 100), would give an upper bound of the form $S(T) \leq \exp(100T)$, indicating that we could only compute accurately to $T = 1$, which would be a gross underestimate. Even in this case with considerable exponential growth, we are far from a worst case Grönwall estimate.

5 Discretization: General Galerkin G^2

In this section we present the general space-time Galerkin least squares stabilized finite element method, referred to as the General Galerkin G^2- method, for the incompressible Navier-Stokes equations (1). This method includes the *streamline diffusion method* on Eulerian space-time meshes, the *characteristic Galerkin method* on Lagrangian space-time meshes with orientation along particle trajectories, and *Arbitrary Lagrangian-Eulerian ALE methods* with different mesh orientation. Further, the least-squares stabilizations present in

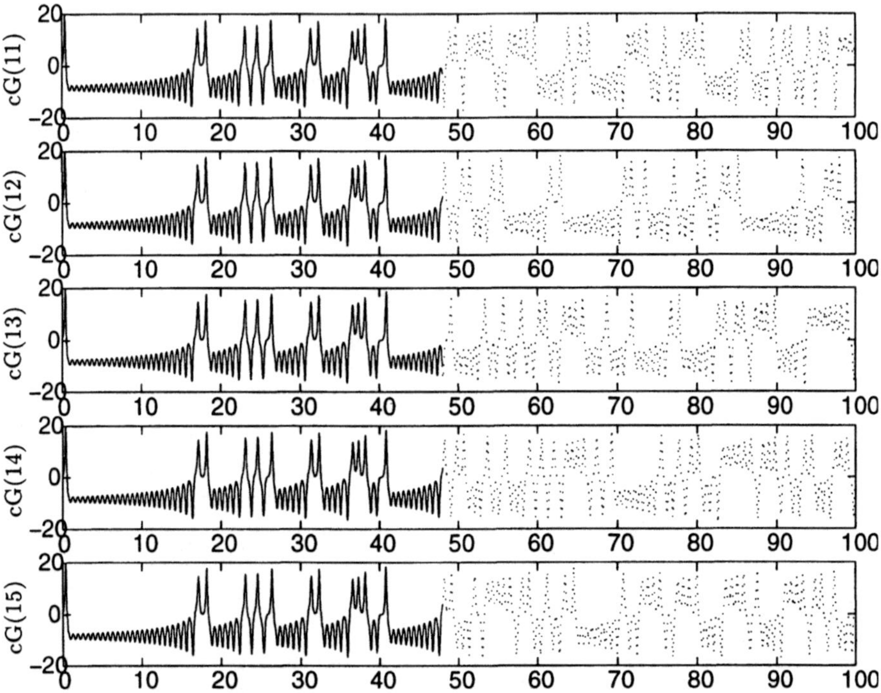

Fig. 2. The x-component of Lorenz solutions on $[0, 50]$ computed with high-order methods using the constant time-step $k = 0.1$.

the G^2-method, does take care of the two difficulties traditionally met in the discretization of the incompressible Navier-Stokes equations, namely

- instabilities from Eulerian discretization of convection terms,
- pressure instabilities in equal order interpolation of velocity and pressure.

Altogether, we are able to present a general flexible methodology for the discretization of the incompressible Navier-Stokes equations applicable to a great variety of flow problems from creeping viscous flow to slightly viscous flow, including free or moving boundaries.

Let $0 = t_0 < t_1 < ... < t_N = T$ be a sequence of discrete time steps with associated time intervals $I_n = (t_{n-1}, t_n]$ of length $k_n = t_n - t_{n-1}$ and space-time slabs $S_n = \Omega \times I_n$, and let $W_n \subset H^1(\Omega)$ be a finite element space consisting of continuous piecewise polynomials of degree p on a mesh $\mathcal{T}_n = \{\kappa\}$ of mesh size $h_n(x)$ with W_{0n} the functions in W_n vanishing on Γ. To define the G^2-method for (1) with homogeneuos Dirichlet boundary conditions for the velocity ($w = 0$), let for a given velocity field β on $S_n =$

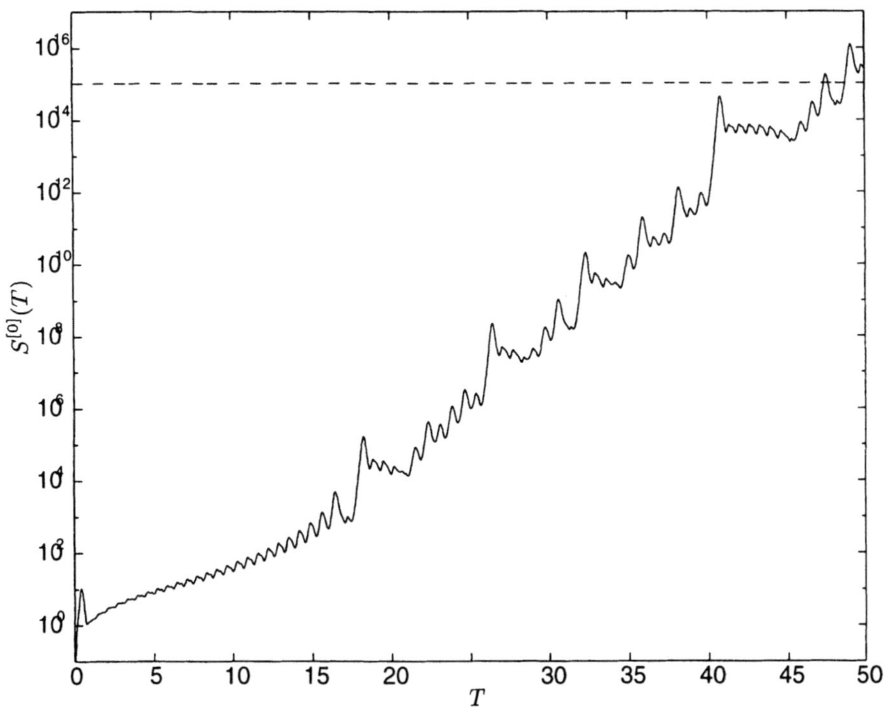

Fig. 3. The stability factor $S(T)$ measuring accumulation of round-off and quadrature errors in the Lorenz system

$\Omega \times I_n$ vanishing on $\Gamma \times I_n$, the particle paths $x(\bar{x}, \bar{t})$ be defined by

$$\begin{aligned} \frac{dx}{d\bar{t}} &= \beta(x, \bar{t}) \quad \bar{t} \in I_n, \\ x(\bar{x}, t_n) &= \bar{x}, \quad \bar{x} \in \Omega, \end{aligned} \tag{15}$$

and introduce the corresponding mapping $F_n^\beta : S_n \to S_n$ defined by $(x,t) = F_n^\beta(\bar{x}, \bar{t}) = (x(\bar{x}, \bar{t}), \bar{t})$, where $x = x(\bar{x}, \bar{t})$ satisfies (15). Define for a given $q \geq 0$, the spaces

$$\bar{V}_n^\beta = \{\bar{v} \in H^1(S_n)^3 : \bar{v}(\bar{x}, \bar{t}) = \sum_{j=0}^{q} (\bar{t} - t_n)^j U_j(\bar{x}),\ U_j \in [W_{0n}]^3\},$$

$$\bar{Q}_n^\beta = \{\bar{q} \in H^1(S_n) : \bar{q}(\bar{x}, \bar{t}) = \sum_{j=0}^{q} (\bar{t}_n - t_n)^j q_j(\bar{x}),\ q_j \in W_n\},$$

together with their analogs in (x,t)-coordinates:

$$V_n^\beta = \{v : \bar{v} \in \bar{V}_n^\beta\}, \qquad Q_n^\beta = \{q : \bar{q} \in \bar{Q}_n^\beta\}, \tag{16}$$

where $v(x,t) = \bar{v}(\bar{x},\bar{t})$ and $q(x,t) = \bar{q}(\bar{x},\bar{t})$. Defining finally $V^\beta \times Q^\beta = \prod_n V_n^\beta \times Q_n^\beta$, we can now formulate the G^2-method as follows: Find $(U,P) \in V^\beta \times Q^\beta$, such that for $n = 1,2,...,N$,

$$\begin{aligned}&(\dot{U} + (U \cdot \nabla)U, v)_n - (P, \operatorname{div} v)_n + (q, \operatorname{div} U)_n + (2\nu\epsilon(U), \epsilon(v))_n \\ &+ (\delta_1 a(U;U,P), a(U;v,q))_n + (\delta_2 \operatorname{div} U, \operatorname{div} v)_n + ([U^{n-1}], v_+^{n-1}) \\ &= (f, v + \delta_1 a(U;v,q))_n \quad \forall (v,q) \in V_n^\beta \times Q_n^\beta,\end{aligned} \tag{17}$$

where $a(w;v,q) = D_{w,t}v + \nabla q - \nu\Delta v$ with the Laplacian defined elementwise, $\delta_1 = \frac{1}{2}(k_n^{-2} + |U|^2 h_n^{-2})^{-1/2}$ in the convection-dominated case $\nu < Uh_n$ and $\delta_1 = \kappa_1 h^2$ otherwise, $\delta_2 = \kappa_2 h$ if $\nu < Uh_n$ and $\delta_2 = \kappa_2 h^2$ otherwise, with κ_1 and κ_2 positive constants of unit size, and

$$(v,w)_n = \int_{I_n} (v,w)dt, \quad (v,w) = \sum_{K \in \mathcal{T}_\backslash} \int_K v \cdot w \, dx,$$

$$(\epsilon(v), \epsilon(w)) = \sum_{i,j=1}^{3} (\epsilon_{ij}(v), \epsilon_{ij}(w)).$$

Further, $[v^n] = v_+^n - v_-^n$ is the jump across the time level t_n with $v_\pm^n$ the limit from $t > t_n/t < t_n$. In the Eulerian *streamline diffusion method* we choose $\beta = 0$, which means that the mesh does not move in time. The *characteristic Galerkin method* is obtained choosing $\beta = U$ (and then $\delta_1 = \kappa_1 h^2$), which means that the mesh moves with the fluid particles. We may also choose β differently which gives various versions of ALE-methods, with the mesh and particle velocity being (partly) different; for example we may move the mesh with the particle velocity at a free boundary, while allowing the mesh to move differently inside the domain.

The variational formulation (17) with $\delta_1 = \delta_2 = 0$ is obtained multiplying the momentum equation by v, integrating over S_n including integration by parts, and adding the incompressibility equation multiplied by q and integrating over S_n. Choosing δ_1 and δ_2 positive as indicated introduces stabilizing least-squares terms. Note that the viscous term $(2\nu\epsilon(U), \epsilon(v))_n$ may alternatively occur in the form $(\nu\nabla U, \nabla v)_n = \sum_{i=1}^3 (\nu\nabla U_i, \nabla v_i)_n$. In the case of Dirichlet boundary conditions the corresponding variational formulations will be equivalent, but not so in the case of Neumann boundary conditions, see below. Note finally that we may write the term $-(P, \operatorname{div} v)$ alternatively in the form $(\nabla P, v)$ if v vanishes on the boundary.

In extreme situations, we may add residual dependent *shock-capturing artificial viscosity*, replacing ν by $\hat{\nu} = \max(\nu, \kappa_3 |R(U,P)| h^2)$, where $R(U,P) = \sum_{i=1}^4 R_i(U,P)$ with

$$\begin{aligned}R_1(U,P) &= |\dot{U} + U \cdot \nabla U + \nabla P - f - \nu\Delta U|, \\ R_2(U,P) &= \nu D_2(U), \\ R_3(U,P) &= |[U^{n-1}]|/k_n \quad \text{on } S_n, \\ R_4(U,P) &= |\operatorname{div} U|,\end{aligned} \tag{18}$$

where $D_2(U)(x,t) = \max_{y\in\partial K}(h_n(x))^{-1}|[\frac{\partial U}{\partial n}(y,t)]|$ for $x \in K$, with $[\cdot]$ the jump across the element edge ∂K, and κ_3 is a positive constant of unit size. Note that $R_1(U,P)$ is defined elementwise and that with piecewise linears in space, the Laplacian ΔU is zero. In the computations presented below, we chose $\kappa_3 = 0$ corresponding to shutting off the artificial viscosity. Note that $R_1(U,P) + R_2(U,P)$ bounds the residual of the momentum equation, with the Laplacian term bounded by the second order difference quotient $D_2(U)$ arising from the jumps of normal derivatives across element boundaries.

The special case of the Stokes equations is of course obtained omitting the nonlinear terms $(U\cdot\nabla)U$ and $(U\cdot\nabla)v$, and setting $\delta_1 = \kappa_1 h^2$, $\delta_2 = \kappa_2 h^2$. This method contains the pressure stabilizing term $(\delta_1\nabla P, \nabla q)$, which corresponds to a weighted Laplacian equation for the pressure in terms of the velocity.

Since in the local Lagrangean coordinates $(\bar{x},\bar{t})$ on each slab S_n with $\beta = U$,

$$\frac{\partial \bar{U}}{\partial \bar{t}} \equiv \frac{\partial}{\partial \bar{t}} U(x(\bar{x},\bar{t}),\bar{t}) = \dot{U} + U\cdot\nabla U,$$

the convection term $U\cdot\nabla U$ effectively dissappears in the characteristic Galerkin method, when expressed in the characteristic coordinates $(\bar{x},\bar{t})$, and thus the discrete equations on each time step effectively correspond to a Stokes problem.

The order of the G^2-method with polynomials of degree p in space/time is generally $p + 1/2$, see [3]. The time stepping method in (17) is dG(q), the discontinuous Galerkin method with piecewise polynomials of order q, which is of order $2q+1$ seen as an ODE-solver, see [23].

5.1 Neumann boundary conditions

If we change to Neumann boundary conditions $\sigma\cdot n = g$ on a part Γ_1 of the boundary Γ, then W_{0n} is chosen to be the functions in W_n vanishing on the remaining Dirichlet part Γ_0 of the boundary, and the right hand side is supplemented with an integral over Γ_1 of $g\cdot v$. As usual this implements the Neumann boundary condition in weak form through the presence of the term $(-P, \operatorname{div} v) + (2\nu\epsilon(U), \epsilon(v)) = (\sigma, \epsilon(v))$ on the left hand side, which when integrated by parts generates an integral over Γ_1 of $(\sigma\cdot n)\cdot v$. If the viscous term appears in the form $(\nu\nabla U, \nabla v)_n$ the corresponding Neumann boundary condition has the form $\nu\frac{\partial u}{\partial n} - pn = 0$, where $\frac{\partial u}{\partial n}$ is the derivative in the unit outward normal direction n.

5.2 Outflow boundary conditions

To simulate an outflow boundary condition we may use a Neumann condition with $g = 0$ corresponding to a zero force at outflow, simulating outflow into a large empty reservoir. The alternative condition $\nu\frac{\partial u}{\partial n} - pn = 0$ acts slightly differently as an approximation of a *transparent* outflow boundary condition.

5.3 The Eulerian cG(1)dG(0) method

We now consider the the G^2-method (17) with $p = 1$, $q = 0$ and $\beta = 0$ for (1), which is the Eulerian cG(1)dG(0) method with continuous piecewise linears in space (cG(1)) and piecewise constants in time (dG(0)) corresponding to the backward Euler method). We then seek an approximate velocity $U(x,t)$ such that $U(x,t)$ is continuous and piecewise linear in x for each t, and $U(x,t)$ is piecewise constant in t for each x. Similarly, we seek an approximate pressure $P(x,t)$ which is continuous piecewise linear in x and piecewise constant in t. More precisely, we seek $U^n \in V_n^0 = W_{0n}^3$ and $P^n \in Q_n^0 = W_n$ for $n = 1, ..., N$, and we define

$$\begin{aligned} U(x,t) &= U^n(x) \quad x \in \Omega, \quad t \in (t_{n-1}, t_n], \\ P(x,t) &= P^n(x) \quad x \in \Omega, \quad t \in (t_{n-1}, t_n]. \end{aligned} \tag{19}$$

We can now write the cG(1)dG(0) method without stabilization as follows: For $n = 1, ..., N$, find $(U^n, P^n) \in V_n^0 \times Q_n^0$ such that

$$\begin{aligned} &(\frac{U^n - U^{n-1}}{k_n}, v) + (U^n \cdot \nabla U^n + \nabla P^n, v) + (\nabla \cdot U^n, q) \\ &+ (\nu \nabla U^n, \nabla v) = (f^n, v) \quad \forall (v,q) \in V_n^0 \times Q_n^0, \end{aligned} \tag{20}$$

The cG(1)dG(0) method with δ_1-stabilization takes the form: For $n = 1, ..., N$, find $(U^n, P^n) \in V_n^0 \times Q_n^0$ such that

$$\begin{aligned} &(\frac{U^n - U^{n-1}}{k_n}, v) + (U^n \cdot \nabla U^n + \nabla P^n, v + \delta_1 (U^n \cdot \nabla v + \nabla q)) + (\nabla \cdot U^n, q) \\ &+ (\nu \nabla U^n, \nabla v) = (f^n, v + \delta_1 (U^n \cdot \nabla v + \nabla q)) \quad \forall (v,q) \in V_n^0 \times Q_n^0, \end{aligned} \tag{21}$$

where $\delta_1 = \frac{1}{2}(k_n^{-2} + |U|^2 h_n^{-2})^{-1/2}$ in the convection-dominated case $\nu < U h_n$. Note that if $k \approx \frac{h}{U}$, which is a natural choice of time step respecting a CFL-condition, then $\delta_1 \approx \frac{h}{U}$. Note that the stabilized form of the cG(1)dG(0) method is obtained by replacing v by $v + \delta_1(U^n \cdot \nabla v + \nabla q)$ in the terms $(U^n \cdot \nabla U^n + \nabla P^n, v)$ and (f^n, v). In principle, we should make the replacement throughtout, but in the present case of the cG(1)dG(0), only the indicated terms get involved because of the low order of the approximations. The perturbation in the stabilized method is of size δ_1, and thus the stabilized method has the same order as the original method (first order in h if $k \sim h$).

Letting v vary in (21) while choosing $q = 0$, we get the following equation (the discrete momentum equation):

$$\begin{aligned} &(\frac{U^n - U^{n-1}}{k_n}, v) + (U^n \cdot \nabla U^n + \nabla P^n, v + \delta_1 U^n \cdot \nabla v) \\ &+ (\nu \nabla U^n, \nabla v) = (f^n, v + \delta_1 U^n \cdot \nabla v) \quad \forall v \in V_n^0, \end{aligned} \tag{22}$$

and letting q vary while setting $v = 0$, we get the following discrete "pressure equation"

$$(\delta_1 \nabla P^n, \nabla q)) = -(\delta_1 U^n \cdot \nabla U^n, \nabla q) - (\nabla \cdot U^n, q) + (\delta f^n, \nabla q) \quad \forall q \in Q_n^0. \tag{23}$$

The cG(1)dG(0) has a backward Euler first order accurate time stepping, and thus in general is too dissipative.

5.4 The Eulerian cG(1)cG(1) method

We now present the a cG(1)cG(1) variant of the above cG(1)dG(0) method using the continuous Galerkin method cG(1) in time instead of dG(0). With cG(1) in time the trial functions are continuous piecewise linear and the test functions piecewise constant. The cG(1)cG(1) variant with δ_1-stabilization reads: For $n = 1, ..., N$, find $(U^n, P^n) \in V_n^0 \times Q_n^0$ such that

$$\begin{aligned}&(\frac{U^n - U^{n-1}}{k_n}, v) + (\hat{U}^n \cdot \nabla \hat{U}^n + \nabla P^n, v + \delta_1(\hat{U}^n \cdot \nabla v + \nabla q)) + (\nabla \cdot \hat{U}^n, q)\\ &+ (\nu \nabla \hat{U}^n, \nabla v) = (f^n, v + \delta_1(\hat{U}^n \cdot \nabla v + \nabla q)) \quad \forall (v, q) \in V_n^0 \times Q_n^0,\end{aligned} \tag{24}$$

where $\hat{U}^n = \frac{1}{2}(U^n + U^{n-1})$. This method corresponds to a second order accurate Crank-Nicolson time-stepping, but the stabilization suffers from an inconsistency up to the term $\delta_1 \dot{u}$ resulting from the piecewise constancy of the test functions. The inconsistency seems to be be acceptable unless $\dot{u}$ is large, and we use cG(1)cG(1) in the computations presented below. The Eulerian cG(1)dG(1)-method would have consistent stabilization, but has two degrees of freedom in time per time step and thus twice as many degrees of freedom.

6 Discrete solvers

The cG(1)cG(1)-method with δ_1-stabilization leads to a system of the following principal form in each step of an outer fixed point iteration with the convection velocity being given from the previous iteration:

$$\begin{aligned}AU^n + k_n BP^n &= k_n F^n,\\ -B^\top U^n + CP^n &= G^n,\end{aligned} \tag{25}$$

where $A = M_n + k_n N_n - k_n \nu \Delta_n$ with M_n a mass matrix, N_n representing a discrete analog of the convection term with frozen velocity from the previous iteration, Δ_n is a discrete Laplacian, B is a discrete gradient, $B^\top$ a discrete divergence, and $C = -\delta_1 \Delta_n$. In the computations presented below we solved this system using a fixed point inner iteration, where we first solve for $P^{n,j+1}$ in terms of $U^{n,j}$ from the equation

$$CP^{n,j+1} = G^n + B^\top U^{n,j}$$

using a multigrid method, and then solve for $U^{n,j+1}$ from the equation

$$AU^{n,j+1} = k_n F^n - k_n BP^{n,j+1}$$

using GMRES. The inner iteration converges if k_n/δ_1 is small enough. Since typically $\delta_1 \approx h_n/U^n$, we need $\frac{U^n k_n}{h_n}$ to be small enough, which is a CFL-like condition.

We may also apply GMRES directly to the equation $AU^n + k_n BP^n = k_n F^n$ with P^n solved in terms of U^n from the equation $B^\top U^n + CP^n = G^n$ using multigrid. The number of GMRES iterations would then depend on the condition number of the matrix $M_n + k_n N_n - k_n \nu \Delta_n + \frac{k_n}{\delta_1} B \Delta_n^{-1} B^\top$, which is bounded with k_n/h_n, $k_n \nu / h_n^2$ and k_n/δ_1.

In both variants the full iterative procedure converges in a few iterations in our typical applications of non-stationary high Reynolds number flow with k_n/h_n and k_n/δ_1 bounded by 1.

7 A posteriori error estimates with stability weights

As an example, we now prove an a posteriori error estimate for (20) following our general methodology. Aiming at error control of $(e(T), \psi)$ with $e = u - U$ and $\psi \in [L_2(\Omega)]^3$ given, we introduce the following linearized dual problem: find $(\varphi, \theta) \in L_2(I; [H_0^1(\Omega)]^3 \times L_2(\Omega)) \equiv W$ such that in $Q = \Omega \times (0, T)$

$$\begin{aligned} -\dot{\varphi} - (u \cdot \nabla)\varphi + \nabla U \cdot \varphi + \nabla \theta - \nu \Delta \varphi &= 0 \quad \text{in } Q \\ \operatorname{div} \varphi &= 0 \quad \text{in } Q \\ \varphi &= 0 \quad \text{on } \Gamma \times I, \\ \varphi(\cdot, T) &= \psi \quad \text{in } \Omega, \end{aligned} \tag{26}$$

where $(\nabla U \cdot \varphi)_j = U_{,j} \cdot \varphi$. Multiplying the first equation by e, integrating over Q together with integration by parts, using that $(u \cdot \nabla)u - (U \cdot \nabla)U = (u \cdot \nabla)e + (e \cdot \nabla)U$, gives

$$(e(T), \psi) = \sum_{n=0}^{N} \{(-\dot{\varphi} - (u \cdot \nabla)\varphi + \nabla U \cdot \varphi, e)_n + (\nabla \theta, e)_n + (\nu \nabla \varphi, \nabla e)_n\}$$

$$= \sum_{n=1}^{N} \{(\varphi, e_t)_n + ((u \cdot \nabla)e, \varphi)_n + ((e \cdot \nabla)U, \varphi)_n$$

$$- (\theta, \operatorname{div} e)_n + (\nu \nabla \varphi, \nabla e)_n - (p - P, \operatorname{div} \varphi)_n\} + \sum_{n=1}^{N} ([U^{n-1}], \varphi_+^{n-1})$$

$$
\begin{aligned}
&= \sum_{n=1}^{N} \{(\dot{u} + u \cdot \nabla u + \nabla p, \varphi)_n + (\nu \nabla u, \nabla \varphi)_n \\
&- (\dot{U} + U \cdot \nabla U + \nabla P, \varphi)_n - (\nu \nabla U, \nabla \varphi)_n \\
&+ (\theta, \operatorname{div} U)_n\} + \sum_{n=1}^{N} ([U^{n-1}], \varphi_+^{n-1}) \\
&= - \sum_{n=1}^{N} \{(\dot{U} + U \cdot \nabla U + \nabla P - f, \varphi - \Phi)_n \\
&- (\nu \nabla U, \nabla(\varphi - \Phi))_n + (\operatorname{div} U, \theta - \Theta)_n\} + \sum_{n=1}^{N} ([U^{n-1}], \varphi_+^{n-1} - \Phi_+^{n-1}).
\end{aligned}
$$

Estimating now the interpolation errors $\varphi - \Phi$ and $\theta - \Theta$, and recalling the definition (18), we obtain an estimate of the form

$$
\begin{aligned}
|(e(T), \psi)| \leq \sum_{i=1}^{3} \int_Q R_i(U)(Ch^m |D^m \varphi| + Ck|\dot{\varphi}|) dx\, dt \\
+ \int_Q R_4(U)(Ch^m |D^m \theta| + Ck|\dot{\theta}|) dx\, dt,
\end{aligned}
\tag{27}
$$

for $m = 1, 2$, where D^m measures derivatives with respect to x of order m, and C represents interpolation constants. To get a concrete a posteriori error estimate, we solve the dual problem numerically and compute approximations of the derivatives of the dual solution involved. With adaptive choice of meshing, we choose $h_n(x)$ and k_n from a principle of equidistribution with the derivatives of the dual solution entering as weights.

8 A posteriori error estimates with stability factors

In the a posteriori error estimate (27), certain derivatives of the the dual solution appear as a weights in a space time integral over the residuals R_i. We may estimate the space-time integral in various ways; for instance using Cauchy's inequality with $L_2(I; (L_2(\Omega))$-norms $\|\cdot\|_I$ with $I = [0, T]$, we obtain an a posteriori error estimate of the form

$$
\begin{aligned}
|(e(T), \psi)| \leq C\|\dot{\varphi}\|_I \sum_{i=1}^{3} \|kR_i(U, P)\|_I \\
+ C\|D^2 \varphi\|_I \sum_{i=1}^{3} \|h^2 R_i(U, P)\|_I \\
+ C\|\dot{\theta}\|_I \|kR_4(U, P)\|_I + C\|D^1 \theta\|_I \|hR_4(U, P)\|_I
\end{aligned}
\tag{28}
$$

with $\|\dot{\varphi}\|_I$, $\|D^2\varphi\|_I$, $\|\dot{\theta}\|_I$ and $\|D^1\theta\|_I$ entering as multiplicative *stability factors*. Equidistribution can now be performed on product quantities such as kR_1 which have a mesh size factor and a residual factor.

9 Computation of lift and drag

Suppose we want to compute an approximation of the quantity

$$N(\sigma(u,p)) = \frac{1}{T}\int_0^T \int_{\Gamma_1} \sum_{i,j=1}^{3} \sigma_{i,j}(u,p) n_j \psi_i \, ds, \tag{29}$$

where $\Gamma = \Gamma_0 \cup \Gamma_1$ is a decomposition of the boundary Γ, and $\psi = (\psi_i)$ is a given function on Γ_1, and (u,p) solves (1). The quantity $N(\sigma)$ may represent the mean value over $[0,T]$ of the drag or lift on a body with boundary Γ_1 immersed in a flow, depending on the choice of ψ with for example $\psi_1 = \psi_2 = 0, \psi_3 = 1$ to give the lift if the x_3 is oriented vertically (and $u_3 \approx 0$). Instead of directly using (29), we may use the following alternative expression with the idea of increasing the precision, see [7],

$$N(\sigma(u,p)) = \frac{1}{T}\int_0^T ((\dot{u} + u\cdot\nabla u, \psi) - (p, \operatorname{div}\psi) + (2\nu\epsilon(u), \epsilon(\psi)))\, dt,$$

where ψ is an extension of the given ψ into Ω with $\psi = 0$ on Γ_0, which follows by integrating by parts in the last two terms, and using the momentum equation for the solution (u,p). We see that the representation does not depend on the particular extension of ψ being used. We are thus led to approximate $N(\sigma(u,p))$ by the quantity

$$N_h(\sigma(U,P)) = \frac{1}{T}\int_0^T ((\dot{U} + U\cdot\nabla U, \Psi) - (P, \operatorname{div}\Psi) + (2\nu\epsilon(U), \epsilon(\Psi)))\, dt$$

where Ψ is a finite element function satisfying $\Psi = \psi$ on Γ_1, assuming ψ is the restriction to Γ_1 of a finite element function and (U,P) is a finite element solution of (1). Again, the discrete momentum equation shows that $N_h(\sigma(U,P))$ is independent of the extension Ψ. Let now (φ,θ) be the solution of the linearized dual problem $\varphi(T) = 0$ and $\varphi(\cdot,t) = \psi$ on Γ_1 and $\varphi(\cdot,t) = 0$ on Γ_0 for $t \in [0,T]$. Reasoning as above we then obtain an a posteriori error estimate for $N(\sigma(u,p)) - N_h(\sigma(U,P))$ of the same form as (28) with corresponding associated stability factors.

10 Gallery of stability factors and dual problems

In this section we present computed stability factors for a set of problems including flow around a bluff body, drag of a bluff body, channel flow with a step down and a channel flow with jets. The dual solution carries information on the growth and propagation of perturbations, and in particular underlies the mesh selection in adaptive methods. We present plots of some dual solutions below. The dual solution may be vastly different depending on the data of the dual problem, with smooth data corresponding to large mean values. We present a selection of stability factors using the $L_1(0,T;L_1(\Omega))$-norm in space-time, denoted by $\|\cdot\|_I$ with $I=[0,T]$, for the dual solution corresponding to $L_\infty(0,T;L_\infty(\Omega))$ for the residuals, see Tab. 1. A variety of combinations of norms in space/time for the residuals and the dual solution are possible. The purpose of computing stability factors is to get a rough measure of relevant stability features. For more precise error estimation, the form of the a posteriori error estimates with (more or less local) stability weights, is advantageous.

We compute on tetrahedral meshes with mesh size $h = 1/32 - 1/64$, and the viscosity ν varies from 10^{-3} to 10^{-4}. The stabilization introduces a numerical vicosity, which may be of size $h^{3/2}$ at best, indicating that we compute with effective viscosities in the range 10^{-2} to 10^{-3}. We note that in this study we have linearized the dual problems at computed approximations of the primal solutions averaged over a regular tetrahedral mesh of size $h = 1/16$, which in the case of a highly irregular primal solution might lead to an under estimation of the stability factors.

Table 1. Stability factors, where $\|\cdot\|_I$ corresponds to the $L_1(I;L_1(\Omega))$-norm

$S_{0,1}$	$S_{0,2}$	$S_{1,1}$	$S_{1,2}$	$S_{1,3}$	$S_{1,4}$
$\|\varphi\|_I$	$\|\theta\|_I$	$\|\nabla\varphi\|_I$	$\|\dot\varphi\|_I$	$\|\nabla\theta\|_I$	$\|\dot\theta\|_I$

10.1 Bluff body

We consider channel flow with no slip walls in a 1×1 rectangular cross section of length 4 containing a cubic body of side length 0.25 with center at $(0.5,0,0)$, where $x=0$ is the inflow boundary. We impose a parabolic inflow condition $u=(16y(1-y)z(1-z),0,0)$, a transparant outflow condition, and we set $\nu = 10^{-3}$. We compute on a regular tetrahedral mesh, with $h=1/32$, using the cG(1)cG(1)-method. We start from $u=0$ at time $t=0$, and we compute to time $T=20$. We consider the problem of computing the space-time average of u_1 over the domain $\omega\times[T-d(\omega),T]$, where $\omega\subset\Omega$ is a cube centered at $(2.5,0.5,0.5)$ with side length $d(\omega)$. To estimate this error the appropriate data to the linearized dual problem is zero final data and a force $\psi=(\chi_\omega/(d(\omega)|\omega|),0,0)$ acting during the time interval $[T-d(\omega),T]$, with $\chi_\omega$the characteristic function of ω and $|\omega|$ the volume of ω. We give stability factors corresponding to a computation starting at $t = 12$ for different $d(\omega)$ in Tab. 2.

The residuals are of order 1, and the product of mesh size and stability factor increases with decreasing $d(\omega)$. This example supports our belief that pointwise quantities are more difficult to compute than mean values. In Fig.5 we note the decrease in the dual solution with (backward) time for $d(\omega)$ when the initial data is convected out of the computational domain, indicating that the error in the mean value of the solution over the time interval $[20-d(\omega), 20]$ is independent of the error in the solution for $t < 15$ in this case.

Table 2. Bluff body: stability factors

$d(\omega)$	$S_{0,1}$	$S_{0,2}$	$S_{1,1}$	$S_{1,2}$	$S_{1,3}$	$S_{1,4}$
1/16	6.0	3.4	121.1	33.1	16.3	57.7
1/8	4.9	3.1	78.6	19.2	12.1	35.8
1/4	3.3	2.6	46.2	11.2	7.8	18.1
1/2	1.7	1.9	20.0	5.2	3.7	8.7

Table 3. Step down: stability factors

$d(\omega)$	$S_{0,1}$	$S_{0,2}$	$S_{1,1}$	$S_{1,2}$	$S_{1,3}$	$S_{1,4}$
1/8	30.9	16.2	836.8	124.0	138.4	278.4
1/4	22.9	4.2	533.4	39.0	48.9	46.8
1/2	10.8	2.4	220.3	10.5	16.1	25.2

10.2 Step down

We consider now a channel with no slip walls and 1×1 rectangular cross section of length 4, and a step down of height and length 0.5. We have a parabolic inflow condition $u = (64(1-y)(y-0.5)z(1-z), 0, 0)$, and we use a transparent outflow condition. We set $\nu = 10^{-3}$, and we compute on a regular tetrahedral mesh, with $h = 1/32$, using the cG(1)cG(1)-method. We start from $u = 0$ at time $t = 0$, and during the start up phase we can follow the formation of the recirculation zone behind the step and also the formation of a corresponding zone where the flow separates from the top boundary.

We consider the problem of computing a space-time average over the time interval $[9, 10]$and a spatial cube $\omega \subset \Omega$ centered at $(1.5, 0.5, 0.5)$with side

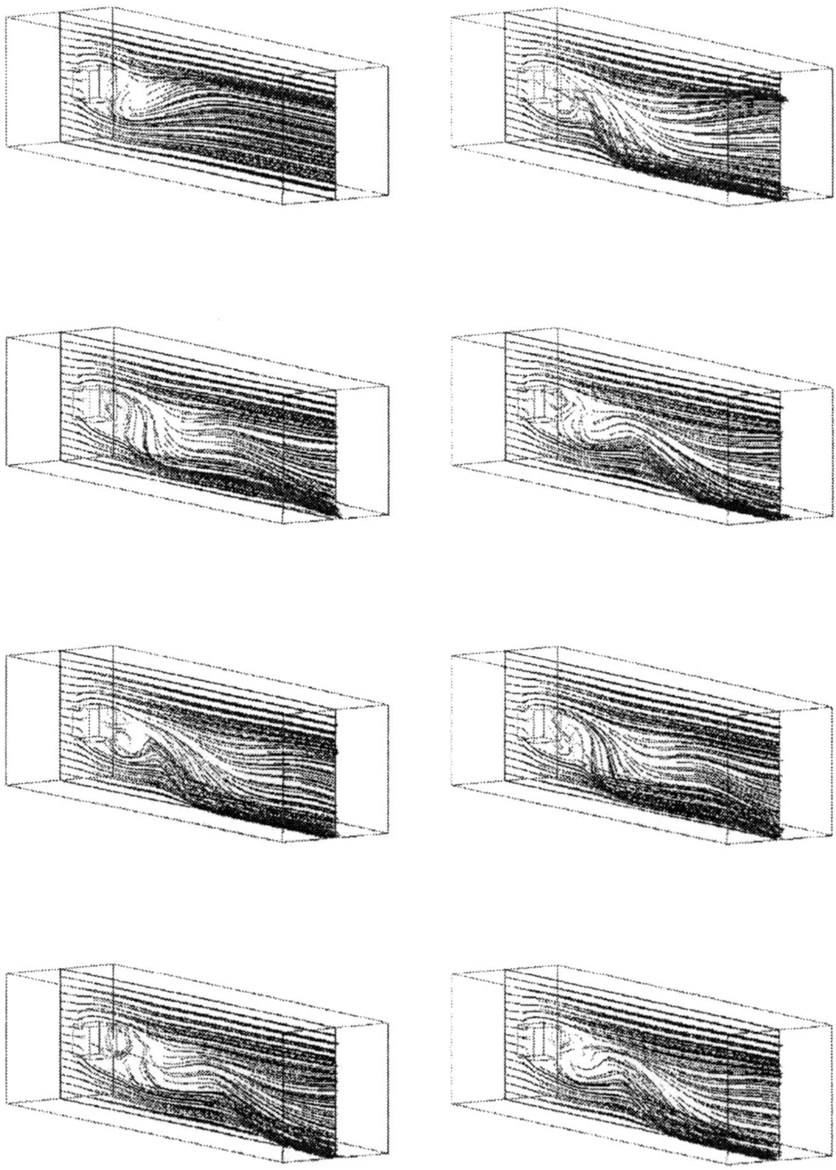

Fig. 4. Time evolution for the bluff body problem with $t = 2, 4, ..., 16$

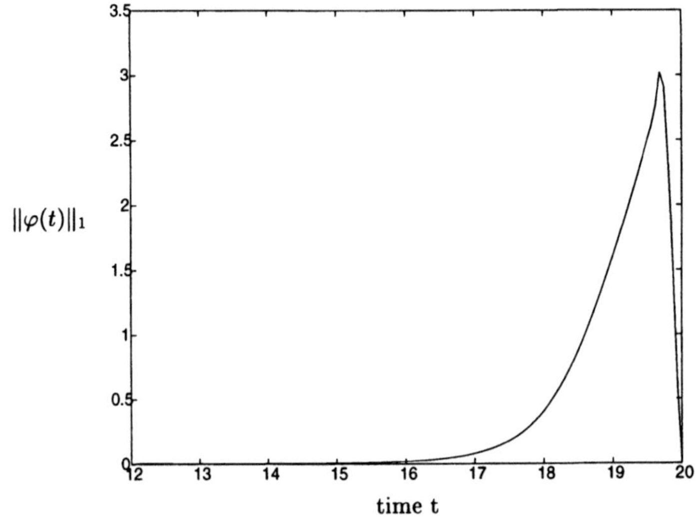

Fig. 5. Bluff body: $\|\varphi(t)\|_1$ for $d(\omega) = 1/4$.

length $d(\omega)$, starting the computation at $t = 5$. The residuals are of order 0.1, and in Tab.3 we present stablity factors corresponding to different $d(\omega)$.

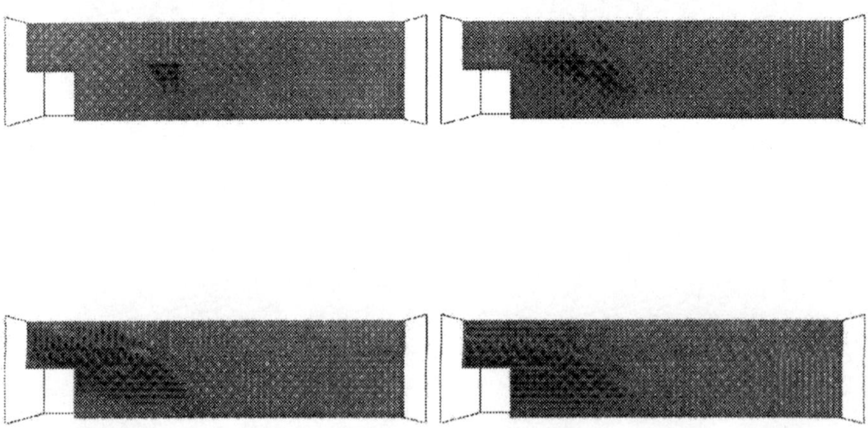

Fig. 6. Dual solution for step down problem for $d(\omega) = 1/8$, and backward time.

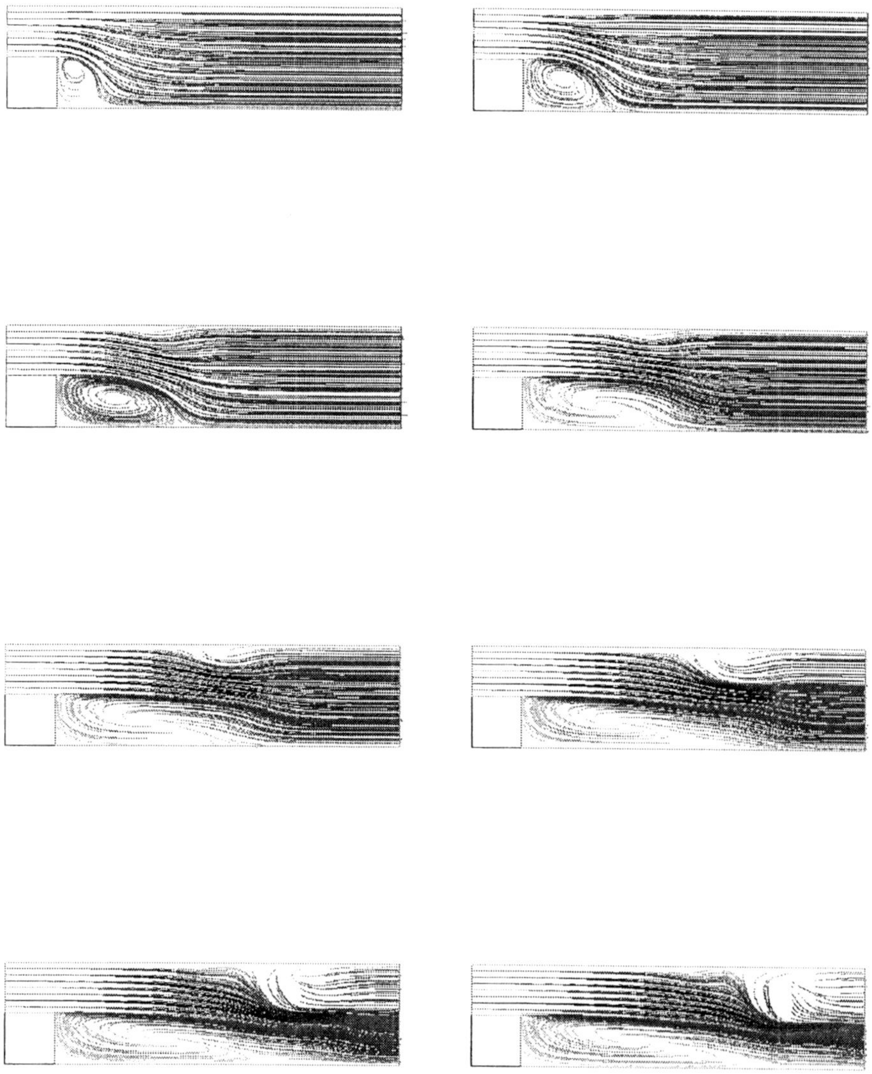

Fig. 7. Time evolution for the step down problem with $t = 1, 2, ..., 8$

10.3 Drag of Bluff body

We reconsider the bluff body problem with now the objective of computing the average of the drag force over a time interval [12, 20]. The corresponding data for the dual problem is a boundary condition $u = (1, 0, 0)$ on the faces of the bluff body. The stability factors for this averaged quantity is several orders of magnitude less than for the more local quantities again supporting the idea that averaged quantities are easier to compute. We also note that the norm of the dual solution after an initial (backwards in time) growth approaches a stable value, see Fig.8. The corresponding dual solution is shown in Fig.9. The dual solution does not go to zero as in the previous cases, indicating (not very surprising) that the computation of a time average is dependent of the quality of the solution during the whole time interval, in contrast to the previous cases in Section 10.1-10.2 where the dependence of the quality of the solution for previous time was decreasing with (backward) time.

Table 4. Bluff body drag: stability factors

$S_{0,1}$	$S_{0,2}$	$S_{1,1}$	$S_{1,2}$	$S_{1,3}$	$S_{1,4}$
0.13	0.11	1.6	0.03	0.43	0.06

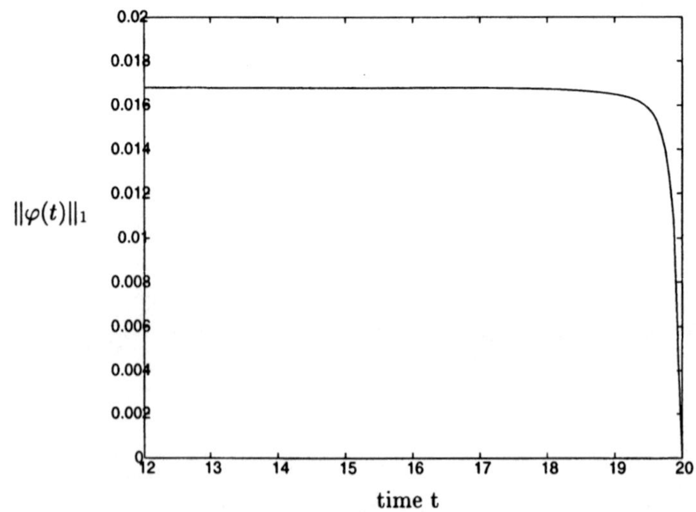

Fig. 8. Bluff body: $\|\varphi(t)\|_1$ for the drag problem.

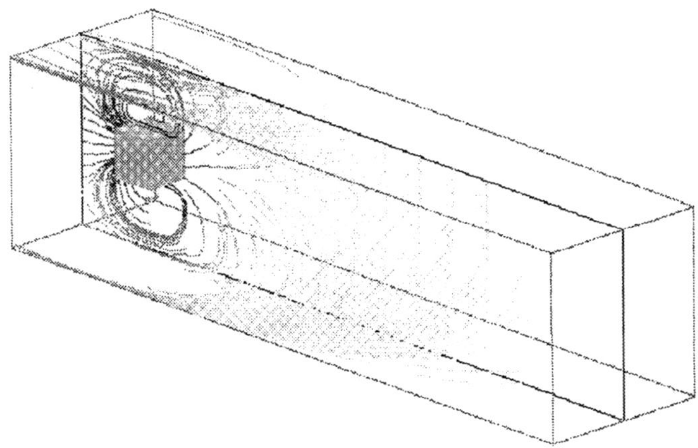

Fig. 9. Dual solution for the Drag problem at $t = 19$

10.4 Jets

We consider a channel with no slip walls and 1×1 rectangular cross section of length 4 with an obstacle with four quadratic holes of size 0.25 at $(0.5, 0, 0)$. We have a parabolic inflow condition and we use a transparant outflow condition. The inflow condition causes the flow to form 4 high velocity jets through the holes, and in the domain behind the obstacle we get a highly irregular flow pattern. We compute on a tetrahedral mesh, with $h = 1/32$, locally refined to $h = 1/64$ for $0.125 \leq x_1 \leq 1.125$, and $\nu = 10^{-4}$, using the cG(1)cG(1)-method. We start from $u = 0$ at time $t = 0$, and we compute to time $t = 10$. The residuals are of order 1, and in In Tab. 5 we consider the case of computing a space-time average over the spatial cube ω with $d(\omega) = 0.125$ for different time intervals $[T_0, 10]$, and we find that we get larger stability factors for the shorter time interval. We then fix the time interval to $[9, 10]$ and we vary the side length $d(\omega)$ of the spatial cube ω, resulting in larger stability factors for smaller $d(\omega)$, see Tab. 6.

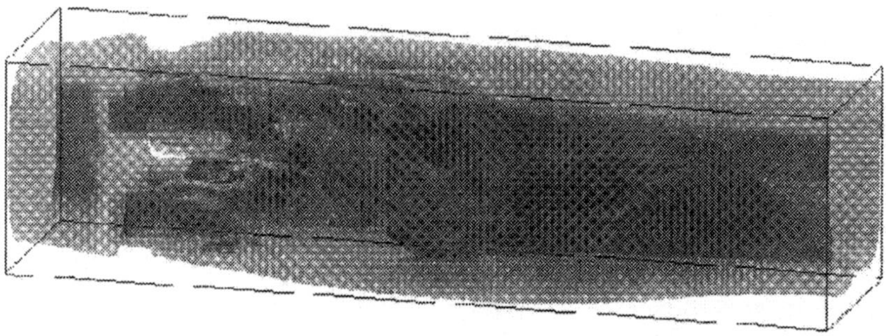

Fig. 10. Jet problem isosurfaces at time $t = 10$.

Table 5. Jets: stability factors corresponding to a space-time average over $\omega \times [T_0, 10]$, with $d(\omega) = 0.125$.

T_0	$S_{0,1}$	$S_{0,2}$	$S_{1,1}$	$S_{1,2}$	$S_{1,3}$	$S_{1,4}$
1/4	4.2	4.2	189.4	28.7	31.7	39.5
1/2	3.0	2.9	101.9	8.6	14.0	12.4

Table 6. Jets: stability factors corresponding to a space-time average over $\omega \times [9, 10]$, with side length $d(\omega)$.

$d(\omega)$	$S_{0,1}$	$S_{0,2}$	$S_{1,1}$	$S_{1,2}$	$S_{1,3}$	$S_{1,4}$
1/16	5.2	5.5	189.8	13.0	21.1	23.1
1/8	5.0	5.4	164.7	12.1	20.0	22.9
1/4	4.1	5.0	115.6	7.8	15.2	19.0
1/2	2.3	4.0	55.9	3.6	8.5	7.9

11 The characteristic Galerkin method for free boundary flow

The characteristic Galerkin method is ideally suited to handle flow problem with free boundaries moving with the fluid, or moving boundaries with prescribed motion: just let the nodes on the fluid boundary move according to (15) with U a computed or prescribed velocity. We present the chG-method for the Navier-Stokes equations for a moving blob of incompressible fluid with zero stress on the entire boundary. Denoting the volume occupied by the fluid at time t by $\Omega(t)$, with $\Omega(0)$ a given initial volume, the equations read: find

(u,p) such that for $t \in (0,T)$,

$$\begin{aligned} \dot{u} + (u \cdot \nabla)u - \operatorname{div}\sigma = f \quad & \text{in} \quad \Omega(t), \\ \operatorname{div} u = 0 \quad & \text{in} \quad \Omega(t), \\ \sigma(u,p) \cdot n = 0 \quad & \text{on} \quad \Gamma(t), \\ u(\cdot,0) = u_0 \quad & \text{in} \quad \Omega(0), \end{aligned} \tag{30}$$

where $\Gamma(t)$ is the boundary of $\Omega(t)$ with outward normal $n = n(t) = (n_j(t))$. In this problem the fluid volume $\Omega(t)$ is unknown, and is the image of the initial volume $\Omega(0)$ given by $\Omega(t) = \{x(\chi,t) : \chi \in \Omega(0)\}$, where $x(\chi,t)$ denote particle paths satisfying $\frac{dx}{dt} = u(x,t), \quad t > 0, \ , x(\bar{x},0) = \bar{x}$.

The characteristic Galerkin method for (30) based on the piecewise polynomial space W_n with no restriction on the boundary, can now be formulated as follows: find $(U,P) \in V^U \times Q^U$ such that

$$\begin{aligned} &(\dot{U} + (U \cdot \nabla U)U, v)_n - (P, \operatorname{div} v)_n + (q, \operatorname{div} U)_n + (2\nu\epsilon(U), \epsilon(v))_n \\ &+\delta_1(a(U;U,P), a(U;v,q))_n + (\delta_2 \operatorname{div} U, \operatorname{div} v)_n + \\ &\delta_3 < \sigma(u,p) \cdot n, \sigma(v,q) \cdot n >_n \\ &= (f, v + \delta_1(a(U;v,q))_n \quad \forall (v,q) \in V_n^U \times Q_n^U, \end{aligned} \tag{31}$$

where

$$\begin{aligned} (v,w)_n &= \int_{I_n} \int_{\Omega_h(t)} v \cdot w dx dt \\ < v, w >_n &= \int_{I_n} \int_{\Gamma_n(t)} v \cdot w ds dt, \end{aligned}$$

where $\Omega_h(t)$ with boundary $\Gamma_h(t)$ is given by $\Omega_h(t) = \{x(\bar{x},t) : \bar{x} \in \Omega(0)\}$ with $x(\bar{x},t)$ satisfying $\frac{dx}{dt} = U(x,t)$ for $t > 0$, and $x(\bar{x},0) = \bar{x} \in \Omega(0)$. In this case $\mathcal{T}_n$, is a triangulation of $\Omega_h(t_{n-1})$. We note that stress zero boundary condition is enforced weakly and that the δ_3-term gives an least squares control of the discrete boundary stress. The integrals are to interpreted as sums of integrals over the elements as usual.

In Fig 11 we present computational results by Thomas Svedberg for a 2d sloshing tank problem using the characteristic Galerkin method.

In the ALE method the mesh nodes on the boundary move with the particle velocity, while the nodes inside the domain move differently, for example according to an elastic string model with the nodes connected with elastic strings which amounts to solving a discrete Laplace equation. In this method interior mesh tangling is avoided.

12 Transition to turbulence in parallel flow

In this section we present aspects of transition to turbulence in parallel flow from [14], including a simple model for transition in pipe flow, and computational results for pipe flow and Couette flow. As a general reference into the

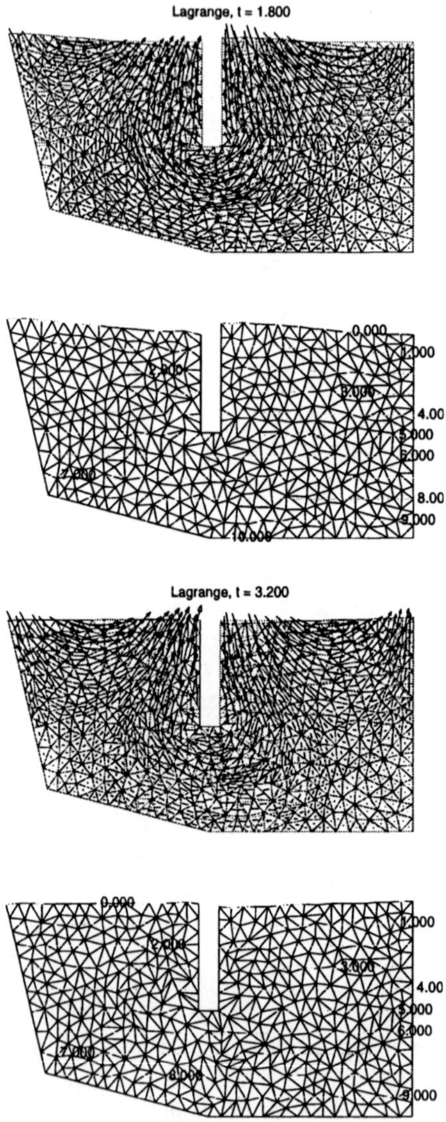

Fig. 11. Sloshing tank computation using the characteristic Galerkin method

large literature on transition, we refer to [28]. Our own early speculations are presented in [18] and [19].

12.1 Laminar pipe flow

We consider the basic case of laminar pipe flow in a pipe directed along the x_1 axis and with cross-section ω, with velocity $u = (u_1, 0, 0)$ and a computed velocity $U = (U_1, 0, 0)$. We assume u and U to depend on (x_2, x_3) and time t, but not on x_1. The associated linearized dual problem (26), takes the following form in this case, assuming also the perturbations to be x_1 independent and using that $\bar{U} = (U_2, U_3) = 0$,

$$\begin{aligned}
-\dot{\varphi}_1 - \nu\Delta\varphi_1 &= 0 && \text{in } \omega \times I,\\
-\dot{\varphi}_2 - \nu\Delta\varphi_2 + \theta_{,2} + U_{1,2}\varphi_1 &= 0 && \text{in } \omega \times I,\\
-\dot{\varphi}_3 - \nu\Delta\varphi_3 + \theta_{,3} + U_{1,3}\varphi_1 &= 0 && \text{in } \omega \times I,\\
\varphi_{2,2} + \varphi_{3,3} &= 0 && \text{in } \omega \times I,\\
\varphi &= 0 && \text{on } \partial\omega \times I,\\
\varphi(\cdot, T) &= \varphi^T && \text{in } \omega,
\end{aligned} \tag{32}$$

with given initial data φ^T. This problem has a decoupling analogous to that found above for the linearized perturbation equations for parallel flow. In the present dual linearized problem, φ_1 is decoupled from the Stokes-like system for $(\bar{\varphi}, \theta)$, where $\bar{\varphi} = (\varphi_2, \varphi_3)$, with $\varphi_1 \bar{\nabla} U_1$ appearing as a driving force. Accordingly, the weak stability factor $S_0(T)$ grows linearly with T or *Re*, indicating that laminar pipe flow is computable for moderately large Reynolds numbers.

12.2 An ODE-model for transition

We consider the following initial value problem for a system of two ordinary differential equations: find $w(t) = (w_1(t), w_2(t))$ such that

$$\begin{aligned}
\dot{w}_1 + \nu w_1 - \lambda w_1 w_2 &= \nu && t > 0,\\
\dot{w}_2 + 2\nu w_2 - \nu w_2 w_1 &= 0 && t > 0,\\
w_1(0) = 1, \quad w_2(0) &= \kappa\nu,
\end{aligned} \tag{33}$$

where ν is a small positive parameter, and λ and κ are positive parameters of moderate size. The system (33) models almost parallel shear flow with w_1 representing the flow velocity in the main direction of the flow, and w_2 the small velocities transversal to the main flow, and the stationary solution $w = (1, 0)$ corresponds to Couette flow between two plates or Poiseuille flow in a pipe. We shall use the model to describe how the small perturbation $\kappa\nu$ of w_2 may cause the base solution $(1, 0)$ to become unstable if $\lambda\kappa$ is larger than some critical value of moderate size.

We shall see that the model (33) contains an essential part of the secret of transition to turbulence in shear flow. The equations for w_1 and w_2 in (33) are coupled through the quadratic terms $\lambda w_1 w_2$ and $\nu w_1 w_2$, and model the following selection of terms from the Navier-Stokes equations

$$\begin{aligned} \dot{u}_1 - \nu\Delta u_1 + \bar{u}\cdot\bar{\nabla}u_1 &= \nu \qquad t > 0, \\ \dot{\bar{u}} - \nu\Delta\bar{u} + u_1\partial\bar{u}/\partial x_1 &= 0 \qquad t > 0, \end{aligned} \tag{34}$$

from the momentum equations for the main flow velocity u_1 and the transversal velocity $\bar{u}$. The nonlinear coupling terms $\bar{u}\cdot\bar{\nabla}u_1$ and $u_1\partial\bar{u}/\partial x_1$ are modeled in the form $\lambda w_1 w_2$ and $\nu w_2 w_1$, corresponding to assuming that $u_{1,i} = -\lambda u_1$, for $i = 2,3$, and $\partial\bar{u}/\partial x_1 = -\nu\bar{u}$, connecting transversal derivatives of u_1 with u_1 through the parameter λ, and the streamwise derivative of the transversal velocity $\bar{u}$ with $\bar{u}$ through the small parameter ν. The relation $\partial\bar{u}/\partial x_1 = -\nu\bar{u}$ models a basic feature of parallel flow with the streamwise variations being small. Since we assume initially that $\bar{u} \approx w_2 \approx \nu$, it corresponds to assuming $\partial\bar{u}/\partial x_1 \sim \nu^2$, which is indeed very small. On the other hand, the assumption that $u_{1,i} = -\lambda u_1$, for $i = 2,3$, with λ of moderate size corresponds to a natural transversal variation of moderate size of the streamwise velocity in a shear flow.

Note that the coupling term $u_{1,1}u_1$ in the equation for u_1 is not modeled in the form of some multiple of w_1^2. This is because (a) assuming $u_{1,1} \approx \nu^2$, with a corresponding very small term $-\nu^2 w_1^2$ in the model, has no destabilizing effect, and (b) assuming $u_{1,1} = -C\nu$ with a corresponding larger term $-C\nu w_1^2$, which may cause exponential growth through self-resonance in w_1, is not realistic. In fact, (b) is more or less the classical scenario based on the 2d Orr-Sommerfeld equations, which require artificially generated perturbation levels in experiments, for example through heavily vibrating ribbons.

In the transition model, we thus seek to build in realistic features of shear flow including realistic perturbation levels. If we assume zero perturbations, then the model reduces to $\dot{w}_1 + \nu w_1 = 0$, $\dot{w}_2 + 2\nu w_1 = 0$, which has no chance of going unstable. If we assume large perturbation levels, then instability may result immediately. However, none of these scenarios occur in reality, and the role of the model is to explain how small but realistic size perturbations, indeed may cause the initially stable base flow to go unstable after some time. Our model builds on the presence of a very small perturbation of order ν^2 of the transversal velocity in the streamwise direction, which naturally may be introduced through the roughness of the pipe. The model does not build on a larger variation of order ν of the streamwise velocity in the streamwise direction, which only seems to be possible with artificially generated perturbations.

The model (33) contains the two basic parameters λ and κ, both of moderate size, λ being related to the transversal geometry of the flow such as pipe cross section, $\kappa\nu$ representing a perturbation level in transversal velocities, and $\kappa\nu^2$ a perturbation level in streamwise derivatives of transversal velocities, including both transversal and streamwise perturbations levels. We will see that if $\lambda\kappa$ is larger than some critical value of moderate size, then transition to instability will take place in the model. This indicates that transition in shear flow builds on a combination of features related the transversal geometry and levels of perturbations in both transversal and streamwise direction. We give below computational results for transition in Couette flow and Poiseuille flow supporting this picture.

The system (33) has two stationary solutions $w = (1,0)$ and $w = (2, \nu/(2\lambda))$, with $(1,0)$ representing the basic Couette or Poiseuille flow. A classical stability analysis based on the eigenvalues of the corresponding linearized system, indicates that $(1,0)$ is stable and $(2, \nu/(2\lambda))$ is unstable. For example, the linear system obtained linearizing around $(1,0)$, takes the form

$$\begin{aligned} \dot\varphi_1 + \nu\varphi_1 - \lambda\varphi_2 &= \nu \qquad t > 0, \\ \dot\varphi_2 + \nu\varphi_2 &= 0 \qquad t > 0, \\ \varphi_1(0) = \varphi_{10}, \quad \varphi_2(0) &= \varphi_{20}, \end{aligned} \tag{35}$$

where the coefficient matrix $A = [\nu \ -\lambda, 0 \ \nu]$ has a double positive eigenvalue ν. The corresponding coefficient matrix linearizing around $(2, \nu/(2\lambda))$, has one positive (stable) and one negative (unstable) eigenvalue. A classical stability analysis shows that $(1,0)$ is stable under sufficiently small perturbations, and that $(2, \nu/(2\lambda))$ is unstable even under small perturbations. As a result $(1,0)$ is unstable under large perturbations bringing the initial value sufficiently close to the unstable solution $(2, \nu/(2\lambda))$. However, the classical eigenvalue stability analysis is unable to explain the intriguing fact that $(1,0)$ may become unstable even under a small perturbation of the initial data $(1,0)$, if we just have patience to wait! We will now present such a scenario of transition, where the stationary solution $(1,0)$ of (33) goes unstable under a small perturbation of initial data of the form $(0, \kappa\nu)$, where κ is a parameter of moderate size, and the scaling with ν makes the perturbation small (since we assume ν to be small). We shall see that if the product $\lambda\kappa$ is above a certain threshold of moderate size, then transition to instability will take place, if we wait over a period of time of length ν^{-1}.

We thus consider the problem (33) with the initial data $(1, \kappa\nu)$ close to $(1,0)$, and we ask if the corresponding solution $\bar w(t)$ may become unstable after some time. We see that $\dot{\bar w}_1(0)/\bar w_1(0) = \lambda\kappa\nu$, while $\dot{\bar w}_2(0)/\bar w_2(0) = -\nu$, which shows that initially $\bar w_1$ grows and $\bar w_2$ decays at rates $\propto \nu$. Now, $\bar w_1$ will continue to grow at that rate as long as $\lambda \bar w_2 > \nu$, and further $\bar w_2$ will start to grow as soon as $\bar w_1 > 2$. Thus, if $\bar w_1$ manages to become larger than 2, before $\bar w_2$ has decayed below ν/λ, then both components will propel each other to infinity, corresponding to instability. We shall see that this will occur if $\lambda\kappa$ is above a certain threshold. We notice that the time scale for significant changes in both $\bar w_1$ and $\bar w_2$ is $\sim \nu^{-1}$, which is a long time since ν is small. The scenario is thus that $\bar w_1$ grows slowly at the rate ν over a long time, and if $\lambda\kappa$ is above the threshold, then $\bar w_1$ may reach the value 2, where also $\bar w_2$ starts to grow after which a blow up follows on a usually somewhat shorter time scale (though still $\propto \nu^{-1}$). This scenario is easy to grasp intuitively, and conforms with the every-day experience of quite sudden blow-up, as a result of an accumulation of small events over a long period.

Solving the linearized equation (35) approximately describing the evolution of $\bar{w} - (1,0)$, we find that

$$\bar{w}_1(t) \approx 1 + \varphi_1 = 1 + \lambda\kappa t\nu \exp -t\nu, \quad \bar{w}_2(t) \approx \varphi_2 = \kappa\nu \exp -t\nu, \qquad (36)$$

which shows the slow growth of $\bar{w}_1$ and slow decay of $\bar{w}_2$ over the long time scale prior to the blow up, occuring if $\lambda\kappa$ is above the threshold. The linear growth in time of φ_1 may be viewed as a consequence of the non-normality of the coefficient matrix A. A classical stability analysis focussing on the double positive eigenvalue ν of $A = [\nu \ -1, 0 \ \nu]$, states that the factor $t \exp -\nu t$ eventually will decay to zero as $t \to \infty$, but misses the substantial transient growth to the level $\propto \nu^{-1}$ after time $\propto \nu^{-1}$ prior to decay. This perturbation growth of size $\propto \nu^{-1}$ is capable of bringing a solution from the point $(1, \kappa\nu)$ very close to $(1,0)$, into a neigborhood of the unstable point $(2, \nu(2\lambda))$ with ensuing blow up.

12.3 Computational transition in Couette and Poiseuille flow

To verify that the predictions of the model (33) applies in the case of nearly parallell flow we present computational results for two test problems: transition to turbulence in Couette and Poiseuille flow in a pipe along the x_1-axis with square cross section 1×1, where we use the cG(1)cG(1) method on a regular tetrahedral mesh with meshsize $h = 1/64$. We set the viscosity $\nu = 1/10000$, which formally would correspond a Reynolds number $Re = 10000$. But due to the stabilized numerical method in the computations we have a numerical viscosity of the order $h^{3/2}$, at best, which means that the effective Reynolds number in the computations is at most $Re \approx h^{-3/2} = 64^{3/2} = 512$, and we consider the computations as *direct numerical simulations* (DNS) of a flow with $Re \approx 200-500$. For both cases we start with a small initial transversal velocity perturbation $\varphi^0 = (0, \varphi_2(x_2,x_3), \varphi_3(x_2,x_3))$ of order 0.1, and we also apply a very small x_1-dependent driving force $f = (0, f_2(x_1), f_3(x_1))$ of order 10^{-3} creating and sustaining a very small streamwise variation of $\bar{u}$, modeling, for example, inperfections in the pipe.

The Couette base flow $u = (u_1, 0, 0)$ has a streamwise velocity profile $u_1 = 2x_2 - 1$, with streamwise velocity ± 1 on the top and bottom, slip side walls, and periodic boundary condition in the streamwise direction. Initially, as in the scenario predicted by the model (33), the streamwise perturbation φ_1 grows linearly through the action of the Taylor-Görtler mechanism, see Fig.12. In Fig.13 we can see the formation of high velocity streaks due to the Taylor-Görtler mechanism, shifting particles with different streamwise velocities causing the formation of high and low velocity streaks. The perturbations φ_2 and φ_3 decreases initially as predicted by the model (33). In the same way, derivatives with respect to x_2 and x_3 grow linearly for u_1, and

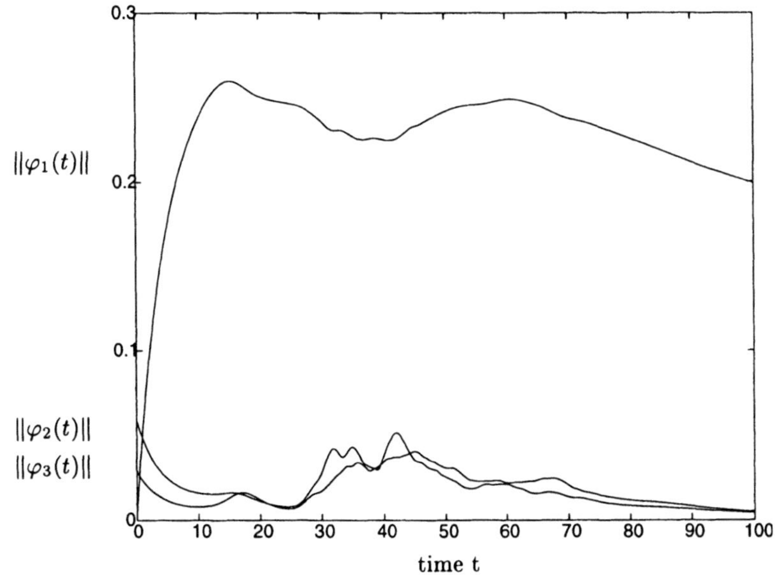

Fig. 12. Couette flow: Perturbations

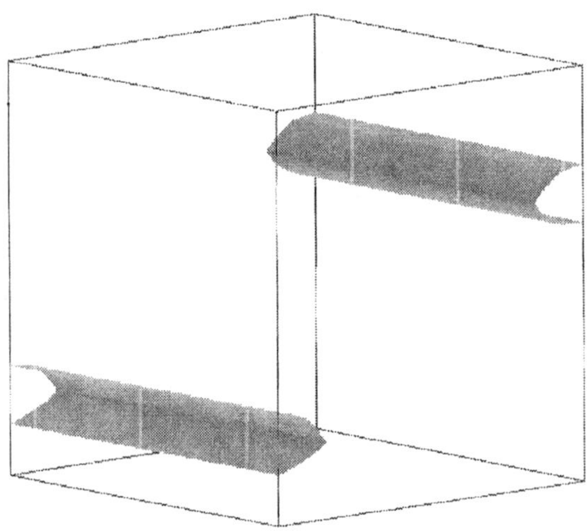

Fig. 13. Couette flow: Isoconcentration surfaces for φ_1 prior to the transistion, corresponding to high speed streaks

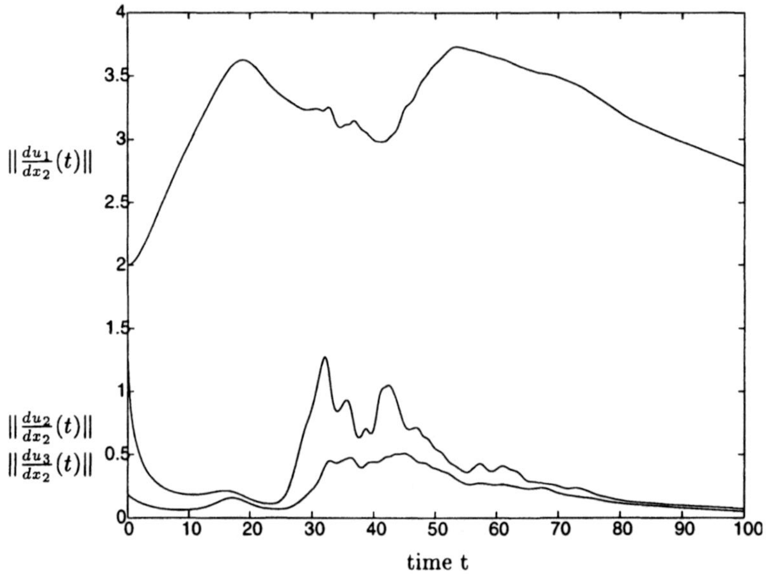

Fig. 14. Couette flow: x_2-derivatives

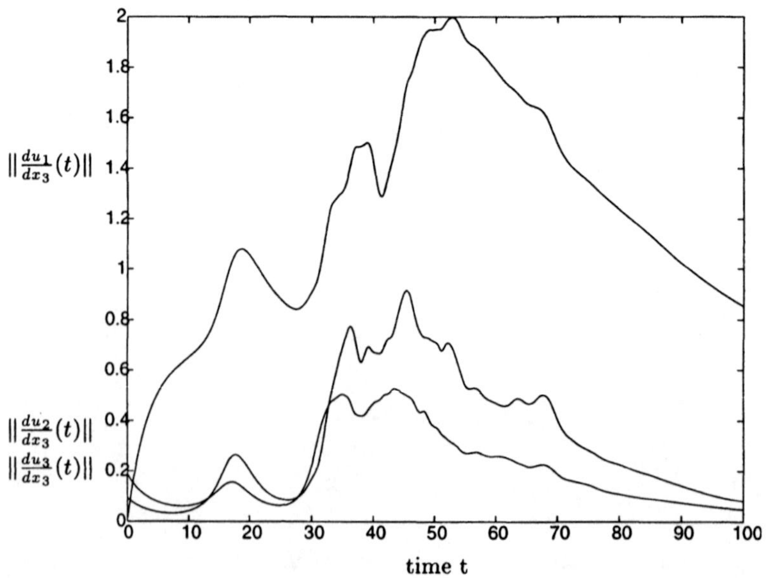

Fig. 15. Couette flow: x_3-derivatives

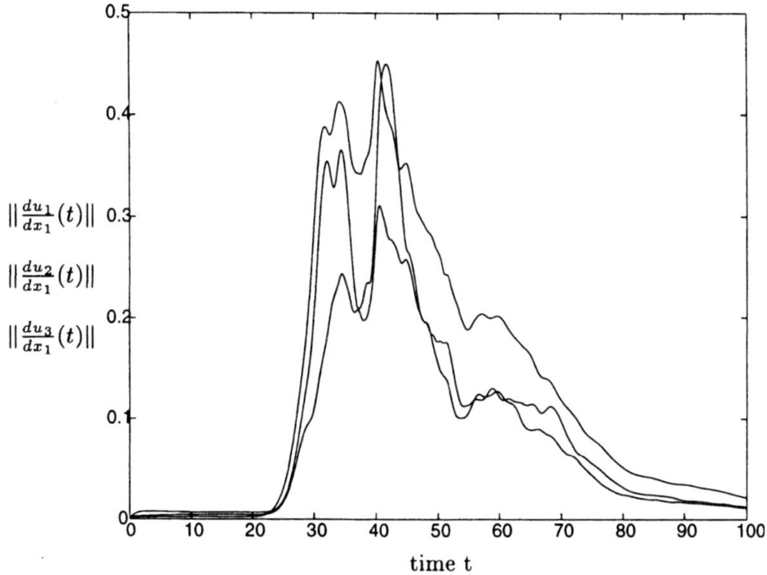

Fig. 16. Couette flow: x_1-derivatives

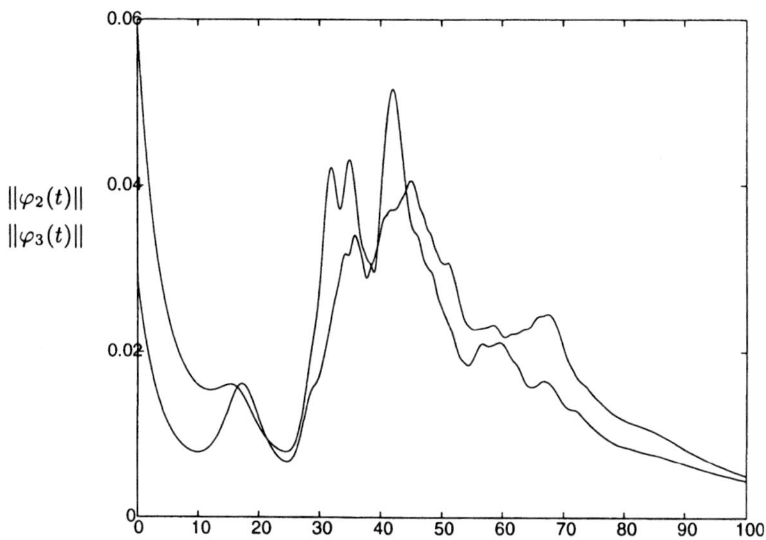

Fig. 17. Couette flow: x_2- and x_3-perturbations

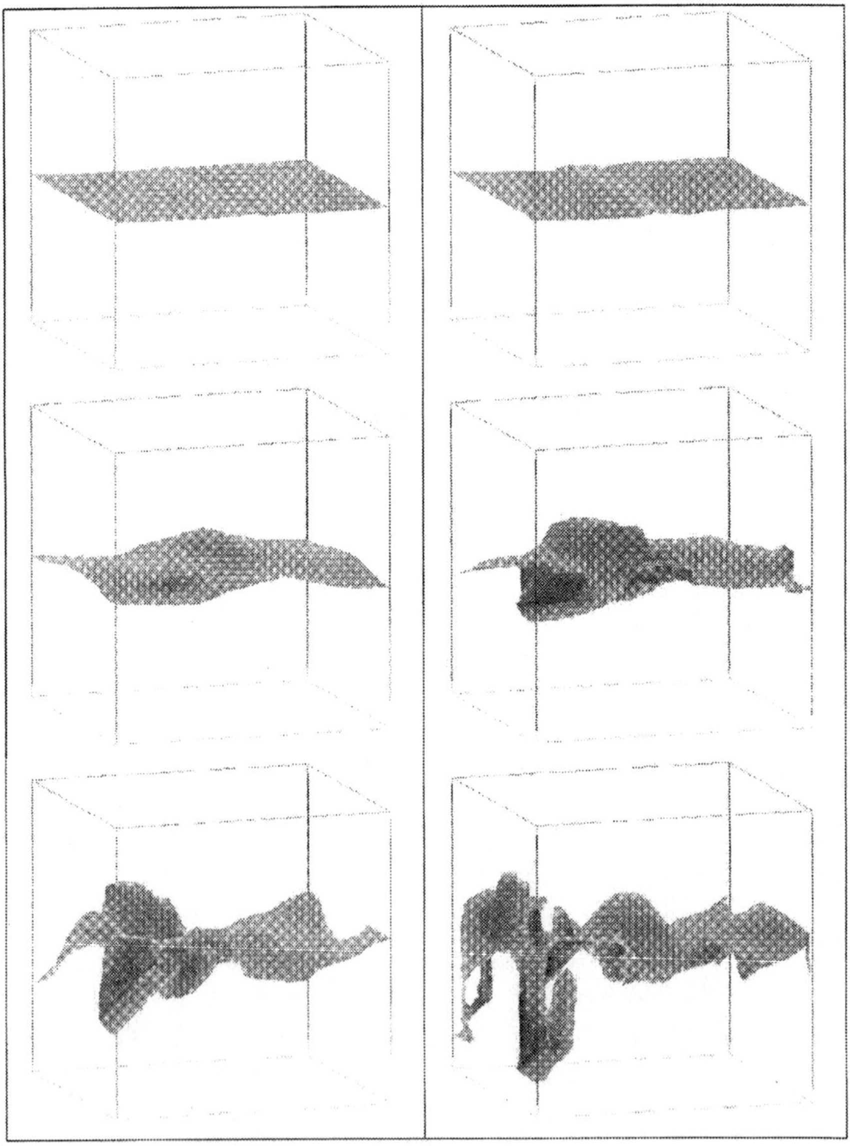

Fig. 18. Streamwise velocity zero isosurfaces in Couette flow for $t = 20, 25, 30, 32, 34, 40$, showing the initial phase of the transition

decreases for u_2 and u_3 (see Figs.14-15). In Fig.16 we see that near $t = 25$ we get a sudden burst where all x_1-derivatives increases by a factor 100 over a short time interval, corresponding to initial transition. We also get a sudden increase in φ_2 and φ_3, see Fig.17, at the same time. A key observation is that the transition is not possible until the perturbation φ_1 and the derivatives $\partial u_1/\partial dx_2$ and $\partial u_1/\partial dx_3$ has reached a certain threshold. Another important observation which is not obvious from studying the global norms is that the perturbations of course vary in space and that the threshold is a local condition that has to be satisfied. In Fig.18 we can follow the initial phase of the transition.

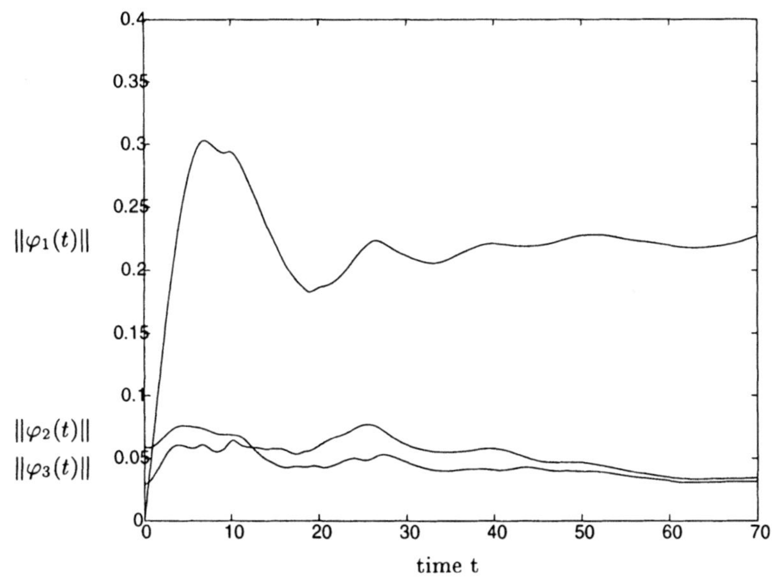

Fig. 19. Poiseuille flow: Perturbations

The Poiseuille base flow has a streamwise velocity profile $u_1(x_2, x_3) = 16y(1-y)z(1-z)$ in a channel with no slip walls and a force term $f = (32(y(1-y) + z(1-z)), 0, 0)$, where we use periodic boundary conditions in the streamwise direction. In Fig.19 we show the perturbation growth in the L_2-norm and note a linear growth in the streamwise perturbation corresponding to the Taylor-Görtler mechanism, whose action is shown in Fig.21 slowly shifting particles with different streamwise velocity transversally with a considerable reorganization of the streamwise velocity from the transversal perturbation. Fig.20 shows the L_2-norm of the x_1-derivatives in the solution as a function of time, with a sudden increase near $t = 6$. Again we note that this increase is not possible until the x_1-perturbation φ_1 is large enough.

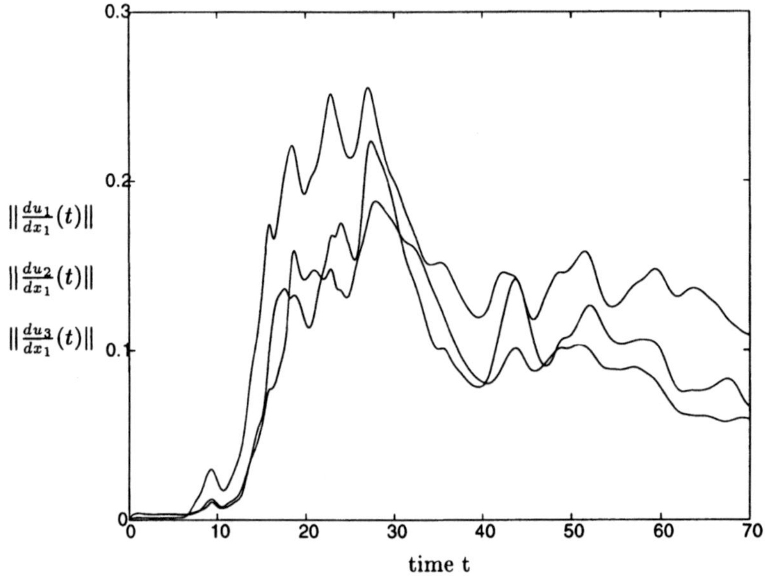

Fig. 20. Poiseuille flow: x_1-derivatives

The computational error in turbulent flow computations is investigated in [15], where the computations in this section are shown to be valid as a DNS if we consider the computations as simulations of flows with a viscosity corresponding to the numerical viscosity.

13 Turbulence modeling: DLES

We now turn to the topic of turbulence modeling. There are many different approaches to this problem, for an overview we refer to [6]. Here we are going to consider the case of *Large Eddy Simulations* (LES), where we compute the larger scales of motion in the flow and model the effect of the finer unresolved scales in a so called subgrid model. The basic problem is to model the divergence of the Reynolds stress tensor $(\tau_{ij}(u))$ with components

$$\tau_{ij}(u) = (u_i u_j)^h - u_i^h u_j^h,$$

where v^h represents a local running average of v in space of size h, in terms of u^h. The classical turbulent viscosity model is the *Smagorinsky model* with the components of τ_{ij} given by

$$\tau_{ij} - \frac{1}{3}\tau_{kk} = -2\nu_T \epsilon_{ij}(u^h),$$

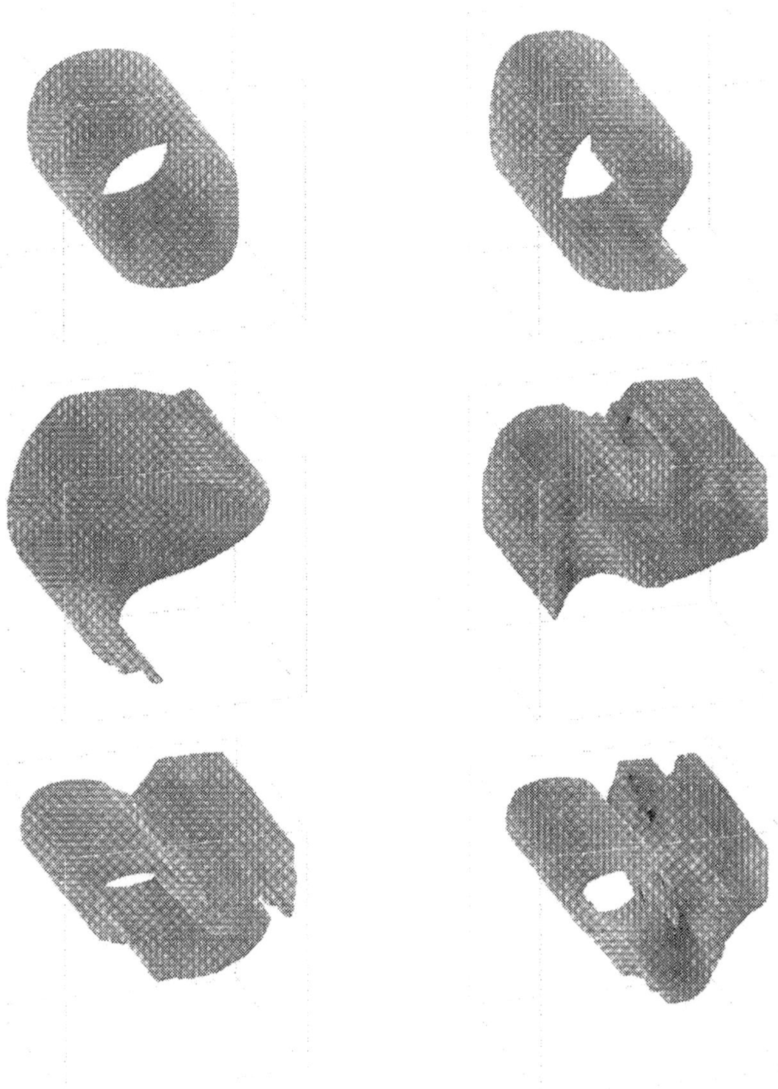

Fig. 21. Streamwise velocity isosurfaces in Poiseuille flow at $t = 1, 3, \ldots, 15$, illustrating the Taylor-Görtler mechanism

with

$$\nu_T = (C_S h)^2 |\epsilon(u^h)|$$

where C_S is the Smagorinsky constant.

In general, the resolved scales in a LES are assumed to lie in the so called inertial range (see [6]), which refers to a range of scales for which the energy spectrum has a simple power law behaviour, corresponding to scale similarity. Different types of scale similarity assumptions also on the Reynolds stresses have been used to motivate various types of subgrid models. For example, in *Dynamic Large eddy Simulation* (DLES) [8] the parameters in a particular model are determined by comparing resolved Reynolds stresses on different scales. In scale similarity models [26] the assumption is that the exact Reynolds stresses are proportional to the resolved Reynolds stresses.

In [10–13] we used an assumption based on the existence of a scale similarity with respect to a *Haar Multiresolution Analysis* (MRA) generated by the hierarchy of successively refined computational meshes. Scale similarity with respect to a Haar MRA has been observed, for example, in experimental aerothermal data [24]. We will base a subgrid model on an ansatz of the form

$$\tau_{ij}(x,t) \approx C(x,t)h^{\mu(x,t)} \tag{37}$$

with the coefficients $C(x,t)$ and $\mu(x,t)$ to be found by fitting the model on coarser scales.

We first discuss a similar subgrid modeling problem for reaction-convection-diffusion problems and present results based on scale similarity. We then discuss results from computation and experiments indicating that turbulent flow has some scale similarity features and thus may be open to the same approach.

13.1 Subgrid modeling in reaction-diffusion-convection problems

In recent years methods of *dynamic subgrid modeling* have been proposed, in particular in turbulence modeling in DLES by Germano (1991). The purpose of a subgrid model is to model the effect of unresolvable scales on resolvable scales corresponding to closure in turbulence modeling. The basic idea in dynamic subgrid modeling is to fit a particular subgrid model based on computed solutions on different resolvable scales, and then extrapolate the model to subgrid scales. In order for such a process based on scale extrapolation to work, it is necessary that the underlying problem has some 'scale similarity', so that the experience gained by fitting the model on a coarse scale with a fine scale solution as reference may be extrapolated to the finer scale. It is conceivable that many problems involving a range of scales from large to small, such as fluid flow at larger Reynolds numbers and flow in heterogeneous porous media, in fact do have such a scale similarity, once the larger scales related to the geometry of the particular problem have been resolved.

In [10] we initiated a study of dynamic subgrid modeling in a stationary convection-diffusion problem with fractal coefficients, where a subgrid model in the form of a corrective force is extrapolated from coarser grids with the finest computational grid as reference. We showed that the quality of the solution on mesh size h with extrapolated subgrid model, is comparable to that of a solution without subgrid model on a scale $h/4$ or smaller. We here present extensions to time-dependent convection-diffusion-reaction systems with 'fractal' solutions, where the fractality again originates from data (see [10–12]). The relevance of studying fractal models is motivated by the abundant number of experimental observations of fractality in turbulent flows, see e.g. [27] and references therein. Extensions to Navier-Stokes equations with the goal of connecting with DLES is under way.

In the setting of a general problem of the form $A(u) = f$ of Section 1.6 with exact solution u, the problem of subgrid modeling can be formulated as follows: For a given function v, let v^h represent a local average on a scale h which represents the finest computational scale. We seek an equation for the average u^h of the exact solution u and by averaging the equation $A(u) = f$ and we obtain $A(u^h) + F_h(u) = f^h$, where $F_h(u) \equiv (A(u))^h - A(u^h)$ has to be modeled in terms of u^h to get a modified equation $\hat{A}(\hat{u}) = A(\hat{u}) + \hat{F}_h(\hat{u}) = f^h = \hat{f}$, where the function $\hat{F}_h$ is the subgrid model and $\hat{F}_h(\hat{u})$is supposed to approximate $F_h(u)$. In dynamic subgrid modelling we seek to model $F_h(u)$ by extrapolation from computing $F_H(U_h)$ where $H > h$ and U_h is a computed solution on the scale h.

13.2 Systems of convection-diffusion-reaction equations

We consider a model problem of the form: Find $u : \Omega \times [0,T] \to R^n$ such that

$$\dot{u} + Lu = \dot{u} - \epsilon \Delta u + \beta \cdot \nabla u = f(u), \quad \forall (x,t) \in \Omega \times (0,T), \tag{38}$$

$$u = u_D, \quad \forall (x,t) \in \Gamma_D \times (0,T), \quad \frac{\partial u}{\partial n} = u_N, \quad \forall (x,t) \in \Gamma_N \times (0,T) \tag{39}$$

$$u(x,0) = u_0(x), \quad \forall x \in \Omega, \tag{40}$$

where $f : R^n \to R^n$ is smooth, $\Omega \subset R^d$ and $\partial\Omega = \Gamma_D \bigcup \Gamma_N$. Typically we will assume that ϵ is small and that the solution u to (38)–(40) contains a range of scales, from very small scales to large scales, induced either by the initial condition $u_0(x)$ or by the differential operator L through β. Assuming we want to find an approximation of u on the scale h, representing the finest spatial computational scale, we define for each fixed t the spatial *running average* u^h of u on the scale h by

$$u^h(x,t) = \frac{1}{h^d} \int_{x_1-h/2}^{x_1+h/2} \dots \int_{x_d-h/2}^{x_d+h/2} u(y,t)\, dy_1 \dots dy_d, \tag{41}$$

where we note that this operator commutes with space and time differentiation. Applying this operator to (38)–(40) we find that the running average u^h satisfies the following equation (modulo boundary effects)

$$\dot{u}^h + L_h u^h = \dot{u}^h + \beta^h \cdot \nabla u^h - \epsilon \Delta u^h = f(u^h) + F_h(u), \quad u^h(x,0) = u_0^h(x), \quad (42)$$

where L_h is a simplified operator on the scale h resulting from approximating β by β^h and the correction term $F_h(u) = (f(u))^h - f(u^h) + L_h u^h - (Lu)^h$ contains the influence of the unresolved scales on u^h. We consider a computational problem without subgrid model of the form

$$\dot{u}_h + L_h u_h = f(u_h), \quad u_h(x,0) = u_0^h(x), \quad (43)$$

and a corresponding problem with subgrid model of the form

$$\dot{\hat{u}}_h + L_h \hat{u}_h = f(\hat{u}_h) + \hat{F}_h(\hat{u}_h), \quad \hat{u}_h(x,0) = u_0^h(x), \quad (44)$$

where $\hat{F}_h(\hat{u}_h)$ should approximate $F_h(u)$. In this note we will consider a subgrid model of the form $\hat{F}_h(\hat{u}_h) = g(F_h(\hat{u}_h), F_{2h}(\hat{u}_h), F_{4h}(\hat{u}_h))$ based on averaging on the coarser scales $2h$ and $4h$, where the function g is derived based on a scale regularity assumption on $F_h(u)$.

13.3 Analysis of $F_h(u)$ using the Haar MRA

In the rest of this section we let $\Omega = [0,1]^2$ and for each $h = 2^{-i}$, with $i = 0, 1, ...,$ we define a corresponding regular quadratic mesh τ^h with elements corresponding to subdomains $\Omega_{i,k}$ with side length h. We denote the space of piecewise constant functions on τ^h by V_i, and the closure of the union of the V_j's is equal to $L_2(\Omega)$. The chain of closed subspaces $V_0 \subset V_1 \subset ... \subset V_j \subset ...$ is denoted a Haar Multi-resolution Analysis of $L_2(\Omega)$.

Each V_j is spanned by the dilates and integer translates of one *scale function* $\Phi \in V_0$, that is, $V_j = \text{span}\{\Phi_{j,k}(x) = 2^j \Phi(2^j x - k)\}$. The functions $\Phi_{j,k}$ form an L_2-orthonormal basis in V_j, and we denote the orthogonal complement of V_j in V_{j+1} by W_j, which is generated by another orthonormal basis (the *wavelets*) $\Psi_{j,k}(x) = 2^j \Psi(2^j x - k)$, where $\Psi \in W_0$ is called the *mother wavelet*. $W_j = W_j^1 \oplus W_j^2 \oplus W_j^3$, where the W_j^ν's represent differences in the horizontal, vertical and diagonal directions respectively. The space $L_2(\Omega)$ can now be represented as the direct sum $L_2(\Omega) = V_0 \oplus W_0^1 \oplus ... \oplus W_0^3 \oplus ... \oplus W_j^1 \oplus ... \oplus W_j^3 \oplus ...$, and each $f \in L_2(\Omega)$ has a unique decomposition $f = f_\Phi \Phi + \sum_{j,k} f_{j,k}^1 \Psi_{j,k}^1 + ... + f_{j,k}^3 \Psi_{j,k}^3 = f_\Phi + \sum_j f_j^1 + ... + f_j^3$, where the f_j^ν's represent the contributions on the different scales 2^{-j}. For the one dimensional Haar MRA in $L_2([0,1])$, the scale function is defined by $\varphi(x) = 1$ for $x \in [0,1]$ and 0 else, and the mother wavelet is defined by $\psi(x) = 1$ for $x \in (0,1/2)$, -1 for $x \in (1/2,1)$ and 0 else. In two dimensions the scale function and the wavelets are tensor products of the one dimensional scale function and wavelets. For the two dimensional Haar MRA in $L_2(\Omega)$ we have the scale function $\Phi(x_1, x_2) = \varphi(x_1)\varphi(x_2)$ and the wavelets $\Psi^1(x_1, x_2) = \varphi(x_1)\psi(x_2)$, $\Psi^2(x_1, x_2) = \psi(x_1)\varphi(x_2)$, $\Psi^3(x_1, x_2) = \psi(x_1)\psi(x_2)$. For $f \in L_2(\Omega)$, we define $[f]^h = f_\varphi + \sum_{j<i} f^1_{j,k}\Psi^1_{j,k} + f^2_{j,k}\Psi^2_{j,k} + f^3_{j,k}\Psi^3_{j,k}$. The linear mapping $L_2 \ni f \to [f]^h \in V_i$ can then be identified with the L_2-projection of f onto V_i, and we note that $[f]^h = \bar{f}^h$, where $\bar{f}^h$ is the piecewise constant function on τ^h that equals f^h in the midpoints of each element in τ^h. If we let $\bar{F}_h(u)$ denote the piecewise constant function on τ^h that equals $F_h(u)$ in the midpoints of the elements of τ^h, we have

$$\bar{F}_h(u) = [f(u)]^h - f([u]^h) - ([\beta \cdot \nabla u]^h - [\beta]^h \cdot [\nabla u]^h),$$

which for second order reaction terms $f(u)$ leads us to model covariances of the form

$$E_h(v, w) = [vw]^h - [v]^h[w]^h, \tag{45}$$

for given functions v and w. The following observation from [11] shows that $E_h(v, w)$ equals the sum of the Haar coefficients corresponding to scales finer than and equal h, scaled by the area of the elements in τ^h: if $x \in \Omega_{i,k}$ then

$$E_h(v, w)(x) = 2^{2i} \sum_{\substack{j \geq i \\ l \,:\, \Omega_{j,l} \subset \Omega_{i,k}}} (v^1_{j,l}w^1_{j,l} + v^2_{j,l}w^2_{j,l} + v^3_{j,l}w^3_{j,l}). \tag{46}$$

13.4 Scale extrapolation using self-similarity

We base our subgrid model on an ansatz of the form of each separate covariance

$$E_h(v, w)(x) \approx C(x)h^{\mu(x)}, \quad x \in \Omega, \tag{47}$$

with the coefficients $C(x)$ and $\mu(x)$ to be extrapolated. The ansatz can be motivated from (46) assuming scale similarity of the Haar coefficients corresponding to fractality. The ansatz leads to the following extrapolation formula: $\hat{F}_h(\hat{u}_h) = \sum \hat{E}_h(\hat{v}_h, \hat{w}_h)$, where

$$\hat{E}_h(\hat{v}_h, \hat{w}_h) = g(E_h(\hat{v}_h, \hat{w}_h), E_{2h}(\hat{v}_h, \hat{w}_h), E_{4h}(\hat{v}_h, \hat{w}_h)) \tag{48}$$

with

$$g(a, b, c) = (1 - (\frac{c - b^{4h}}{b^{4h} - a^{4h}})^{-n}) \frac{b^{4h} - a^{4h}}{\frac{c - b^{4h}}{b^{4h} - a^{4h}} - 1} \tag{49}$$

and $2^{-(n+i)}$ is the finest scale present in the exact solution.

13.5 Applications

In all examples we will have $h = 2^{-5}$. We construct two dimensional fractal data as sums of local tensor products of the one dimensional fractal *Weierstrass function* $W_{\gamma,\delta}(x) = \gamma \sum_{j=0}^{N} 2^{-j\delta} \sin(2^j \cdot 2\pi x)$, where we let $\gamma = \delta = 0.1$ in all examples.

Volterra-Lotka (VL) We consider a reaction dominated problem of the form

$$\dot{u}_1 - \epsilon \Delta u_1 = u_1(1-u_2), \quad \dot{u}_2 - \epsilon \Delta u_2 + \beta \cdot \nabla u_2 = u_2(u_1 - 1),$$
$$\frac{\partial u}{\partial n}|_{\partial \Omega} = 0, \quad u(x,0) = (W^{2D}_{\gamma,\delta}(x), 1),$$

where $\epsilon = 10^{-6}$, which corresponds to the classical Volterra-Lotka system with small diffusion and convection in one component. We have $F_h(u) = (-(u_1u_2)^h + u_1^h u_2^h, (u_1u_2)^h - u_1^h u_2^h)$, and with subgrid model we have that $\hat{F}_h(\hat{u}_h) = (-\hat{E}_h(\hat{u}_1, \hat{u}_2), \hat{E}_h(\hat{u}_1, \hat{u}_2))$, with $n = 4$ (the 2-logarithm of the reference scale minus the computational scale). For these problems we use a central difference-Crank-Nicolson scheme for the midpoints of the elements, where we let these midpoint values represent a piecewise constant approximation over the elements, and the reference scale is 2^{-9} in the computation of the error. The solutions are oscillating and both u_1 and u_2 are fractal for $t > 0$, even though $u_2(x,0) = 1$. We want to approximate u^h, and the errors $\|u^h - U_h\|$ are shown in Tab.7-9, where U_h is the solution without model, without model but computed on the finer scale $h/2$ and then projected onto the scale h, with the subgrid model (49), and with a simplified model $\hat{F}_h(\hat{u}_h) = F_{2h}(\hat{u}_h, \hat{u}_h)$ corresponding to a standard scale similarity model. We first let $\beta = 0$ and compute to $T = 2$, then we let β be a rotational mixing of order h: $\beta = h\ (\sin(\pi x_1)\cos(\pi x_2), -\cos(\pi x_1)\sin(\pi x_2))$ and here we only compute to $T = 1$, since after $T = 1$ the subgrid scales are dominated by the convective streaks due to β. We study the error for each component individually, and for $\beta = 0$ we find that the solution with the subgrid model is the best for both components, even though the modeling errors are smaller in u_2 since $u_2(x,0)$ is constant. For $\beta \neq 0$ the solution with the subgrid model is best for u_1 but, because of the convection, subgrid scales in u_2 do not develop and the solutions with subgrid models does not differ significantly from the solution without subgrid model. The solution on $h/2$ is better since the discretization error is then reduced.

Table 7. Error in $u_1$in L_1-norm for VL with $\beta = 0$

subgrid model	$t = 0.5$	$t = 1.0$	$t = 1.5$	$t = 2.0$
no model	$1.9 \cdot 10^{-3}$	$7.2 \cdot 10^{-3}$	$1.1 \cdot 10^{-2}$	$1.1 \cdot 10^{-2}$
no model on $h/2$	$1.4 \cdot 10^{-3}$	$4.6 \cdot 10^{-3}$	$7.0 \cdot 10^{-3}$	$6.8 \cdot 10^{-3}$
$\tilde{F}_h(\tilde{u}_h) = F_{2h}(\tilde{u}_h, \tilde{u}_h)$	$1.4 \cdot 10^{-3}$	$4.8 \cdot 10^{-3}$	$6.5 \cdot 10^{-3}$	$6.3 \cdot 10^{-3}$
formula (49)	$1.0 \cdot 10^{-3}$	$2.8 \cdot 10^{-3}$	$3.8 \cdot 10^{-3}$	$4.2 \cdot 10^{-3}$

Table 8. Error in $u_2$in L_1-norm for VL with $\beta = 0$

subgrid model	$t = 0.5$	$t = 1.0$	$t = 1.5$	$t = 2.0$
no model	$1.4 \cdot 10^{-4}$	$3.4 \cdot 10^{-4}$	$2.0 \cdot 10^{-4}$	$4.4 \cdot 10^{-4}$
no model on $h/2$	$1.0 \cdot 10^{-4}$	$2.1 \cdot 10^{-4}$	$1.2 \cdot 10^{-4}$	$2.9 \cdot 10^{-4}$
$\tilde{F}_h(\tilde{u}_h) = F_{2h}(\tilde{u}_h, \tilde{u}_h)$	$9.0 \cdot 10^{-5}$	$2.0 \cdot 10^{-4}$	$1.5 \cdot 10^{-4}$	$2.8 \cdot 10^{-4}$
formula (49)	$7.0 \cdot 10^{-5}$	$1.4 \cdot 10^{-4}$	$1.5 \cdot 10^{-4}$	$2.4 \cdot 10^{-4}$

Table 9. Error in $u_1$in L_1-norm for VL with $\beta \neq 0$

subgrid model	$t = 0.5$	$t = 1.0$	$t = 1.5$	$t = 2.0$
no model	$2.5 \cdot 10^{-5}$	$1.5 \cdot 10^{-4}$	$3.4 \cdot 10^{-4}$	$6.4 \cdot 10^{-4}$
no model on $h/2$	$2.5 \cdot 10^{-5}$	$1.1 \cdot 10^{-4}$	$1.9 \cdot 10^{-4}$	$3.3 \cdot 10^{-4}$
$\tilde{F}_h(\tilde{u}_h) = F_{2h}(\tilde{u}_h, \tilde{u}_h)$	$2.5 \cdot 10^{-5}$	$1.1 \cdot 10^{-4}$	$2.1 \cdot 10^{-4}$	$5.0 \cdot 10^{-4}$
formula (49)	$2.0 \cdot 10^{-5}$	$6.0 \cdot 10^{-5}$	$2.0 \cdot 10^{-4}$	$5.6 \cdot 10^{-4}$

Fractal convection (FC) We now consider a convection dominated problem of the form

$$\dot{u} + \beta \cdot \nabla u - \epsilon \Delta u = 1, \quad u|_{x_1=0, x_2=0} = 0, \quad \frac{\partial u}{\partial n}|_{x_1=1, x_2=1} = 0, \quad u(x, 0) = 0, \tag{50}$$

for $\beta = (W^{2D}_{\gamma,\delta}, W^{2D}_{\gamma,\delta})$ and $\epsilon = 10^{-3}$, which we solve by a Streamline Diffusion cG(1)cG(1) method [2] with bilinear, quadratic elements. In the computation of the error the reference scale is 2^{-8}. The solution is in this case relatively smooth since the fractal field β is now only acting on the derivatives of the solution. We have that $F_h(u) = \beta^h \cdot (\nabla u)^h - (\beta \cdot \nabla u)^h$, and the error in the solution with the subgrid model is smaller than in the solution without subgrid model.

Table 10. Error in L_1-norm for FC

subgrid model	$t = 0.25$	$t = 0.5$	$t = 0.75$	$t = 1.0$
no model	$7.2 \cdot 10^{-3}$	$1.2 \cdot 10^{-2}$	$1.1 \cdot 10^{-2}$	$1.1 \cdot 10^{-2}$
formula (49)	$5.8 \cdot 10^{-3}$	$7.2 \cdot 10^{-3}$	$6.2 \cdot 10^{-3}$	$6.2 \cdot 10^{-3}$

13.6 Do we have scale similarity in turbulent flows?

The above discussion is based on an assumption of scale similarity of covariances of individual solution components, with respect to the Haar basis induced by a mesh hierarchy. In 3d we get a corresponding Haar basis induced by the tetrahedral mesh hierarchy used in the above computations of transition to turbulence, see Fig.22. In Fig.23 we have plotted the sum of Haar coefficients $\sum_N$ on three different scales $N = 2h, 4h, 8h$, of the Reynolds stress component τ_{11}^{8h} ($h = 1/64$) for the Couette flow from Section 12.3 at $t = 30$, which has started its transition to turbulence but is not fully developed (the component τ_{11} is chosen since the streamwise velocity dominates the Couette flow). The plot shows 60 elements in the coarsest mesh corresponding to the scale $8h$, and we see that the decrease in the Haar coefficients is reasonable regular and we consider this as some evidence of scale similarity of the Reynolds stresses for this flow, although this test is not satisfactory because the turbulence is not fully developed and the computational mesh is too coarse to represent a fully developed turbulent flow. For experimental observations of fractality in turbulent flows we refer to e.g. [24] or [27] and references therein.

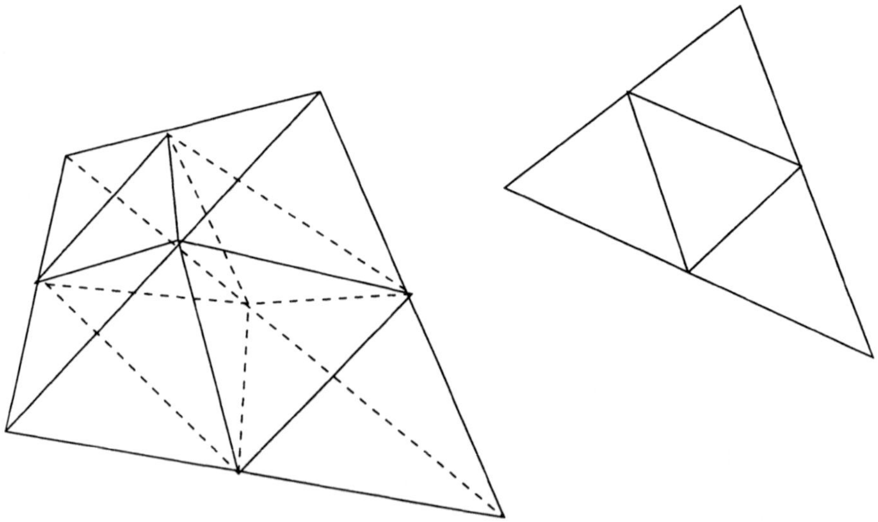

Fig. 22. Haar MRA generated from successively refining unstructured tetrahedral and triangular meshes, in 3d and 2d respectively.

13.7 A posteriori error estimation for turbulent flow

In deriving a posteriori error estimates for turbulent flows we have to take into account both the numerical error from discretization and the modeling

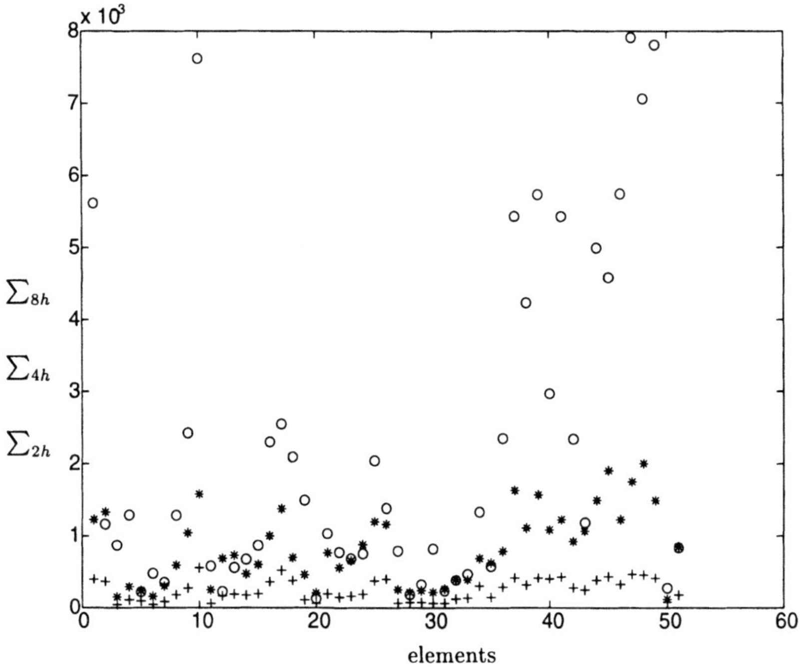

Fig. 23. Couettte flow at $t = 21$: sum of Haar coefficients $\sum_N$ for $\tau_{11}^{8h}(U_h)$ on scales $N = 8h$ ('o'), $N = 4h$ ('$*$'), and $N = 2h$ ('$+$')

error from subgrid modeling. Adaptivity and error control for turbulent flow computations, including subgrid modeling, is investigated in [15], and here we explain the basic steps in the setting of a general subgrid modeling problem $A(u) = f$, where u is the exact solution. Due to subgrid scales in u we can not compute a pointwise accurate approximation to u. Instead we aim for a pointwise accurate approximation of u^h, where u^h represent a local average of u on a scale h, which represents the finest computational scale. We seek an equation for the average u^h of the exact solution u and by averaging the equation $A(u) = f$ we obtain $A(u^h) + F_h(u) = f^h$, where $F_h(u) \equiv (A(u))^h - A(u^h)$ has to be modeled in terms of u^h to get a modified equation $\hat{A}(\hat{u}) = A(\hat{u}) + \hat{F}_h(\hat{u}) = f^h = \hat{f}$, where $\hat{u}$ is an approximation of u^h and $\hat{F}_h(\hat{u}) \approx F_h(u)$ represents a subgrid model.

We then solve the Galerkin equation: find $U_h \in V_h$ such that $(A(U_h) + \hat{F}_h(U_h), v) = f^h$, for all $v \in V_h$, where V_h is a finite dimensional subspace, on a mesh of size h, of the Hilbert space V with inner product $(\cdot,\cdot)$ and norm $\|\cdot\|$. U_h is now an approximation of u^h and the error $e = u^h - U_h$ is therefore the appropriate error to study, not $u - U_h$. This means that the appropriate linearized dual problem to solve is $(A'(u^h, U_h)w, \varphi) = (w, \psi)$, for all $w \in V$, where $A'(u^h, U_h)e = A(u^h) - A(U_h)$, that is we linearize the dual problem at u^h and not at the exact solution u. When we solve the dual problem numerically we do not have access to u^h and we will therefore have to use U_h instead of u^h. The linearization error is then $u^h - U_h$, which we anticipate to be pointwise small. If we on the other hand linearize at the exact solution u we get a linearization error $u - U_h$, which is not pointwise small. We get the following error representation

$$\begin{aligned}(e, \psi) &= (u^h - U_h, \psi) = (A'(\hat{u}^h, U_h)e, \varphi) = (A(u^h) - A(U_h), \varphi)\\ &= (f^h - F_h(u) - A(U_h), \varphi) = (f^h - A(U_h) - \hat{F}_h(U_h), \varphi)\\ &+ (\hat{F}_h(U_h) - F_h(u), \varphi) = (\hat{R}(U_h), \varphi) + (\hat{F}_h(U_h) - F_h(u), \varphi),\end{aligned}$$

where $R(U_h) = f^h - A(U_h) - \hat{F}_h(U_h)$ is a computable numerical residual related to the discretization error in solving the equation $A(\hat{u}) + \hat{F}_h(\hat{u}) = f^h$, and $\hat{F}_h(U_h) - F_h(u)$ is a modeling residual related to the error in the subgrid model $\hat{F}_h$. We note that the linearized dual problem is independent of both $F_h(u)$ and $\hat{F}_h$, and thus contain the stability properties of the average u^h.

Even if we do not use a turbulence model in computing U_h we will have to estimate $F_h(u)$, possibly by using a turbulence model. Using a turbulence model $\hat{F}_h$ in computing U_h, we instead need to estimate $F_h(u) - \hat{F}(U_h)$.

14 Multi-adaptivity: individual time-steps

We finally present the multi-adaptive Galerkin methods mcG(q) and mdG(q) with individual time steps for different solution components developed in [23]. We consider an initial value problem of the form

$$\begin{aligned} u_t + f(t,u) &= 0 \qquad \text{for } 0 < t < T,\\ u(0) &= u_0, \end{aligned} \tag{51}$$

where $f(t,\cdot) : \mathbb{R}^N \to \mathbb{R}^N$ is a given mapping, $u_t = \frac{du}{dt}$, and T is a final time. To formulate the mcG(q)-method, we partition the interval $(0,T)$ individually for the different components with individual time-intervals $\{I_{ij}\}_j$ and time-steps $\{k_{ij}\}_j$ for every individual component $U_i(t)$. The mcG(q)-method for (51) reads: Find $U \in V$ with $U(0) = u_0$, such that

$$\int_0^T (\dot{U}, v)\, dt + \int_0^T (f(U,\cdot), v)\, dt = 0 \;\forall v \in W,$$

where

$$\begin{aligned} V &= \{v \in C([0,T]) : v_i|_{I_{ij}} \in \mathcal{P}^{q_{ij}}(I_{ij}),\ j = 1,\ldots,M_i,\ i = 1,\ldots,N\},\\ W &= \{v : v_i|_{I_{ij}} \in \mathcal{P}^{q_{ij}-1}(I_{ij}),\ j = 1,\ldots,M_i,\ i = 1,\ldots,N\}, \end{aligned}$$

and where $\mathcal{P}^q(I)$ denotes the linear space of polynomials of degree $\leq q$ on I. The trial functions in V are thus continuous piecewise polynomials, locally of degree q_{ij}, and the test functions in W are discontinuous piecewise polynomials that are locally of degree $q_{ij} - 1$. The mdG(q) method is similar with discontinuous trial functions of degree q_{ij} and discontinuous test functions also of degree q_{ij}.

14.1 Computability and predictability of the Solar System

We consider the problem of computability and predictability of the Solar System, with the objective of using a multi-adaptive solver with individual time steps for individual components, see [23] where also applications to heat flow and the Burgers equation are presented. We expect multi-adaptivity to be of importance also in CFD.

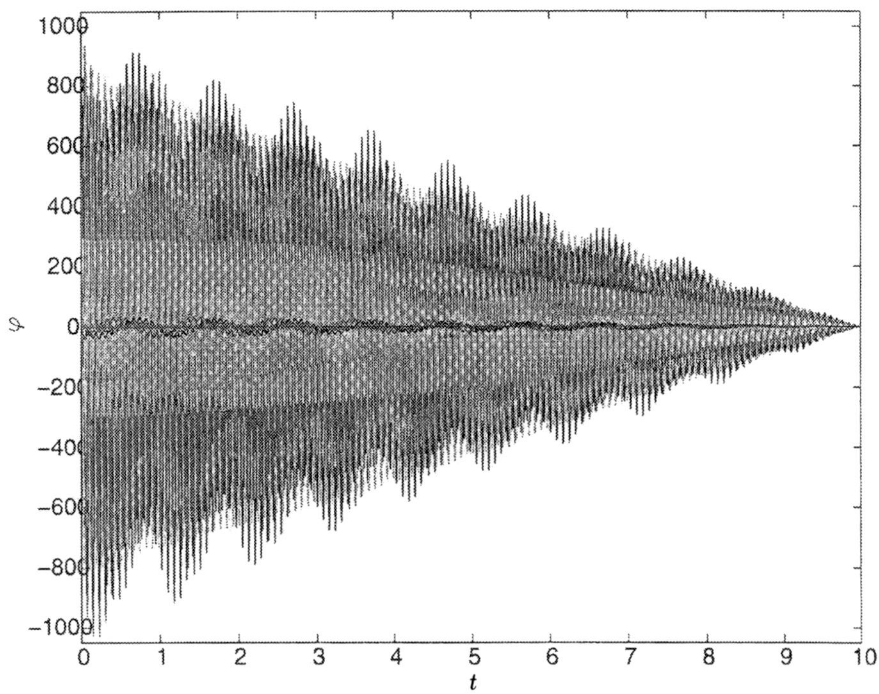

Fig. 24. Part of the dual of the Solar System with data chosen for control of the error in position of the Moon at final time.

We thus consider the Solar System, including the Sun, the Moon, and the nine planets, which is a particular n-body problem of fundamental importance:

$$m_i \ddot{x}_i = \sum_{j \neq i} \frac{G m_i m_j}{|x_j - x_i|^3}(x_j - x_i), \tag{52}$$

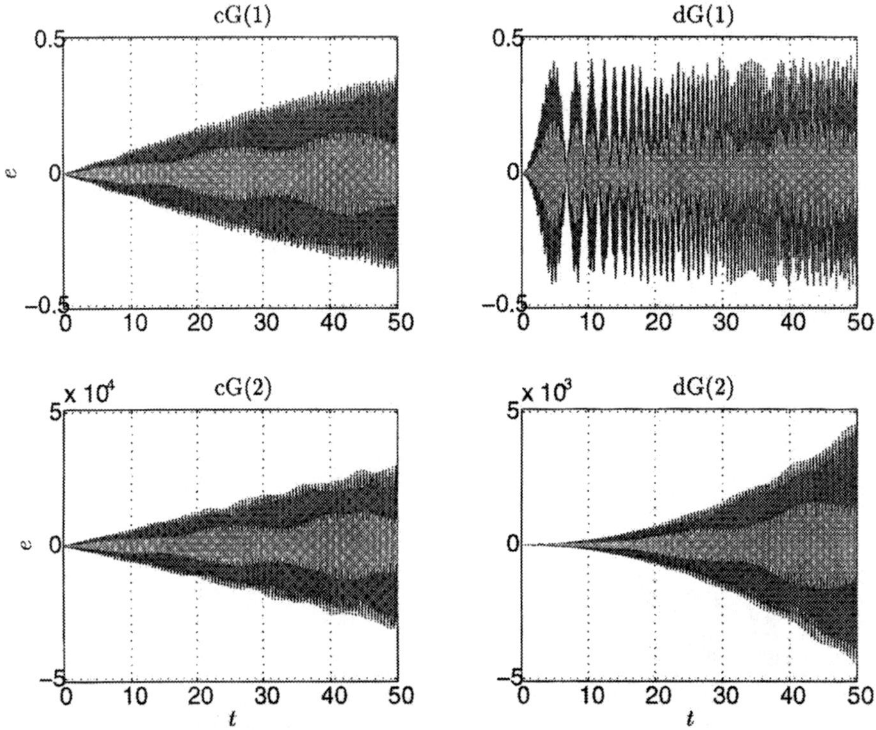

Fig. 25. The growth of the error over 50 years for the Earth-Moon-Sun system as described in the text.

where $x_i(t) = (x_i^1(t), x_i^2(t), x_i^3(t))$ denotes the position of body i at time t, m_i is the mass of body i, and G is the gravitational constant.

As initial conditions we take the values at 00.00 GMT on January 1:st 2000, obtained from the US Naval Observatory with initial velocities obtained by fitting a high-degree polynomial to the values of December 1999. The initial data should be correct to five or more digits, which is similar to the available precision for the masses of the planets. We normalize length and time to have the space coordinates per astronomical unit, AU, which is (approximately) the mean distance between the Sun and Earth, the time coordinates per year, and the masses per solar mass. With this normalization, the gravitational constant is $4\pi^2$.

Investigating the *predictability* of the Solar System, the question is how far we can accurately compute the solution, given the precision in initial data. In order to predict the accumulation rate of errors, we solve the dual problem and compute stability factors. Assuming the initial data is correct to five or more digits, we find that the Solar System is computable on the order of 500 years. Including also the Moon, we cannot compute more than a

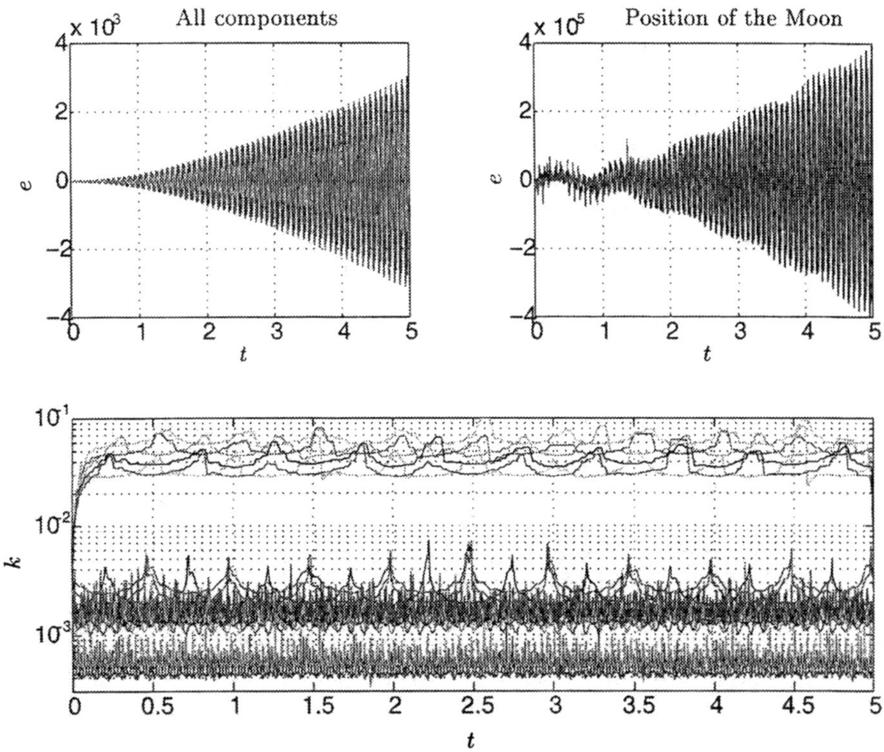

Fig. 26. The growth of the error over 5 years for the Earth-Moon-Sun system computed with the mcG(2) method, together with the multi-adaptive time-steps.

few years. The dual solution grows linearly backward in time, see Figure 24, and so errors in initial data grow linearly with time. For every extra digit of increased precision, we thus reach ten times further.

We now touch briefly the fundamental question of the *computability* of the Solar System. Assuming correct initial data and model, we compute the trajectories for Earth, the Moon and the Sun over a time interval of 50 years using different methods. Since errors in initial data grow linearly, we expect stability factors for quadrature and round off to grow quadratically.

In Figure 25 we plot the errors for the 18 components of the solution, computed with time step $k = 0.001$ with $cG(1)$, $cG(2)$, $dG(1)$ and $dG(2)$. We see that the error seems to grow linearly for $cG(q)$, which is in accordance with earlier observations [21] for periodic Hamiltonian systems, recalling that the $cG(q)$ conserves energy, [23]. The stability factors, however, grow quadratically and thus overestimate the error growth for this particular problem.

Examining the solutions obtained with the $dG(1)$ and $dG(2)$, we see that for these methods the error indeed grows quadratically. For the $dG(1)$ solution, the error reaches a maximum level of ~ 0.5 for the velocity components of the Moon. The error in position for the Moon is much smaller. This means that the Moon is still in orbit around Earth, the position of which is still very accurate, but the position relative to Earth is incorrect and thus also the velocity. The error thus grows quadratically until it reaches a limit. This effect is also visible for the error of the $cG(1)$ solution; the linear growth flattens out as the error reaches the limit. Notice also that even if the higher-order $dG(2)$ performs better on a short time-interval, it will be outrun on a long enough interval by the $cG(1)$ method with its linear accumulation of errors (for this particular problem).

Solving with the multi-adaptive method mcG(2), see Figure 26, the error grows quadratically because mcG(2) is not fully energy conserving. Computing in double precision the limit of computability seems to be $T \sim 10^6$.

15 Summary

We have presented aspects of computability of the non-stationary incompressible Navier Stokes equations with medium large Reynolds numbers. We have shown that computational solution of dual linearized problems is feasible and produces information on the growth and propagation of computational errors, which in particular may be used in adaptive error control and evaluation of computability of different mean values. We have also studied transition to turbulence computationally and discussed basic aspects of turbulence modeling based on scale similarity. Needless to say, these notes are preliminary with many open ends, which we hope to follow up in future work.

References

1. R. Becker and R. Rannacher. (2001) An optimal control approach to a posteriori error estimation in finite element methods. Acta Numerica, pp 1-102.
2. K. Eriksson, D. Estep, P. Hansbo and C. Johnson. (1995) Introduction to Adaptive Methods for Differential Equations. Acta Numerica, pp 105–158.
3. K. Eriksson, D. Estep, P. Hansbo and C. Johnson. (1996) *Computational Differential Equations*, Cambridge University Press.
4. K. Eriksson, D. Estep and C. Johnson. (2001) *Applied Mathematics: Body and Soul*, Springer.
5. K. Eriksson, D. Estep, P. Hansbo, C. Johnson. *Advanced Computational Differential Equations*, to appear 2002.
6. T.B Gatski, M. Y. Hussaini and J.L. Lumley. (1996) *Simulation and Modeling of Turbulent Flow*, Oxford Univ. Press.
7. M. Giles, M. Larson, M. Levenstam and E. Süli. (1997) Adaptive error control for finite element approximations of the lift and drag coefficients in viscous flow, Technical Report NA-76/06, Oxford Univ. Computing Laboratory.
8. M.Germano, U.Poimelli, P.Moin and W.Cabot. (1991) A dynamic subgrid scale eddy-viscosity model. Phys. Fluids A 3, 1760.

9. J. Heywood, Remarks on the possible global regularity of solutions of the three-dimensional Navier-Stokes equations. Preprint, Univ of British Columbia.
10. J. Hoffman, C. Johnson and S. Bertoluzza. (1999) Dynamic Subgrid Modeling I. Preprint 2000-04a, Chalmers Finite Element Center, to appear in Comp. Meth. Appl. Mech. Engrng.
11. J. Hoffman. (2000) Dynamic Subgrid Modeling II. Preprint 2000-04b, Chalmers Finite Element Center.
12. J. Hoffman. (2001) Dynamic Subgrid Modeling for Convection-Diffusion Equations with Fractal Coefficients. *Multiscale and Multiresolution Methods*, Springer Lecture Series in Engineering, Springer Verlag.
13. J. Hoffman. (2001) Dynamic Subgrid Modeling for Convection-Diffusion-Reaction Systems with Fractal solutions. to appear in International Journal for Numerical Methods in Fluids.
14. J. Hoffman and C. Johnson. (2002) A computational study of transition to turbulence in shear flow. Preprint 2002-04, Chalmers Finite Element Center.
15. J. Hoffman. (2002) Adaptive finite element methods for turbulent flow. Preprint, Chalmers Finite Element Center.
16. D. Estep and C. Johnson. (1998) The pointwise computability of the Lorenz system. M^3AS 8, pp 1277-1306.
17. C. Johnson, Finite Element Methods for Flow Problems, in Unstructured Grid Methods for Advection Dominated Flows, AGARD Report 787, 1992.
18. C. Johnson, R. Rannacher and M. Boman (1995) On transition to turbulence and error control in CFD, Preprint 95-06, SFB 359, Univ. of Heidelberg.
19. C. Johnson and R. Rannacher. (1995) Numerics and hydrodynamic stability: Towards error control in CFD, SIAM J. Numer. Anal., 32, pp 1058-1079.
20. C. Johnson and R. Rannacher. (1994) On error control in CFD, Proc. Int. Workshop "Numerical Methods for the Navier-Stokes equations", Heidelberg Oct. 25-28, 1993, NNFM, Vol. 47, pp. 25-28, Vieweg, Braunschweig.
21. M.G. Larson. (2000) Error Growth and A Posteriori Error Estimates for Conservative Galerkin Approximations of Periodic Orbits in Hamiltonian Systems M^3AS 10, pp 31-46.
22. J. Leray. (1934) Sur le Mouvement d'un Liquide Visqueux Emplissent l'Espace, Acta Math. J 63, pp 193-248.
23. A. Logg. (2001) Multi-Adaptive Galerkin Methods for ODEs I and II. Submitted to SIAM J. Sci. Comput.
24. G.Papanicolau, K.Solna. (2001) Wavelet based estimation of local Kolmogorov turbulence. *Long-range Dependence Theory and Applications.* Birkhauser.
25. R. Rannacher. (1999) Finite element methods for the incompressible Navier Stokes equations, Preprint Intsitute of Applied Mathematics, Univ. of Heidelberg.
26. F.Sarghini, U.Piomelli, E.Balaras. (1999) Scale-similar models for large-eddy simulations. *Phys. Fluids* **11**, *1596.*
27. A.Scotti, C.Meneveau. Fractal dimension of velocity signal in high-Reynolds-number hydrodynamic turbulence. *Phys. Review E* **51**, *5594*, 1995.
28. P. Schmid and D. Henningson. (2001) *Stability and Transition in Shear Flows*, Applied Mathematical Sciences 142, Springer.
29. G. Wagner and W.K. Liu. (1999) Turbulence simulation and multiple scale subgrid models, to appear in Computational Mechanics.

A General Lagrangian Formulation for the Computation of *A Posteriori* Finite Element Bounds

Anthony T. Patera and Jaume Peraire

Massachusetts Institute of Technology, Cambridge MA 02139, USA

Abstract. We present our general framework for implicit *a posteriori* finite element procedures to compute upper and lower bounds for functional outputs of finite element solutions. The approach is based on a finite element domain decomposition technique and the construction of an augmented Lagrangian, in which the objective is a "quadratic" energy re-formulation of the desired output, and the constraints are the finite element equilibrium equations and inter-subdomain continuity requirements. Bounds for the output, on a suitably refined "truth" mesh, are then obtained by appealing to a dual min-max relaxation, evaluated for Lagrange multipliers computed on a coarse "working" mesh. For coercive problems, the computed bounds are uniformly valid regardless of the size of the underlying coarse discretization. An extension of the basic method to deal with non-coercive problems is also presented. In this case, the bounds computed can only be proven to be asymmmptotic although in practice, they are also found to be valid for very coarse meshes. Adaptive mesh procedures can be incorporated naturally into the bounding framework to produce optimized meshes that meet a target bound gap. The methods presented are illustrated with several application examples.

Contents

1 Introduction

Engineering applications require the prediction of certain quantities of interest, or outputs – such as compliance, stresses, flow rates and heat transfer. These outputs are functionals of field variables – such as displacements, velocity field, temperature – which, in turn, are solutions of partial differential equations that need to be approximated numerically. Engineering decisions are thus based on approximations to the desired outputs that are generated from computed approximations to the field variables. It is often difficult to determine *a priori* the size of the discretization that would yield the outputs at a required level of accuracy. In fact, it is well understood that for a given field solution, different outputs can be predicted at varying levels of accuracy. The situation is further complicated in a real design, or optimization, setting where one may require multiple appeals to the PDE solver to evaluate solutions for different values of the design parameters. In this case, using a conservative fine discretization for every solution, in an attempt to guarantee a prescribed accuracy, may prove prohibitive.

In these notes, we present a summary of our work on *a posteriori* error estimation. We propose a finite element *a posteriori* method to compute inexpensive upper and lower bounds for engineering outputs of interest. The proposed method requires the solution of non-linear local Neumann sub-problems and is therefore referred to as an implicit method [2]. When compared to the simpler more inexpensive *a posteriori* explicit methods which only require residual evaluations (e.g. [3]), implicit methods offer the potential for quantitative constant-free bounds. The original implicit methods [9], [1], [4] were developed, for linear self-adjoint problems, to provide bounds for the energy norm of the error. In reality however, it is not the error in the energy norm which is of interest, but the error in the quantities on which an engineering decision will be based, such as stresses, deflection etc. The first attempts at developing *a posteriori* error estimation for functional outputs were carried out in the context of explicit error estimation [5]. It turns out that the error in the quantities of interest can be related to a weighted norm of the residual where the specific weights are determined by the solution of an adjoint problem. Unfortunately, the explicit nature of the procedure means that these estimates contain unknown constants, which severely limit its quantitative value.

The development of an implicit procedure yielding *a posteriori* constant-free bounds for linear-functional outputs of partial differential equations was presented in [15], [16], [18]. The method is applicable to elliptic coercive problems including non-symmetric terms. The approach is based on a finite element domain decomposition technique and the construction of an augmented Lagrangian, in which the objective is a "quadratic" energy re-formulation of the desired output, and the constraints are the finite element equilibrium equations and inter-subdomain continuity requirements. Bounds for the output, on a suitably refined "truth" mesh, are then obtained by appealing to a dual min-max relaxation, evaluated for Lagrange multipliers computed on a coarse "working" mesh. The focus of these notes is on the computation of bounds, but the method presented provides naturally a local indicator that can be used to drive a mesh adaptive procedure [18].

The framework presented extends naturally to the computation of bounds for reduced basis methods. This extension is treated in detail in [20].

1.1 Motivation example

We consider an example that was introduced and analyzed in [13]. Figure 1 shows a stack of heat generating units, such as, for example, computer chips. These units are supported by a material with low thermal conductivity and surrounded by air. A possible question a designer may want to ask is : for a given Q and a given geometry, is the average temperature $\bar{\theta}$ on Γ, within the "acceptable" range $[\bar{\theta}_{min}, \bar{\theta}_{max}]$?

An appropriate mathematical model to describe the relevant physical phenomena is given by the incompressible Navier-Stokes equations complemented with the energy equation to account for density variations due to temperature changes via the Boussinesq approximation. The relevant design parameters for this problem are the Prandtl and Grashof numbers [10], and the geometry. An output of engineering interest is the average temperature $\bar{\theta}$ over the boundary of the top unit.

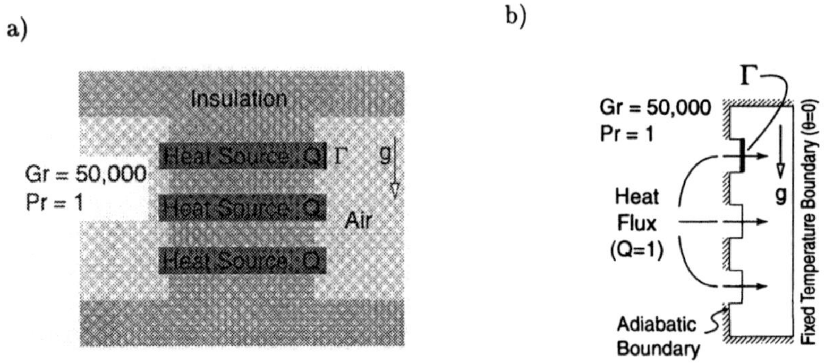

Fig. 1. Illustrative example showing three heat generating units supported by an insulating material and surrounded by air: (a) problem description, and (b) computational domain.

Given a suitable mathematical model we can now use a standard finite element solver to predict the flow and temperature distributions in an attempt to answer the original design question. One of the first decisions we must make is about the discretization to be employed. Figure 2 shows a coarse "working" triangulation $\mathcal{T}_H$, leading to an inexpensive but uncertain result $\bar{\theta}_H$, and a very fine "truth" triangulation $\mathcal{T}_h$, leading to an accurate but expensive prediction $\bar{\theta}_h$. The fine triangulation is sufficiently conservative so that we confidently assume that $\bar{\theta}_h \approx \bar{\theta}$.

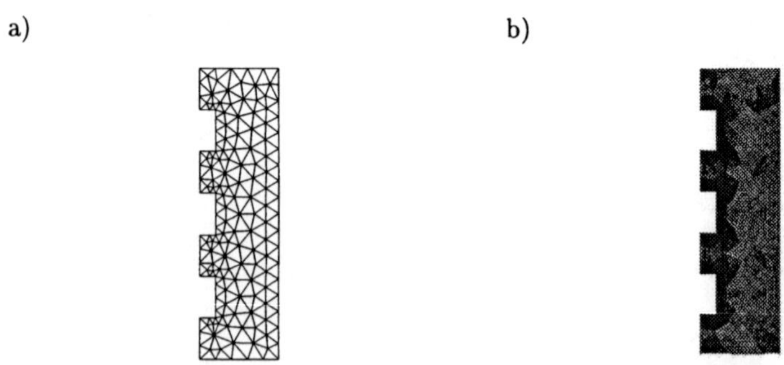

Fig. 2. Finite Element meshes: (a) "working" mesh, and (b) "truth" mesh.

The use of rigorous *a posteriori* bounds can help determine a suitable discretization. The proposed procedure allows us to calculate upper and lower bounds for the output of interest computed on the fine mesh with a computational cost which is roughly proportional to that of the coarse mesh. Of course if the initial coarse mesh happens to be inadequate to resolve the features of interest the computed upper and lower bounds will most likely be very far apart. The finer the mesh used to compute the solution the smaller the bound gap we can expect.

In figure 3 we illustrate the results obtained by our *a posteriori* error estimation approach. On the initial mesh, shown on the left, the bound procedure predicts that the solution $\bar{\theta}^{init.}$ lies inside the interval $[0.253, 0.297] = 0.275 \pm 0.022(8\%)$. If the bound gap is deemed to be too large, the error estimation procedure can be used to drive a mesh adaption algorithm. The adapted mesh together with the solution computed is shown on the right, and for this mesh, the bound procedure predicts that the temperature $\bar{\theta}^{adapt.}$ is within the interval $[0.2745, 0.2795] = 0.277 \pm 0.0025(1\%)$.

a) b)

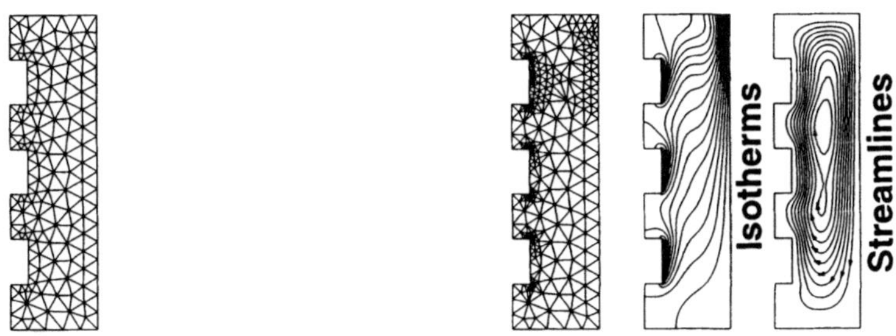

Fig. 3. (a) Initial mesh, and (b) adapted mesh and computed solution

In the following sections we develop the methods which are used to produce the results described above. First, we introduce the general formulation in a relatively abstract setting. We then present, in the second part, efficient relaxations to compute accurate bounds for finite element solutions of partial differential equations.

2 Model Problem

2.1 Continuous Problem

In order to facilitate the description of the methods to be presented we will introduce a simple model problem. We consider the following convection-diffusion problem

$$-\nu \nabla^2 u + U \cdot \nabla u = f^I, \quad \text{for all} \quad x \in \Omega \tag{1}$$

and

$$u = 0 \quad \text{on} \quad \Gamma .$$

Here, u is the unknown function, U is a given velocity field satisfying $\nabla \cdot U = 0$, ν is a diffusivity constant and f^I is a known forcing term which can be a function of x. We also consider our output of interest to be of the form

$$s = \int_\Omega f^O u \, d\Omega,$$

where f^O is a given function defined over Ω.

In order to introduce the weak form of this problem we define $X(\Omega) \equiv H_0^1(\Omega)$ (or simply X) to be the space of solutions and variations. The weak form of the problem is then: find $u \in X$ such that, $\forall v \in X$,

$$\int_\Omega \nu \nabla u \cdot \nabla v \, d\Omega + \int_\Omega U \cdot \nabla u \, v \, d\Omega = \int_\Omega f^I v \, d\Omega, \tag{2}$$

or, in shorthand form: find $u \in X$ such that

$$a(u, v) = l(v), \qquad \forall v \in X,$$

where $a : X \times X \mapsto \mathbb{R}$ and $l : X \mapsto \mathbb{R}$ are given by

$$a(w, v) = \int_\Omega \nu \nabla w \cdot \nabla v \, d\Omega + \int_\Omega U \cdot \nabla w \, v \, d\Omega$$
$$l(v) = \int_\Omega f^I v \, d\Omega$$

It is relatively straightforward to show that all solutions to the strong statement (1) will also be a solution of the weak form (2). To do that, we simply select an arbitrary function $v \in X$, multiply it by the differential equation, and integrate over Ω

$$\int_\Omega \{-\nu \nabla^2 u + U \cdot \nabla u - f^I\} \, v \, d\Omega = 0.$$

Given that v belongs to X, we can always integrate by parts the term involving the second derivative to obtain the weak form.

To show that the solution of the weak statement (2) is also a solution of the original strong form (1), requires reversing the steps followed above. This will only be possible when the solution to our weak problem is sufficiently regular so that the integration by parts can be performed. This will not be the case in general. In summary, we note that the weak statement places weaker regularity requirements on the solution than the original strong form.

2.2 Galerkin Approximation

Given a finite dimensional subspace $\tilde{X} \subset X$, $dim\{\tilde{X}\} = n$, the Galerkin approximation $\tilde{u} \in \tilde{X}$ satisfies

$$a(\tilde{u}, v) = l(v), \quad \forall v \in \tilde{X}.$$

Let the functions $\phi_i, i = 1, \ldots, n$ form a basis of $\tilde{X}$

$$\tilde{X} = span\{\phi_1, \ldots, \phi_n\}.$$

By writing

$$\tilde{u} = \sum_{i=1}^{n} \tilde{u}_i \phi_i(x), \qquad v = \sum_{i=1}^{n} v_i \phi_i(x),$$

we can explicitly carry out the spatial integrations and generate a linear set of equations for the coefficients of $\tilde{u}$, $\tilde{u}_i$. This can be expressed in vector form as

$$\underline{v}^T \tilde{A} \underline{\tilde{u}} = \underline{v}^T \underline{\tilde{F}}, \quad \forall \underline{v} \in \mathbb{R}^n,$$

or equivalently,

$$\tilde{A}\, \underline{\tilde{u}} = \underline{\tilde{F}},$$

where $\tilde{A} \in \mathbb{R}^{n \times n}$ and $\underline{\tilde{u}}, \underline{\tilde{F}} \in \mathbb{R}^n$.

Finally, an approximation to the output of interest, $\tilde{s}$, is obtained as

$$s \approx \tilde{s} = \int_{\Omega} f^O \tilde{u}\, d\Omega.$$

Given an approximation to s, one may ask how good is this approximation. The aim of our *a posteriori* bounding procedures is to answer this question by computing approximations in the $\tilde{X}$ space which can yield guaranteed upper and lower bounds for s.

3 Function Minimization

In this section we present a brief review of some basic concepts in function minimization. The results are presented in the context of general finite or infinite dimensional spaces. For clarity purposes, we use the vector-like notation so that all spaces can be readily interpreted as $\mathbb{R}^n$. That is, for $w, v \in X$ with $dim\{X\}$ *finite* or *infinite* we write $a(w, v)$"="$v^T Aw$ and $l(v)$"="$v^T F$. The term $v^T Aw$ can be interpreted as a vector multiplying a matrix times a vector when the dimension of X is finite. But we will also assume that, when the dimension of X happens to be infinite in the sense of the usual distributional operator and associated duality pairing, it has the meaning of $a(w, v)$. Regarding the function minimization procedures described below we refer to references [22,6] for additional details.

3.1 Unconstrained Minimization

We consider the minimization of an objective function $f : Y \mapsto \mathbb{R}$ defined over a space Y (e.g. $Y \equiv \mathbb{R}^n$)

$$f^* = \min_{v \in Y} f(v)$$

Here, $f^* \equiv f(v^*)$ denotes the minimum which is attained by the minimizer v^*. A more precise statement would replace minimum (maximum) by infimum (supremum), since in general we can not guarantee that the infimum (supremum) will belong to Y. This subtlety will not be important for our discussions and in these notes we will simply use min and max.

Upper Bound for f^* An upper bound, f^+, for the minimum f^*, can be easily computed by setting $f^+ = f(\bar{v})$ for any $\bar{v} \in Y$,

$$f^* = \min_{v \in Y} f(v) \equiv f(v^*) \leq f(\bar{v}) \equiv f^+.$$

For f sufficiently regular, we expect a reasonable bound if $\bar{v} \approx v^*$.

For illustration purposes we consider the case in which f is a linear quadratic function of $v \in \mathbb{R}^n$ of the form

$$f(v) = \frac{1}{2} v^T A v - b^T v$$

with $A \in \mathbb{R}^n \times \mathbb{R}^n$ is symmetric positive definite, and $b \in \mathbb{R}^n$, as illustrated in figure 4. The stationary point condition implies that

$$A v^* = b,$$

and hence,

$$f^* = \min_{v \in Y} f(v) = f(v^*) = -\frac{1}{2} v^{*T} A v^*.$$

In order to show that the stationary point is a minimum we need to show that $f(v^*) < f(\bar{v}), \quad \forall \bar{v} \in Y$. To this end we consider $\bar{v} = v^* + \epsilon$ for $\epsilon \in \mathbb{R}^n$ small (in some appropriate norm), and write

$$\begin{aligned}
f(\bar{v}) &= \frac{1}{2}(v^* + \epsilon)^T A (v^* + \epsilon) - (v^* + \epsilon)^T b \\
&= \frac{1}{2} v^{*T} A v^* - v^{*T} b \\
&+ \frac{1}{2}\epsilon^T A v^* + \frac{1}{2} v^{*T} A \epsilon - \epsilon^T b \\
&+ \frac{1}{2}\epsilon^T A \epsilon \\
&= f(v^*) + \underbrace{\epsilon^T (A v^* - b)}_{=0} + \frac{1}{2}\epsilon^T A \epsilon
\end{aligned}$$

or,

$$f^+ = f(\bar{v}) = f(v^*) + \frac{1}{2}\epsilon^T A \epsilon,$$

which implies that the error in the objective function $f^+ - f^*$ is of order $\mathcal{O}(\epsilon^2)$.

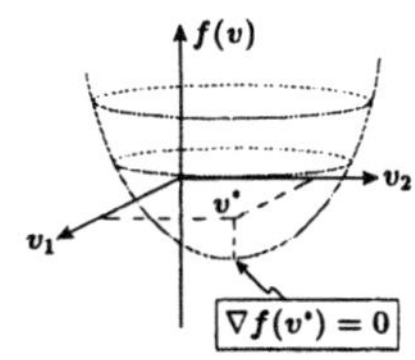

Fig. 4. Illustration of a linear-quadratic objective function in $\mathbb{R}^2$

We see that in the case of unconstrained minimization, upper bounds are trivially obtained. The computation of lower bounds is less straightforward and, for this purpose, we will need some additional tools which are considered in the following section.

3.2 Equality Constrained Minimization and Duality

We now consider the minimization of an objective function $f : Y \mapsto \mathbb{R}$ subject to equality constraints, which we write as $g(v) = 0$, where $g : Y \mapsto \Lambda$. In the finite dimensional case we will typically have $Y \equiv \mathbb{R}^n$, and $\Lambda \equiv \mathbb{R}^m$. We write

$$f^* = \min_{\substack{v \in Y \\ g(v)=0}} f(v).$$

We note that one obvious way of obtaining lower bounds in the constrained problem is to simply remove the constraint and perform the minimization over a larger space. Unfortunately, this approach is often too simplistic and does not yield useful bounds.

A convenient way to address these type of constrained minimization problems is to introduce a Lagrangian $\mathcal{L}(v, \lambda) : Y \times \Lambda \to \mathbb{R}$, as

$$\mathcal{L}(v, \lambda) = f(v) + \lambda^T g(v)$$

where λ are the Lagrange multipliers, and write

$$f^* = \min_{v \in Y} \max_{\lambda \in \Lambda} \mathcal{L}(v, \lambda). \tag{3}$$

The proof of the above statement is straightforward. Performing first the maximization with respect to lambda, we note that

$$\max_{\lambda \in \Lambda} \mathcal{L}(v,\lambda) = \begin{cases} f(v) \text{ if } g(v) = 0 \\ +\infty \text{ if } g(v) \neq 0 \end{cases}$$

since, by construction, $\mathcal{L}(v,\lambda)$ is a linear function of λ. Hence, (3) follows after taking the minimum over $v \in Y$.

By maximizing the Lagrangian with respect to λ, and minimizing with respect to v, we can write

$$\mathcal{L}(v,\lambda^*(v)) \equiv \max_{\lambda \in \Lambda} \mathcal{L}(v,\lambda),$$
$$\mathcal{L}(v^*(\lambda),\lambda) \equiv \min_{v \in Y} \mathcal{L}(v,\lambda),$$

and consequently,

$$\mathcal{L}(v,\lambda^*(v)) \geq \mathcal{L}(v,\lambda) \geq \mathcal{L}(v^*(\lambda),\lambda), \quad \forall v,\lambda.$$

Since this expression is applicable to all v and λ, we can now take the minimum with respect to v and the maximum with respect to λ to prove the following duality statement

$$\min_{v \in Y} \max_{\lambda \in \Lambda} \mathcal{L}(v,\lambda) \geq \max_{\lambda \in \Lambda} \min_{v \in Y} \mathcal{L}(v,\lambda). \tag{4}$$

We can define the duality gap as

$$\delta = \min_{v \in Y} \max_{\lambda \in \Lambda} \mathcal{L}(v,\lambda) - \max_{\lambda \in \Lambda} \min_{v \in Y} \mathcal{L}(v,\lambda).$$

For linear-quadratic objective functions and linear constraints [22], and more generally, for convex functions with convex constraints [6], the duality gap can be shown to be zero. For more general functions, $\delta \geq 0$.

Although duality can be invoked to calculate lower bounds for s, the computation of the max − min appearing in the above expression is often just as complicated as solving the original min − max problem. Below we consider some relaxations that can be used to compute good lower bounds at a much reduced cost.

Lower Bound for f^* Combining equation (3) with duality we can write,

$$f^* = \min_{v \in Y} \max_{\lambda \in \Lambda} \mathcal{L}(v,\lambda) \geq \max_{\lambda \in \Lambda} \min_{v \in Y} \mathcal{L}(v,\lambda) \geq \min_{v \in Y} \mathcal{L}(v,\bar{\lambda}), \quad \forall \bar{\lambda} \in \Lambda, \tag{5}$$

and we simply define $f^- \equiv \min_{v \in Y} \mathcal{L}(v,\bar{\lambda})$. In addition, we would expect accurate bounds when $\bar{\lambda} \approx \lambda^*$.

For illustration purposes, we consider a constrained minimization problem involving a linear-quadratic functional, with linear constraints of the form

$$f(v) = \frac{1}{2} v^T A v - b^T v$$
$$g(v) = Bv - c$$

with $A \in \mathbb{R}^n \times \mathbb{R}^n$ is symmetric positive definite, $B \in \mathbb{R}^m \times \mathbb{R}^n$, $b \in \mathbb{R}^n$, and $c \in \mathbb{R}^m$.

The Lagrangian takes the form

$$\mathcal{L}(v, \lambda) = \frac{1}{2} v^T A v - b^T v + \lambda^T (Bv - c),$$

and the constrained minimization problem can be written as,

$$\begin{aligned} f^* &= \min_{v \in \mathbb{R}^n} \max_{\lambda \in \mathbb{R}^m} \mathcal{L}(v, \lambda) \\ &= \max_{\lambda \in \mathbb{R}^m} \min_{v \in \mathbb{R}^n} \mathcal{L}(v, \lambda) \equiv \mathcal{L}(v^*, \lambda^*). \end{aligned}$$

The last equality follows from the fact that the duality gap for this problem is zero [6].

The stationary point condition can be written as

$$\begin{aligned} Av^* - b + B^T \lambda^* &= 0 \\ Bv^* - c &= 0, \end{aligned}$$

and it is not difficult to show that the point (v^*, λ^*) does indeed correspond to a saddle point.

The value of f^* can thus be expressed in terms of (v^*, λ^*) as

$$f^* = -\frac{1}{2} v^{*T} A v^* - \lambda^{*T} c$$

If we now consider an approximate multiplier $\bar{\lambda} = \lambda^* + \epsilon$, for $\epsilon \in \mathbb{R}^m$ small (in some appropriate norm), and use (5), we have that,

$$f^* \geq \min_{v \in \mathbb{R}^n} \mathcal{L}(v, \bar{\lambda}) \equiv f^-$$

The solution $\hat{v}$ satisfying the above problem will be determined by solving

$$A\hat{v} - b + B^T \bar{\lambda} = 0.$$

Since $\bar{\lambda} = \lambda^* + \epsilon$, we have that

$$\hat{v} = -\ A^{-1}(b - B^T(\lambda^* + \epsilon)) = v^* - A^{-1} B^T \epsilon.$$

Finally,

$$f^- = \frac{1}{2} \hat{v}^T A \hat{v} - b^T \hat{v} + (\lambda^{*T} + \epsilon^T)(B\hat{v} - c),$$

or,

$$\begin{aligned} f^- &= -\frac{1}{2}\, v^{*T} A v^* - \lambda^{*T} c \\ &\quad -\frac{1}{2}\, \epsilon^T B A^{-1} B^T \epsilon \\ &\quad +\epsilon^T (B v^* - c) \\ &= f^* - \frac{1}{2}\, \epsilon^T B A^{-1} B^T \epsilon. \end{aligned}$$

This expression shows that the error in approximating the solution is quadratic in the error in approximating the Lagrange multiplier, ϵ. e.g. $f^* - f^- = \mathcal{O}(\epsilon^2)$.

4 Abstract Bound Formulations

4.1 "Energy" Output

In this section, we extend the bounding strategies presented above to outputs which are functionals of solutions of our abstract problem

$$Au = F, \tag{6}$$

where $A : Y \mapsto Y'$ is a symmetric positive definite operator, $u \in Y$, and $F \in Y'$. Here, Y' stands for the dual space of Y. We recall that for $\mathbb{R}^n$, $Y \equiv Y'$.

To simplify the presentation we start by considering a particular output s, which is given simply as

$$s = u^T F.$$

By introducing the linear-quadratic functional $J : Y \mapsto \mathbb{R}$,

$$J(v) = \frac{1}{2} v^T A v - v^T F,$$

problem (6) can be re-formulated as an unconstrained minimization problem

$$J(u) = \min_{v \in Y} J(v),$$

where the minimizer $v^* \equiv u$.

At the minimum, we will have that

$$J(u) = \frac{1}{2} u^T A u - u^T F = -\frac{1}{2} u^T A u = -\frac{1}{2} u^T F,$$

and therefore our "energy" output can be easily related to $J(u)$

$$s = u^T F = u^T A u = -2J(u). \tag{7}$$

In some situations, it will prove advantageous to work with the difference between an approximation to the solution, $\tilde{u}$, and the exact solution, u, rather than with u itself. Generally, the approximation $\tilde{u}$ will be a Galerkin solution computed on a subspace $\tilde{Y} \subset Y$. That is, $\tilde{u}$ will be the solution to a problem $\tilde{A}\,\tilde{u} = \tilde{F}$.

Given an approximation $\tilde{u} \in \tilde{Y} \subset Y$ to u, $\tilde{u} \approx u$, we define the error $e \in Y$ as

$$e = u - \tilde{u}.$$

We can also write down the problem satisfied by the error as

$$A\,e = F - A\tilde{u} \equiv R, \tag{8}$$

and express the output in terms of the approximate output and the error

$$s = \tilde{u}^T F + e^T F = \tilde{s} + e^T F.$$

Analogously, we can introduce the linear quadratic functional $J^{err} : Y \mapsto \mathbb{R}$,

$$J^{err}(v) = \frac{1}{2} v^T A v - v^T R,$$

and characterize the solution e of problem (8) as the minimizer, v^*, of $J^{err}(v)$. That is,

$$J^{err}(e) = \min_{v \in Y} J^{err}(v).$$

At the minimum

$$J^{err}(e) = \frac{1}{2} e^T A e - e^T R = -\frac{1}{2} e^T A e,$$

and since $e^T F = e^T R + e^T A\tilde{u} = e^T A e + R^T \tilde{u}$ we have that

$$s = \tilde{s} + \tilde{u}^T R + e^T A e = \tilde{s} + \tilde{u}^T R - 2J^{err}(e). \tag{9}$$

We note that, in such cases where the approximation $\tilde{u}$ is a Galerkin approximation carried out in a subspace $\tilde{Y} \subset Y$, the residual R will be orthogonal to all the elements of $\tilde{Y}$, and in particular $\tilde{u}^T R = 0$.

Lower Bounds Given that

$$J(\bar{v}) \geq \min_{v \in Y} J(v), \qquad \forall \bar{v} \in Y,$$

we can replace u by any $\bar{v} \in Y$ in (7) to obtain a lower bound for s,

$$s^- \equiv -2J(\bar{v}) \leq -2J(u) \equiv s.$$

Alternatively, a lower bound can also be obtained by replacing e for any $\bar{v} \in Y$ in (9),

$$s^- \equiv \tilde{s} + \tilde{u}^T R - 2J^{err}(\bar{v}) \leq \tilde{s} + \tilde{u}^T R - 2J^{err}(e) \equiv s \tag{10}$$

It is worth noting that, in the case when $\tilde{u}$ corresponds to a Galerkin approximation, the term $\tilde{u}^T R$ is zero, and the choice $\bar{v} = 0$ in 10 gives $s^- \equiv \tilde{s}$, thus showing that the value of the output computed using the Galerkin approximaion is indeed a lower bound for the true output.

Upper Bounds Since the output of interest is $-2J(u)$, upper bounds for the output of interest will be readily obtained if we can compute lower bounds for $J(u)$ (or $J^{err}(e)$).

We will consider here three possible approaches.

Approach I : Relax Y

If we can construct $\hat{Y} \supset Y$, we can write

$$J(\hat{u}) = \min_{\hat{v} \in \hat{Y}} J(\hat{v}) \leq J(u).$$

Although simple, this approach is unlikely to yield useful bounds in practice and often we will find that $J(\hat{u}) \to -\infty$. A more useful approach will often result from relaxing Y, and the same time introducing some additional constraints thus formulating the problem as a constrained minimization problem.

Let $B : \hat{Y} \mapsto \Lambda$ be such that

$$Y = \{\hat{v} \in \hat{Y} \,|\, B\hat{v} = 0\}. \tag{11}$$

For instance, when $\hat{v} \in \hat{Y} \equiv \mathbb{R}^{\hat{n}}$, we can think of $B : \mathbb{R}^{\hat{n}} \mapsto \mathbb{R}^m$ as a rectangular matrix where m, is the number of constraints.

We can write,

$$\begin{aligned} J(u) = \min_{v \in Y} J(v) = & \min_{\substack{\hat{v} \in \hat{Y} \\ B\hat{v} = 0}} J(\hat{v}) = \min_{\hat{v} \in \hat{Y}} \max_{\lambda \in \Lambda} J(\hat{v}) + \lambda^T B\hat{v} \\ & \geq \max_{\lambda \in \Lambda} \min_{\hat{v} \in \hat{Y}} \mathcal{L}(\hat{v}, \lambda) \geq \min_{\hat{v} \in \hat{Y}} \mathcal{L}(\hat{v}, \bar{\lambda}), \quad \forall \bar{\lambda} \in \Lambda \end{aligned}$$

where the Lagrangian is given (implicitly) by $\mathcal{L}(\hat{v}, \lambda) = J(\hat{v}) + \lambda^T B\hat{v}$.

Now, we can easily obtain an upper bound, s^+, for s by setting $s^+ \equiv -2\min_{\hat{v} \in \hat{Y}} \mathcal{L}(\hat{v}, \bar{\lambda})$ since

$$s \equiv -2J(u) \leq -2\min_{\hat{v} \in \hat{Y}} \mathcal{L}(\hat{v}, \bar{\lambda}) \equiv s^+$$

Analogously, we can define for the error formulation,

$$\mathcal{L}^{err}(\hat{v}, \lambda) = J^{err}(\hat{v}) + \lambda^T B\hat{v},$$

and note that

$$J^{err}(e) \geq \min_{\hat{v} \in \hat{Y}} \mathcal{L}^{err}(\hat{v}, \bar{\lambda}), \quad \forall \bar{\lambda} \in \Lambda.$$

Finally, for any $\bar{\lambda}$),

$$\begin{aligned} s &= \tilde{s} + \tilde{u}^T R - 2J^{err}(e) \\ &\leq \tilde{s} + \tilde{u}^T R - 2 \min_{\hat{v} \in \hat{Y}} \mathcal{L}^{err}(\hat{v}, \bar{\lambda}) \equiv s^+. \end{aligned}$$

As an example of a space and its relaxation, we consider $Y \equiv \mathbb{R}^n$ to the space of grid functions over a one dimensional grid with n nodes; $\hat{Y} \equiv \mathbb{R}^{2n-2}$ is the space of grid functions over the same grid, but now the functions are allowed to be multi-valued at the interior grid nodes as shown in figure 5. Clearly, $Y \subset \hat{Y}$.

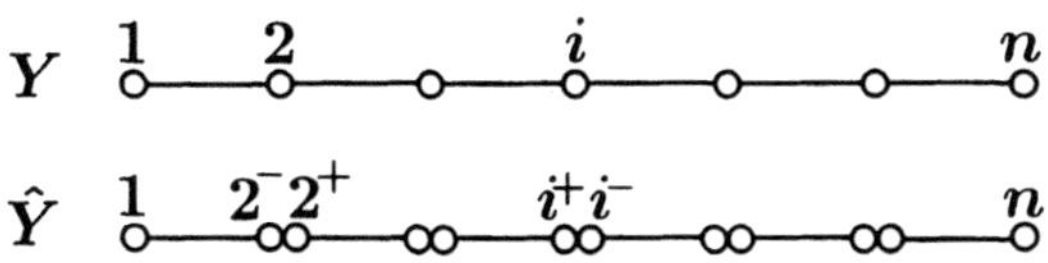

Fig. 5. A space Y consisting of all the grid functions over a one dimensional grid, of size n, and an enlarged space $\hat{Y}$, where for each grid node i, functions are allowed to take different values at i^+ and i^-.

In addition, one could define constraints that would to identify all the members of $\hat{Y}$ that belong to Y. In this case the constraints would be simply $\hat{v}_{i^+} - \hat{v}_{i^-} = 0$, for $i = 2, \ldots n-1$.

Approach II : Use a modified form A_{mod}

Let A_{mod} (SPD) be such that:

$$v^T A_{mod}\, v \leq v^T A v, \quad \forall v \in Y.$$

Then we can introduce a modified functional,

$$J_{mod}(v) = \frac{1}{2} v^T A_{mod}\, v - v^T F \leq J(v), \quad \forall v \in Y,$$

and,

$$\min_{v \in Y} J_{mod}(v) = J_{mod}(u_{mod}) = -\frac{1}{2} u_{mod}^T A_{mod}\, u_{mod}.$$

Here, u_{mod}, satisfies the stationary point condition $A_{mod}\, u_{mod} = F$. In addition, since

$$J_{mod}(u_{mod}) \leq J_{mod}(u) \leq J(u),$$

we have that

$$s \equiv -2J(u) \leq -2J_{mod}(u_{mod}) \equiv s^+.$$

A graphical "proof" is given in figure 6.

This approach is useful in the context of the reduced basis method [20]. As one may expect, good bounds rely on a good choice for A_{mod}, which is not generally obvious. A much more effective approach, which is less sensitive to

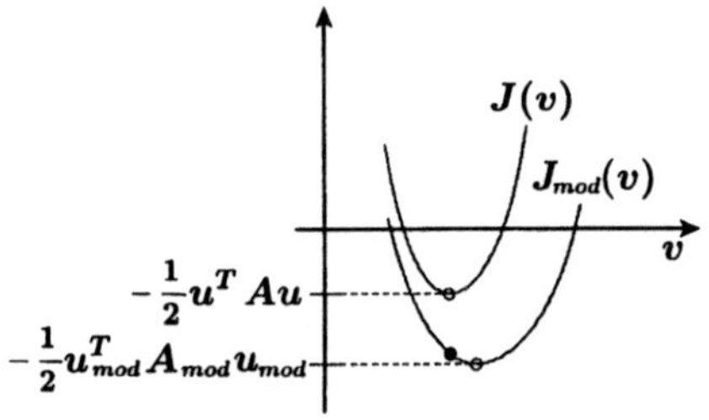

Fig. 6. One dimensional illustration of the modified functional $J_{mod}(v)$.

the choice of A_{mod}, is obtained if we work with the error rather than u. Here, we define

$$J_{mod}^{err}(v) = \frac{1}{2} v^T A_{mod}\, v - v^T R.$$

Then

$$J_{mod}^{err}(e_{mod}) \leq J^{err}(e),$$

where e_{mod} satisfies the stationary point condition $A_{mod}\, e_{mod} = R$ Finally, we have

$$s \equiv \tilde{s} + \tilde{u}^T R - 2J^{err}(e) \leq \tilde{s} + \tilde{u}^T R - 2J_{mod}^{err}(e_{mod}) \equiv s^+.$$

Approach III : Additive Decomposition
Here, we assume that A and F can be decomposed as

$$A = \sum_{k=1}^{K} A_k, \qquad F = \sum_{k=1}^{K} F_k,$$

where $A_k : Y \mapsto Y'$ and is symmetric positive semi-definite for all k. Using this decomposition we can define $J_k(v) = \frac{1}{2}(v^T A_k v) - v^T F_k$ and write

$$J(v) = \frac{1}{2} v^T A v - v^T F = \frac{1}{2} v^T (\sum_{k=1}^{K} A_k) v - v^T (\sum_{k=1}^{K} F_k)$$
$$= \sum_{k=1}^{K} (\frac{1}{2} v^T A_k v - v^T F_k) = \sum_{k=1}^{K} J_k(v).$$

Now,

$$J(u) = \min_{v \in Y} J(v) = \min_{v \in Y} \sum_{k=1}^{K} J_k(v) \geq \sum_{k=1}^{K} \min_{v \in Y} J_k(v),$$

and thus

$$s = -2J(u) \leq -2 \sum_{k=1}^{K} \min_{v \in Y} J_k(v) \equiv s^+.$$

Similarly,

$$s \leq \tilde{s} + \tilde{u}^T R - 2 \sum_{k=1}^{K} \min_{v \in Y} J_k^{err}(v) \equiv s^+.$$

4.2 More general outputs

The bounds developed in the previous section were applicable to symmetric positive-definite operators and to the particular case of the so called "energy" (or "compliance") output. We consider now the more general case where A may be non-symmetric, and the output is any (bounded) linear functional. However, we will still require that its symmetric part of A, A^S, be positive definite.

We consider the problem

$$Au = F,$$

and attempt to develop bounds for the scalar output

$$s = u^T L^O.$$

here $A : Y \mapsto Y'$, $A^S = \frac{1}{2}(A + A^T)$ is symmetric positive definite, $u \in Y$, and $F, L^O \in Y'$.

We introduce the following functionals

$$\mathcal{E}(v) \equiv v^T A^S v - v^T F,$$

and

$$\mathcal{E}^{err}(v) \equiv v^T A^S v - v^T R,$$

and re-write our output s as

$$\pm s = \pm u^T L^O + \kappa \mathcal{E}(u), \quad \forall \kappa \in \mathbb{R}^+$$

The parameter κ is any positive number and, at this stage, it is kept fixed. We shall see below how we can optimize our computed bounds with respect to κ, thus obtaining sharper bounds.

We now write,

$$\pm s = \min_{\substack{v \in Y \\ Av = F}} \left(\pm v^T L^O + \kappa \mathcal{E}(v) \right) = \min_{v \in Y} \max_{\lambda \in Y'} \mathcal{L}^{\pm}(v, \lambda).$$

It is clear that the first minimization is a trivial statement, since there is only one member of Y, namely u, which satisfies the constraint. Nevertheless, we can view this problem as a minimization problem with equality constraints, similar to that encountered in section 3.2, and define a corresponding augmented Lagrangian

$$\mathcal{L}^{\pm}(v, \lambda) = \pm v^T L^O + \kappa \mathcal{E}(v) + \lambda^T (Av - F).$$

Lower (Upper) Bounds Using the duality result (4), we have

$$\pm s = \min_{v \in Y} \max_{\lambda \in Y'} \mathcal{L}^{\pm}(v, \lambda) \geq \max_{\lambda \in Y'} \min_{v \in Y} \mathcal{L}^{\pm}(v, \lambda).$$

Clearly, for any $\bar{\lambda} \in Y'$,

$$s \geq \min_{v \in Y} \mathcal{L}^{+}(v, \bar{\lambda}) \equiv s^{-},$$
$$-s \geq \min_{v \in Y} \mathcal{L}^{-}(v, \bar{\lambda}) \equiv -s^{+},$$

which implies that $s^{-} \leq s \leq s^{+}$

Similarly, if we are given an approximation $\tilde{u} \approx u$, $\tilde{u} \in \tilde{Y} \subset Y$,

$$\pm s = \pm \tilde{s} + \min_{v \in Y} \max_{\lambda \in Y} \mathcal{L}^{\pm err}(v, \lambda).$$

The corresponding Lagrangian will be

$$\mathcal{L}^{\pm err}(v, \lambda) = \pm v^T L^O + \kappa \mathcal{E}^{err}(v) + \lambda^T (A\, v - R),$$

and, for any $\bar{\lambda} \in Y'$,

$$s \geq \tilde{s} + \min_{v \in Y} \mathcal{L}^{+err}(v, \bar{\lambda}) \equiv s^{-},$$
$$-s \geq \tilde{s} + \min_{v \in Y} \mathcal{L}^{-err}(v, \bar{\lambda}) \equiv -s^{+}.$$

We note that in order to compute bounds, in principle, we need to solve the following minimization problems,

$$\min_{v \in Y} \mathcal{L}^{\pm}(v, \bar{\lambda}), \tag{12}$$

or,

$$\min_{v \in Y} \mathcal{L}^{\pm err}(v, \bar{\lambda}). \tag{13}$$

However, it is clear that if instead we compute lower bounds for the above minimums, we will also obtain lower bounds to $\pm s$. The computation of lower bounds for (12) and (13) is an unconstrained minimization problem to which we can apply either of the three approaches discussed in section 4.1.

5 Symmetric Coercive Partial Differential Equations : Bounds for the Energy Norm

The abstract bound results presented in the previous section will now be used to yield bounds for the energy norm of the solution of a symmetric coercive partial differential equation. As we shall see, this will be the basis for computing bounds for more general equations and outputs. We will use Laplace's equation to illustrate the presentation,

$$-\nabla^2 u = f^I, \quad \text{in } \Omega, \qquad \begin{aligned} u|_{\Gamma_D} &= 0, \\ \nabla u \cdot n|_{\Gamma_N} &= h^I. \end{aligned} \tag{14}$$

The output of interest is the energy of the solution

$$s = \int_\Omega |\nabla u|^2 \, d\Omega.$$

It is well known that the solution, u, to the above problem (14), can be characterized as the function that minimizes the following problem

$$J(u) = \min_{v \in X} J(v), \tag{15}$$

where,

$$J(v) \equiv \frac{1}{2} \int_\Omega |\nabla v|^2 \, d\Omega - \int_\Omega f^I v \, d\Omega - \int_{\Gamma_N} h^I v \, ds.$$

The field variable u is defined over a domain $\Omega \subset \mathbb{R}^2$ with boundary $\Gamma = \Gamma_D \bigcup \Gamma_N$; u must reside in $X = \{v \in H^1(\Omega) \mid v = 0 \text{ on } \Gamma_D\}$. Here $H^1(\Omega)$ is the usual Sobolev space; Γ_D is the portion of the domain boundary on which the essential (Dirichlet) boundary data, is imposed; and Γ_N is the portion of Γ on which natural (normal derivative) boundary conditions are imposed.

The minimizer in (15), satisfies the stationary conditions which can be expressed in the following variational statement: find $u \in X$ such that

$$a(u, v) = l(v), \qquad \forall v \in X.$$

Here, X is the space of variations which consists of the functions in $H^1(\Omega)$ satisfying $v|_{\Gamma_D} = 0$, and

$$a(w, v) \equiv \int_\Omega \nabla w \cdot \nabla v \, d\Omega, \qquad l(v) \equiv \int_\Omega f^I v \, d\Omega + \int_{\Gamma_N} h^I v \, ds.$$

Note that, the form $a(\cdot, \cdot)$ is symmetric and positive definite. That is $a(v, v) > 0, \quad \forall v \in X$ (v non zero). Finally,

$$J(u) = -\frac{1}{2} \int_\Omega |\nabla u|^2 \, d\Omega$$

and,

$$s = -2 \, J(u).$$

5.1 Meshes and Spaces

Two triangulations of the computational domain Ω are considered: the working or design H-mesh, $\mathcal{T}_H$, consisting of K_H elements T_H; and the "truth" h-mesh, $\mathcal{T}_h$, consisting of K_h elements T_h. We require that $\mathcal{T}_h$ is a refinement of $\mathcal{T}_H$; the geometric requirements on $\mathcal{T}_H$ are discussed in the context of our refinement strategy.

To each of these meshes we associate regular piecewise–linear continuous finite element subspaces,

$$X_H = \{v \in X \mid v|_{T_H} \in \mathbf{P}_1(T_H),\ \forall T_H \in \mathcal{T}_H\},$$
$$X_h = \{v \in X \mid v|_{T_h} \in \mathbf{P}_1(T_h),\ \forall T_h \in \mathcal{T}_h\},$$

where $\mathbf{P}_1(T)$ denotes the space of linear polynomials over T. Note that, by construction, $X_H \subset X_h \subset X$.

We shall assume that, for all practical purposes, the triangulation $\mathcal{T}_h$ is conservatively fine so that any difference between s and the output computed from the h-mesh solution, s_h, is sufficiently small so that $s \approx s_h$. Then, our objective will be to compute bounds for s_h using only H-mesh global solves.

Here, we will only consider "truth" meshes that are obtained by h-refining a given coarse triangulation $\mathcal{T}_H$. However, the algorithms described in these notes can be extended to "truth" meshes obtained by p-refinement of the coarse grid triangulation with minimal changes.

The algorithms to be presented require that our spaces and forms be expressed as sums of contributions over the H-elements T_H. Towards this end, we introduce the subdomain local spaces $Z_H(T_H)$ and $Z_h(T_H)$,

$$Z_H(T_H) = \mathbf{P}_1(T_H), \quad \forall T_H \in \mathcal{T}_H,$$
$$Z_h(T_H) = \{v|_{T_h} \in \mathbf{P}_1(T_h),\ \forall T_h \in \mathcal{R}_{T_H}\} \cap H^1(T_H), \quad \forall T_H \in \mathcal{T}_H,$$

where $\mathcal{R}_{T_H}$, denotes the set of h-mesh elements contained in T_H. The global representation of $Z_H(T_H)$ and $Z_h(T_H)$ are the "broken" spaces $\hat{X}_H$ and $\hat{X}_h$,

$$\hat{X}_H = \{v \in L_2(\Omega) \mid v|_{T_H} \in Z_H(T_H),\ \forall T_H \in \mathcal{T}_H\},$$
$$\hat{X}_h = \{v \in L_2(\Omega) \mid v|_{T_H} \in Z_h(T_H),\ \forall T_H \in \mathcal{T}_H\},$$

where $L_2(\Omega)$ is the space of square-integrable functions over Ω. The different meshes and associated finite element spaces are illustrated in figure 7.

Finally, the bilinear — $a(\cdot,\cdot)$ — and linear — $l(\cdot)$, $l^O(\cdot)$ — forms are extended to accept discontinuous functions in the "broken" spaces by redefining these forms as a sum of H-element contributions. For instance, $a(\cdot,\cdot)$ is now written as

$$a(v,w) = \sum_{T_H \in \mathcal{T}_H} a_{T_H}(v|_{T_H}, w|_{T_H}), \quad \forall v, w \in \hat{X}_h;$$

similar expressions are obtained for l, and l^O.

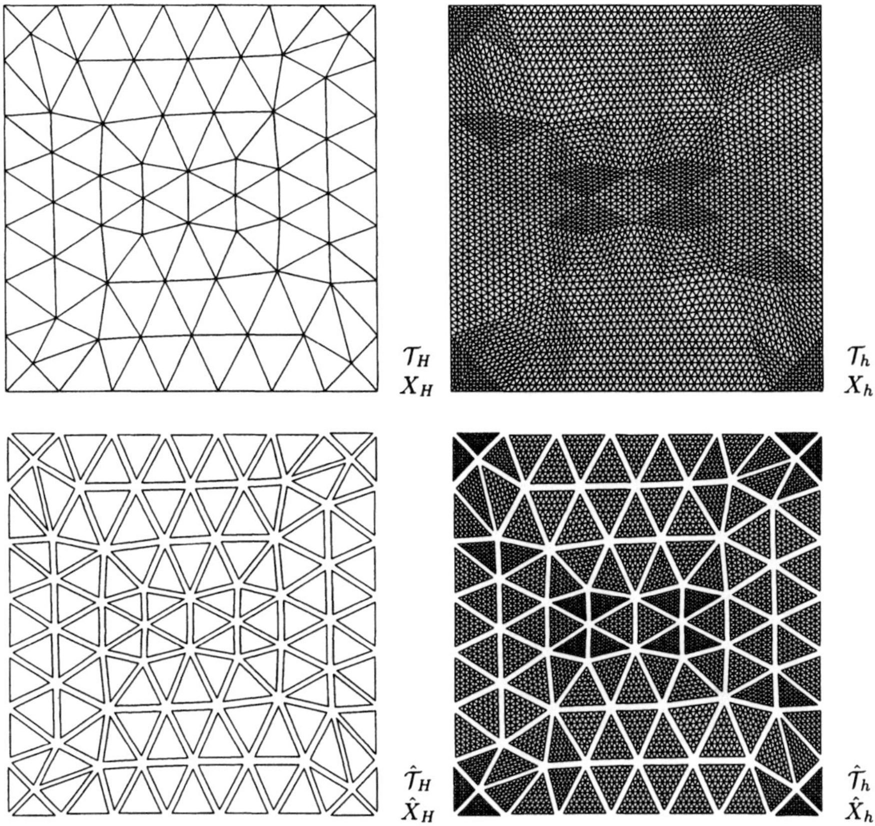

Fig. 7. Illustration of the different meshes and finite element spaces used in the bounds formulation.

5.2 Continuity constraint

Let $\mathcal{E}(\mathcal{T}_H)$ (respectively, $\mathcal{E}(\mathcal{T}_h)$) denote the set of open edges in the triangulation $\mathcal{T}_H$ (respectively, $\mathcal{T}_h$). We can then introduce the spaces of functions over the element edges $\gamma_H \in \mathcal{E}(\mathcal{T}_H)$.

$$Q_H = \{q|_{\gamma_H} \in \mathbf{P}_1(\gamma_H),\ \forall \gamma_H \in \mathcal{E}(\mathcal{T}_H) \mid q|_{\Gamma_N} = 0\}$$
$$Q_h = \{q|_{\gamma_h} \in \mathbf{P}_1(\gamma_h),\ \forall \gamma_h \in \mathcal{E}(\mathcal{T}_h) \cap \mathcal{E}(\mathcal{T}_H) \mid q|_{\Gamma_N} = 0\}$$

It follows that $Q_H \subset Q_h \subset H^{-1/2}(\mathcal{E}(\mathcal{T}_H))$; the functions in these spaces can, of course, be discontinuous [7]. Now, we introduce the "jump" bilinear form

$b : \hat{X}_h \times Q_h \mapsto \mathbb{R}$

$$b(v,q) = \sum_{\gamma_H \in \mathcal{E}(\mathcal{T}_H)} \int_{\gamma_H} [v]_{\gamma_H} \, q|_{\gamma_H} \, ds,$$

where $[v]_{\gamma_H}$ is the *jump* in v across γ_H when γ_H is an interior edge, and the *trace* of v on γ_H when γ_H is on the boundary Γ. The form $b(\cdot,\cdot)$ can be used to enforce continuity on functions in $\hat{X}_H$ and $\hat{X}_h$; in particular,

$$X_H \equiv \{\hat{v} \in \hat{X}_H \mid b(\hat{v},q) = 0, \ \forall q \in Q_H\},$$

$$X_h \equiv \{\hat{v} \in \hat{X}_h \mid b(\hat{v},q) = 0, \ \forall q \in Q_h\}.$$

Note that $b(\cdot,\cdot)$ places no restriction on v on natural boundaries.

5.3 Lower Bound for s_h

Lower bounds for s are trivially computed. Since

$$J(u_h) = \min_{v \in X_h} J(v),$$

and $X_H \subset X_h$, we have

$$J(u_H) = \min_{v \in X_H} J(v) \geq J(u_h).$$

Therefore,

$$s^- \equiv s_H = -2J(u_H) \leq -2J(u_h) = s_h$$

and thus, a lower bound is trivially obtained.

5.4 Upper Bound for s_h

It is clear that in order to obtain an upper bound for s_h, the minimization needs to be carried out over a space larger than X_h so that a lower bound for $J(u_h)$ is obtained. To this end, we consider the "broken" space $\hat{X}_h$. We start by formulating the following trivial minimization statement

$$J(u_h) = \min_{v \in X_h} J(v) = \min_{\begin{cases} \hat{v} \in \hat{X}_h \text{ subject to} \\ b(\hat{v},q) = 0, \forall q \in Q_h \end{cases}} J(\hat{v}).$$

The above constrained minimization problem suggests the formation of the linear-quadratic Lagrangian $\mathcal{L}_h : \hat{X}_h \times Q_h \to \mathbb{R}$

$$\mathcal{L}_h(\hat{v},q) = J(\hat{v}) + b(\hat{v},q),$$

which allows us to express $J(u_h)$ as

$$J(u_h) = \min_{\hat{v} \in \hat{X}_h} \max_{q \in Q_h} \mathcal{L}_h(\hat{v},q).$$

Relaxation From duality results presented in (4), it follows that

$$\begin{aligned} J(u_h) &= \min_{\hat{v}\in\hat{X}_h} \max_{q\in Q_h} \mathcal{L}_h(\hat{v},q) \\ &\geq \max_{q\in Q_h} \min_{\hat{v}\in\hat{X}_h} \mathcal{L}_h(\hat{v},q) \\ &\geq \min_{\hat{v}\in\hat{X}_h} \mathcal{L}_h(\hat{v},\bar{q}) = J^-(\bar{q}), \quad \forall \bar{q}\in Q_h. \end{aligned}$$

Thus, the upper bound s^+ is given by

$$s^+ \equiv -2J^-(\bar{q}) \geq -2J(u_h) = s_h$$

for any $\bar{q} \in Q_h$. It is important to note that, once a suitable $\bar{q}$ has been determined, the minimization required to compute $J^-(\bar{q})$ is carried out over functions which are discontinuous, and therefore can be effected independently for each macro-element T_H.

Hybrid Flux Calculation We know that, at the saddle point (u_h, p_h),

$$J(u_h) = \min_{v\in X_h} J(v) \min_{\hat{v}\in\hat{X}_H} \max_{q\in Q_H} \mathcal{L}(\hat{v},q) = \mathcal{L}(u_h,p_h).$$

We refer to q as the hybrid flux, and for this particular problem it can be identified with an approximation to the normal derivative of u on the edges.

We consider here the problem of determining H-mesh approximations to the hybrid flux p_h, $\bar{q}$, that will yield sharp bounds. For this purpose, we look for the saddle-point of the Lagrangian in the subspaces $\hat{X}_H \subset \hat{X}_h$ and $Q_H \subset Q_h$. That is we solve,

$$J(u_H) = \min_{v\in X_H} J(v) = \min_{\hat{v}\in\hat{X}_H} \max_{q\in Q_H} \mathcal{L}(\hat{v},q) = \mathcal{L}(u_H,p_H), \tag{16}$$

and set,

$$\bar{q} \equiv p_H.$$

The stationary conditions corresponding to (16), are

$$\begin{aligned} a(u_H,\hat{v}) - l(\hat{v}) + b(\hat{v},p_H) &= 0, \qquad \forall \hat{v}\in\hat{X}_H \\ b(u_H,q) &= 0, \qquad \forall q\in Q_H. \end{aligned}$$

The second equation forces $u_H \in X_H$. The first equation must be satisfied for all $\hat{v} \in \hat{X}_H$, and thus, for all $v \in X_H \subset \hat{X}_H$. Therefore, u_H can be obtained by solving the original problem

$$a(u_H,v) = l(v), \quad \forall v \in X_H,$$

which yields u_H. Now we can solve for p_H,

$$b(\hat{v}, p_H) = l(\hat{v}) - a(u_H, \hat{v}), \quad \forall \hat{v} \in \hat{X}. \tag{17}$$

The above equations represent a solvable but indeterminate system. The computation of an acceptable compatible solution is known as the *equilibration problem*. We follow here the approach proposed in [9,4,1], which involves the solution of an indeterminate system at each vertex of the $\mathcal{T}_H$ grid, the size of which is given by the number of edges that meet at the vertex. Details of this procedure can be found in [2].

There exist other approaches to obtaining implicit error bounds which circumvent the equilibration problem, typically at the expense of either some additional computational cost, or some limitations in applicability. We refer the reader to the partition-of-unity flux-free approach presented in [11], and also the method described in [4].

Local Problems Once suitable hybrid fluxes have been computed, we proceed to the solution of the local minimization problems. The minimizers, $\hat{u}$, will satisfy the following stationary condition: find $\hat{u}_h \in \hat{X}_h$ such that

$$a(\hat{u}_h, \hat{v}) = l(\hat{v}) - b(\hat{v}, p_H), \qquad \forall \hat{v} \in \hat{X}_h.$$

This involves the solution of K_H independent *local* Neumman sub-problems, which are *semi-definite*, but solvable provided $\hat{X}_H$ and $\hat{X}_h$ share the same singular modes. In this case, this is guaranteed by the equilibration procedure.

Finally,

$$s^+ \equiv -2J^-(p_H) = a(\hat{u}_h, \hat{u}_h) \geq a(u, u) \equiv s_h \tag{18}$$

since

$$\min_{\hat{v} \in \hat{X}_h} \mathcal{L}_h(\hat{v}, p_H) = \mathcal{L}_h(\hat{u}_h, p_H) = \frac{1}{2} a(\hat{u}_h, \hat{u}_h) - l(\hat{u}_h) + b(\hat{u}_h, p_H) = -\frac{1}{2} a(\hat{u}_h, \hat{u}_h).$$

5.5 Energy Norm of the Error

For a given Galerkin solution $u_H \in X_H$, the above approach allows us to compute an upper bound of the error, s^{err}, in the energy norm by simply taking $s^{err+} = s^+ - s_H$, where $s_H = -2J(u_H)$. An alternative is to compute directly an upper bound for the energy norm of the error. This procedure is outlined below using an error formulation. As it turns out, the two approaches yield the same result for this particular problem.

Let $u_H \in X_H$ be such that

$$a(u_H, v) = l(v), \qquad \forall v \in X_H.$$

The error, e_h, is then, defined as $e_h = u_h - u_H \in X_h$, and the output of interest is $s_h^{err} = a(e_h, e_h)$. The error equation is given by

$$a(e_h, v) = R(v), \qquad \forall v \in X_h, \tag{19}$$

where the residual $R : X_h \mapsto \mathbb{R}$, is $R(v) = l(v) - a(u_H, v)$.

An alternative minimization formulation of (19), is obtained by introducing $J^{err}(v) = \frac{1}{2}a(v,v) - R(v)$,

$$J^{err}(e_h) = \min_{v \in X_h} J^{err}(v).$$

At the minimum,

$$J^{err}(e_h) = -\frac{1}{2}a(e_h, e_h)$$

and

$$s_h^{err} = -2J^{err}(e_h).$$

In order to compute an upper bound for s_h^{err}, we need a lower bound for $J^{err}(v)$ and this can be done following an identical path to that described previously for $J(u)$. Below we simply highlight the main steps.

Upper Bound for s^{err} Consider,

$$J^{err}(e_h) = \min_{v \in X_h} J^{err}(v) = \min_{\hat{v} \in \hat{X}_h} \max_{q \in Q_h} \mathcal{L}^{err}(\hat{v}, q),$$

where,

$$\mathcal{L}^{err}(\hat{v}, q) = J^{err}(\hat{v}) + b(\hat{v}, q).$$

Using duality (4) plus a coarse grid approximation (17) to the hybrid fluxes, we write

$$J^{err}(e_h) \geq \min_{\hat{v} \in \hat{X}_h} \mathcal{L}^{err}(\hat{v}, p_H).$$

The local problems: find $\hat{e}_h \in X_h$ such that

$$a(\hat{e}_h, \hat{v}) = R(\hat{v}) + b(\hat{v}, p_H), \quad \forall \hat{v} \in X_h,$$

are then solved to yield

$$J^{err}(e_h) \geq J^{err}(\hat{e}_h) = -\frac{1}{2}a(\hat{e}_h, \hat{e}_h),$$

and finally,

$$s^{err+} \equiv a(\hat{e}_h, \hat{e}_h) \geq a(e, e) \equiv s_h^{err}.$$

We point out that the upper bound s^{err+}, can be expressed as the sum of H-mesh elemental contributions

$$a(\hat{e}_h, \hat{e}_h) = \sum_{T_h \in \mathcal{T}_H} a_{T_H}(\hat{e}_h|_{T_H}, \hat{e}_h|_{T_H}).$$

In addition, since each of the elemental contributions is positive, we can identify the contribution of each element to the final bound thus providing a procedure for adaptivity. This concept will be further exploited in the next section.

6 Coercive Partial Differential Equations : Bounds for Functional Outputs

The procedure developed in the previous section can now be extended to account for linear functional outputs. We will use an error formulation for the purposes of describing the approach but the formulation in terms of the total unknown follows analogous steps.

Let us consider a linear second order coercive (not necessarily symmetric) problem written in variational form as : find $u \in X^D(\Omega)$ such that

$$a(u, v) = l(v), \quad \forall v \in X.$$

The output of interest $s \in \mathbb{R}$, will be a linear functional of the solution which we write as,

$$s = l^O(u).$$

Here, $X = \{v \in H^1(\Omega) \mid v = 0 \text{ on } \Gamma_D\}$, $X^D = \{v + u_D \mid v \in X\}$, $u_D|_{\Gamma_D} = g_D$, $\Omega \subset \mathbb{R}^2$ and $\Gamma = \Gamma_D \bigcup \Gamma_N$; Γ_D is the portion of the domain boundary on which the essential (Dirichlet) boundary data, g_D, is imposed; u_D is any function in $H^1(\Omega)$ satisfying the essential boundary conditions, that is, $u_D|_{\Gamma_D} = g_D$; and Γ_N is the portion of Γ on which natural (normal derivative) boundary conditions are imposed. Also, $a : H^1(\Omega) \times H^1(\Omega) \mapsto \mathbb{R}$ is a continuous bi-linear form, and $l : H^1(\Omega) \mapsto \mathbb{R}$ (respectively, $l^O : H^1(\Omega) \mapsto \mathbb{R}$) is a continuous linear (respectively, affine) functional.

The symmetric (coercive) part of $a(\cdot, \cdot)$, $a^S : X \times X \mapsto \mathbb{R}$, is given by

$$a^S(v, w) = \frac{1}{2}(a(v, w) + a(w, v)), \quad \forall v, w \in X,$$

and norm induced by it is represented as $|||\cdot|||$,

$$|||v|||^2 = a^S(v, v), \quad \forall v \in X.$$

6.1 $\mathcal{T}_H$ and $\mathcal{T}_h$ Problems

We consider the Galerkin finite element solution on $\mathcal{T}_H$, $u_H \in X_H^D$,

$$a(u_H, v) = l(v), \quad \forall v \in X_H \quad \text{and,} \quad s_H = l^O(u_H).$$

Analogously, on $\mathcal{T}_h$, $u_h \in X_h^D$, satisfies

$$a(u_h, v) = l(v), \quad \forall v \in X_h \quad \text{and,} \quad s_h = l^O(u_h).$$

In the smooth case, $|s - s_H| \sim \mathcal{O}(H^2)$, and $|s - s_h| \sim \mathcal{O}(h^2)$, where H and h are the diameters of $\mathcal{T}_H$ and $\mathcal{T}_h$, respectively. This is proved by standard Aubin-Nitsche duality methods.

It will be assumed that the calculation of u_H, and hence s_H, can be performed at moderate cost. However, the h-mesh will typically be much finer than the H-mesh, in order to ensure that $|s - s_h|$ is suitably small; the computation of u_h and s_h will thus, most likely, be impractical. Our objective is, therefore, to devise a procedure that will yield sharp upper and lower bounds for s_h in an affordable manner.

6.2 Lagrangian Formulation

We start by first deriving the so called "energy" equality which will be used to convexify our problem. Noting that $u_h - u_H \in X_h$, we have that

$$a(u_h, u_h - u_H) - l(u_h - u_H) = 0.$$

Using the bi-linearity of $a(\cdot,\cdot)$, we can write

$$a^S(u_h - u_H, u_h - u_H) - l(u_h - u_H) + a(u_H, u_h - u_H) = 0,$$

or,

$$a^S(e_h, e_h) - R(e_h) = 0.$$

Here $e_h = u_h - u_H \in X_h$ and $R(v) \equiv l(v) - a(u_H, v), \quad \forall v \in \hat{X}_h$. Note that $a^S(e_h, e_h)$ is simply $|||e_h|||^2$.

Consider now the following functional space $\mathcal{S} \subset \hat{X}_h$,

$$\mathcal{S} = \{v \in \hat{X}_h \mid a(v, w) = R(w), \ \forall w \in X_h; \ b(v, q) = 0, \ \forall q \in Q_h\}.$$

We observe that the second constraint enforces continuity and the homogeneous essential conditions, that is, $v \in X_h$, and the first constraint then forces $v = e_h$. Therefore, the space $\mathcal{S}$ consists of a single function, e_h. The following trivial minimization statement can then be written,

$$\pm s_h = \min_{v \in \mathcal{S}} \ \kappa(a^S(v, v) - R(v)) \ \pm \ l^O(u_H + v), \quad \forall \kappa \in \mathbb{R}^+,$$

where κ is a non-negative parameter that will be used later to optimize the computed bounds.

The above constrained minimization problem suggests the formation of the linear-quadratic Lagrangian $\mathcal{L} : \hat{X}_h \times X_h \times Q_h \mapsto \mathbb{R}$,

$$\begin{aligned} \mathcal{L}^\pm(v, \theta, q) = & \ \kappa \ (a^S(v, v) - R(v)) \pm l^O(u_H + v) \\ & + a(u_H + v, \theta) - l(\theta) + b(v, q) \end{aligned}$$

which allows us to express $\pm s_h$ as

$$\pm s_h = \min_{v \in \hat{X}_h} \ \max_{\theta \in X_h, \ q \in Q_h} \ \mathcal{L}^\pm(v, \theta, q).$$

We know from section 3.2 that, at the saddle–point $(e_h, \psi_h^\pm, p_h^\pm)$,

$$\mathcal{L}^\pm(e_h, \psi_h^\pm, p_h^\pm) = \min_{v \in \hat{X}_h} \ [\max_{\theta \in X_h, \ q \in Q_h} \ \mathcal{L}^\pm(v, \theta, q)].$$

We shall refer to θ as the adjoint (dual) variable, and q as the hybrid flux.

Duality From linear-quadratic duality theory, [22] (see also section 4), it follows that, for any $\bar{\theta}_h^{\pm} \in X_h$ and any $\bar{q}_h^{\pm} \in Q_h$,

$$\begin{aligned}
\pm s_h &= \max_{\theta \in X_h,\ q \in Q_h} \ \min_{v \in \hat{X}_h} \mathcal{L}^{\pm}(v, \theta, q) \\
&\geq \min_{v \in \hat{X}_h} \ \mathcal{L}^{\pm}(v, \bar{\theta}_h^{\pm}, \bar{q}_h^{\pm}),
\end{aligned}$$

with equality for $(\bar{\theta}_h^{\pm}, \bar{q}_h^{\pm}) = (\psi_h^{\pm}, p_h^{\pm})$. Upper and lower bounds for the output s_h, s^+ and s^-, respectively, are now readily constructed as

$$s^- \equiv \min_{v \in \hat{X}_h} \mathcal{L}^+(v, \bar{\theta}_h^+, \bar{q}_h^+) \leq s_h \leq - \min_{v \in \hat{X}_h} \mathcal{L}^-(v, \bar{\theta}_h^-, \bar{q}_h^-) \equiv s^+. \tag{20}$$

We note that this expression holds for any functions $\bar{\theta}_h^{\pm} \in X_h$ and $\bar{q}_h^{\pm} \in Q_h$. In particular, we will choose these functions to be members of $X_H \subset X_h$ and $Q_H \subset Q_h$.

Adjoint and Hybrid Fluxes We consider here the problem of determining H-mesh approximations to the adjoint $\psi_h^{\pm}$, $\bar{\theta}_h^{\pm}$, and the hybrid flux $p_h^{\pm}$, $\bar{q}_h^{\pm}$, that will yield sharp bounds. Towards this end, we look for the saddle point of the Lagrangian in the subspaces $\hat{X}_H \subset \hat{X}_h$, $X_H \subset X_h$, and $Q_H \subset Q_h$. In particular we find $(e_H^{\pm}, \psi_H^{\pm}, p_H^{\pm}) \in \hat{X}_H \times X_H \times Q_H$ such that,

$$\mathcal{L}^{\pm}(e_H, \psi_H^{\pm}, p_H^{\pm}) = \min_{v \in \hat{X}_H} [\max_{\theta \in X_H,\ q \in Q_H} \mathcal{L}^{\pm}(v, \theta, q)],$$

and set $\bar{\theta}_h^{\pm} = \psi_H^{\pm}$ and $\bar{q}_h^{\pm} = p_H^{\pm}$. The stationary conditions are obtained by requiring that the variations of $\mathcal{L}^{\pm}$ with respect to v, θ, and q vanish. Thus, we find $e_H^{\pm} \in \hat{X}_H$, $\psi_H^{\pm} \in X_H$, and $p_H^{\pm} \in Q_H$ such that

$$\begin{aligned}
b(w, p_H^{\pm}) &= -\{\kappa\ (2a^S(e_H^{\pm}, w) - R(w)) \pm l^O(w) + a(w, \psi_H^{\pm})\}, \quad \forall w \in \hat{X}_H, \\
a(u_H + e_H^{\pm}, w) &= l(w), \qquad \forall w \in X_H, \\
b(e_H^{\pm}, q) &= 0, \qquad \forall q \in Q_H.
\end{aligned}$$

First, the third equation forces $e_H^{\pm} \in X_H$, which combined with the second and the primal H-mesh problem implies that $e_H^{\pm} = 0$. Next, the first equation must be satisfied for all $w \in \hat{X}_H$, and thus, for all $w \in X_H \subset \hat{X}_H$. The adjoint, $\psi_H^{\pm} \in X_H$, can therefore be determined (precisely as in Aubin–Nitsche procedures [5]) as

$$a(w, \psi_H) = -l^O(w), \quad \forall w \in X_H,$$

with $\psi_H^{\pm} = \pm\psi_H$. Note that, in deriving this equation, we exploit, first, $R(w) = 0$, $\forall w \in X_H$, and second $b(w, p_H^{\pm}) = 0$, $\forall w \in X_H$. Finally, we can determine the hybrid fluxes $p_H^{\pm} \in Q_H$, such that

$$b(w, p_H^{\pm}) = \kappa R(w) \mp l^O(w) - a(w, \psi_H^{\pm}), \quad \forall w \in \hat{X}_H.$$

We set $p_H^{\pm} = \pm p_{0H} + \kappa p_{1H}$ and find $p_{0H} \in Q_H$ and $p_{1H} \in Q_H$ such that

$$b(w, p_{0H}) = -l^O(w) - a(w, \psi_H), \quad \forall w \in \hat{X}_H,$$
$$b(w, p_{1H}) = R(w) \; (= l(w) - a(u_H, w)), \quad \forall w \in \hat{X}_H.$$

Note that, p_{0H} *equilibrates* the H-mesh adjoint problem, whereas p_{1H} *equilibrates* the original H-mesh primal solution. The above equations represent a solvable but indeterminate system which is solved using analogous procedures to those mentioned in section 5.4.

Local Problems We now turn to the solution of the minimization problems: find $\hat{e}_h^{\pm} \in \hat{X}_h$ such that

$$\mathcal{L}^{\pm}(\hat{e}_h^{\pm}, \psi_H^{\pm}, p_H^{\pm}) = \min_{v \in \hat{X}_h} \mathcal{L}^{\pm}(v, \psi_H^{\pm}, p_H^{\pm}).$$

The corresponding stationary condition is ,

$$2\kappa a^S(\hat{e}_h^{\pm}, w) = \kappa R(w) \mp l^O(w) - a(w, \psi_H^{\pm}) - b(w, p_H^{\pm}), \quad \forall w \in \hat{X}_h,$$

which we recognize as $2K_H$ independent T_H–local symmetric Neumann subdomain problems. Note that, although a^S is coercive over $X_h \times X_h$, it is typically only semi-definite over $\hat{X}_h \times \hat{X}_h$; however, thanks to the equilibration process, if the singular modes of $\hat{X}_h$ are shared by $\hat{X}_H$, the above system will be solvable, and our bounds finite. The equations for $\hat{e}_h^+$ and $\hat{e}_h^-$ can be rearranged into the following problems for $\hat{e}_{0h}$ and $\hat{e}_{1h}$,

$$2a^S(\hat{e}_{0h}, w) = -l^O(w) - a(w, \psi_H) - b(w, p_{0H}), \quad \forall w \in \hat{X}_h,$$
$$2a^S(\hat{e}_{1h}, w) = l(w) - a(u_H, w) - b(w, p_{1H}), \quad \forall w \in \hat{X}_h,$$

where $p_H^{\pm} = \pm p_{0H} + \kappa p_{1H}$ and $\hat{e}_h^{\pm} = \pm \frac{1}{\kappa} \hat{e}_{0h} + \hat{e}_{1h}$.

Note that $\hat{e}_{0h}$ and $\hat{e}_{1h}$ are related to the error in the adjoint ψ_H, and the error in the solution u_H, respectively.

6.3 Bounds Calculation

To evaluate the bounds, it remains only to insert the computed values for $\hat{e}_h^{\pm}$, $\psi_H^{\pm}$, and $p_H^{\pm}$ into our Lagrangian.

$$s^+(\kappa) \equiv -\mathcal{L}^-(\hat{e}_h^-, \psi_H^-, p_H^-) = l^O(u_H) + \kappa a^S(\hat{e}_h^-, \hat{e}_h^-)$$
$$s^-(\kappa) \equiv \mathcal{L}^+(\hat{e}_h^+, \psi_H^+, p_H^+) = l^O(u_H) - \kappa a^S(\hat{e}_h^+, \hat{e}_h^+)$$

where the arguments of s^+ and s^- indicate that the quality of the bounds depends, of course, on the working mesh $\mathcal{T}_H$, but also on the parameter κ. Note that our bounds for s_h are also bounds for s_H. Note also that, for the "compliance" case — $a(\cdot,\cdot)$ symmetric, $l^O = l$, homogeneous essential conditions, and $\kappa = 1$ — $\psi_H^+ = -u_H$ and $p_H^+ = 0$; it follows that $\hat{e}_h^+ = 0$, and thus $s^- = l^O(u_H)$.

We can also express our bounds in terms of $\hat{e}_{0h}$, $\hat{e}_{1h}$,

$$s^+(\kappa) = l^O(u_H) - 2a^S(\hat{e}_{0h}, \hat{e}_{1h}) + \frac{1}{\kappa}a^S(\hat{e}_{0h}, \hat{e}_{0h}) + \kappa a^S(\hat{e}_{1h}, \hat{e}_{1h})$$
$$s^-(\kappa) = l^O(u_H) - 2a^S(\hat{e}_{0h}, \hat{e}_{1h}) - \frac{1}{\kappa}a^S(\hat{e}_{0h}, \hat{e}_{0h}) - \kappa a^S(\hat{e}_{1h}, \hat{e}_{1h}),$$

which we recall is valid for any positive value of the parameter κ. The choice of κ will affect the (half) bound gap,

$$\Delta(\kappa) \equiv \frac{1}{2}(s^+ - s^-) = \frac{1}{\kappa}a^S(\hat{e}_{0h}, \hat{e}_{0h}) + \kappa a^S(\hat{e}_{1h}, \hat{e}_{1h}),$$

but will not affect the output predictor,

$$s_{pre} \equiv \frac{1}{2}(s^+ + s^-) = l^O(u_H) - 2a^S(\hat{e}_{0h}, \hat{e}_{1h}).$$

Note that $|s_h - s_{pre}| \le \Delta$.

Optimum κ Since $\hat{e}_{0h}$ and $\hat{e}_{1h}$ do not depend on the choice of κ, we can readily find that κ which minimizes the bound gap, and hence renders the lower and upper bounds as sharp as possible. This procedure gives

$$\kappa^* = \sqrt{\frac{a^S(\hat{e}_{0h}, \hat{e}_{0h})}{a^S(\hat{e}_{1h}, \hat{e}_{1h})}}.$$

The optimized bounds are

$$s^+(\kappa^*) = l^O(u_H) - 2a^S(\hat{e}_{0h}, \hat{e}_{1h}) + 2\,|||\hat{e}_{0h}|||\;|||\hat{e}_{1h}|||, \tag{21}$$
$$s^-(\kappa^*) = l^O(u_H) - 2a^S(\hat{e}_{0h}, \hat{e}_{1h}) - 2\,|||\hat{e}_{0h}|||\;|||\hat{e}_{1h}|||, \tag{22}$$

and the optimized (half) bound gap is

$$\Delta(\kappa^*) \equiv \frac{1}{2}(s^+(\kappa^*) - s^-(\kappa^*)) = 2\,|||\hat{e}_{0h}|||\;|||\hat{e}_{1h}|||.$$

Note that, in the energy output case of section 5, $\kappa^* = 1$. We thus see that our procedure is a *bound formulation* of the classical Aubin–Nitsche result; an implicit counterpart to existing explicit schemes for linear functional outputs [3,5]. In the smooth case we expect that $|||\hat{e}_{0h}|||$ and $|||\hat{e}_{1h}|||$ will each be

$\mathcal{O}(H)$, and thus $\Delta(\mathcal{T}_H) \sim \mathcal{O}(H^2)$ will vanish at the same rate as $|s - s_H|$ as $H \to 0$.

For the special situation in which $\kappa = 1$ and $l^O(u_H + v) \equiv R(v)$, we see that our upper bound corresponds to an upper bound for the error in the energy norm, and thus in this case our method is a (rather trivial) extension of the method of [1] to nonsymmetric operators. (The lower bound is zero, consistent with our compliance result.) Note that, for both the symmetric and nonsymmetric problems, $\psi_H^- = 0$, however only for the symmetric problem is $\psi_h^- = 0$. It follows that only for the symmetric problem is e_h the minimizer of $\mathcal{L}^{\pm}(v, \psi_H^- = 0, \cdot\,)$ ($= 2J(v)$ of [1] *in this particular case*) over all $v \in X_h$. This highlights the distinction between our *energy functional*, $a^S(v,v) - R(v) \pm l^O(u_H + v)$, and a variational principle: the former is only required to *evaluate* $l^O(u_h)$ for $v = e_h$, while the latter must attain its *minimum* $l^O(u_h)$ at $v = e_h$. We are able to consider more general objective functions since the task of enforcing $v = e_h$ has been moved to a constraint.

Finally, we briefly summarize computational complexity. On $\mathcal{T}_H$, we require two global calculations, one for u_H and one for ψ_H, and two (local) hybrid flux equilibrations. On $\mathcal{T}_h$ we need only invert $2K_H$ independent T_H-local *symmetric* Neumann operators. It follows from the superlinearity of most general solution algorithms that the effort to compute our bounds will be much less than the effort to directly compute u_h and s_h. Note that for each additional output of interest, we require only *one* additional global calculation on $\mathcal{T}_H$, and *one* additional equilibration, and K_H additional T_H-local Neumann problems.

7 Adaptivity

The procedure presented in the previous section to calculate bounds can be implemented on any pair of meshes, $\mathcal{T}_H$ and $\mathcal{T}_h$, which satisfy the requirement that $\mathcal{T}_h$ is a refinement of $\mathcal{T}_H$. Since the cost of computing the bounds is essentially a function of the number of elements K_H in $\mathcal{T}_H$, it is desirable to construct *optimized* triangulations that maximize the bound accuracy (minimize the bound gap) for a given number of degrees-of-freedom. In this section we present an algorithm for generating such optimized grids in an adaptive iterative manner.

We first note that the bound gap can be expressed as a sum of elemental contributions,

$$\Delta = \sum_{T_H \in \mathcal{T}_H} \Delta_{T_H},$$

$$\Delta_{T_H} = \frac{1}{\kappa^*} a^S_{T_H}(\hat{e}_{0h}, \hat{e}_{0h}) + \kappa^* a^S_{T_H}(\hat{e}_{1h}, \hat{e}_{1h}) \geq 0.$$

Note that $\Delta_{T_H}(\mathcal{T}_H)$ is non-negative and can thus be directly interpreted as the *contribution* to the bound gap from element T_H.

We can now describe our iterative adaptive strategy. Starting from an initial grid $\mathcal{T}_H^0$ with bound gap Δ^0, we generate a sequence of triangulations $\{\mathcal{T}_H^k,\ k = 1, 2, \ldots\}$, with corresponding bound gaps $\{\Delta^k,\ k = 1, 2, \ldots\}$, such that each triangulation $\mathcal{T}_H^k$ is a refinement of the preceding triangulation $\mathcal{T}_H^{k-1}$. The approach does not guarantee that $\Delta^k \leq \Delta^{k-1}$ for any particular k, but it does ensure that, for a sufficiently large k, $\Delta^k \leq \Delta^{targ}$, where $\Delta^{targ} > 0$ is a specified positive gap target.

In order to identify the elements in $\mathcal{T}_H^{k-1}$ that need to be refined, we first calculate the largest elemental contribution Δ_{max}^{k-1} to the bound gap Δ^{k-1},

$$\Delta_{max}^{k-1} = \max_{T_H \in \mathcal{T}_H^{k-1}} \Delta_{T_H}^{k-1},$$

and refine all those elements $T_H \in \mathcal{T}_H^{k-1}$ for which

$$\Delta_{T_H}^{k-1} \geq \alpha \Delta_{max}^{k-1}, \qquad 0 < \alpha < 1.$$

Here, α is a parameter which controls the fraction of elements to be refined at each iteration. At present, this parameter is specified *a priori*, and is independent of k. More sophisticated strategies for choosing α can, of course, be devised; for instance, we can determine α^k such that, based on *a priori* scalings, a given target gap $\Delta^{targ,k}$ should be satisfied by the refined grid $\mathcal{T}_H^k$.

The refinement process is terminated when a selected accuracy goal is achieved. In the applications that follow, the accuracy measure is the relative (half) bound gap r^k,

$$r^k = \frac{\Delta^k}{s_{pre}^k}.$$

The iteration is thus halted when $r^k \leq r^{obj}$ where r^{obj} is the prescribed accuracy.

7.1 Grid Refinement Strategy

The triangulation $\mathcal{T}_H^k$ is generated from $\mathcal{T}_H^{k-1}$ by refinement of the selected triangles T_H such that $\Delta_{T_H}^{k-1} > \alpha \Delta_{max}^{k-1}$. Each triangle that satisfies this condition is subdivided into *four* similar triangles. This refinement strategy leads to the generation of *hanging nodes* whenever only one of the two triangles sharing an edge is refined. Since an element T_H with a hanging node — a node on ∂T_H which is *not* a vertex of T_H — requires special computational treatment, we permit only one hanging node per element. We thus, first, refine all those elements satisfying the above threshold equation, and then, in addition, recursively refine those elements that have more than one hanging node. At the conclusion of the process, only two types of elements (having either zero or one hanging node) are present. This approach has the benefit that fairly smooth grids are generated, and that only one type of *special* element needs to be developed.

The unknowns corresponding to the hanging nodes appearing as a result of the grid refinement process are constrained to be the equal to the average of the unknown at the two end nodes of the associated edge. Details about this procedure can be found in [18].

7.2 Mesh Adaptivity Results

Linear Elasticity Results are presented for the two-dimensional generalized planar stress linear elasticity equations. Although here the solution field u is the displacement vector, the theory presented for the scalar problem can be trivially extended to handle this case. The particular problem considered is that of one-fourth of a symmetrically loaded plate with two rectangular holes.

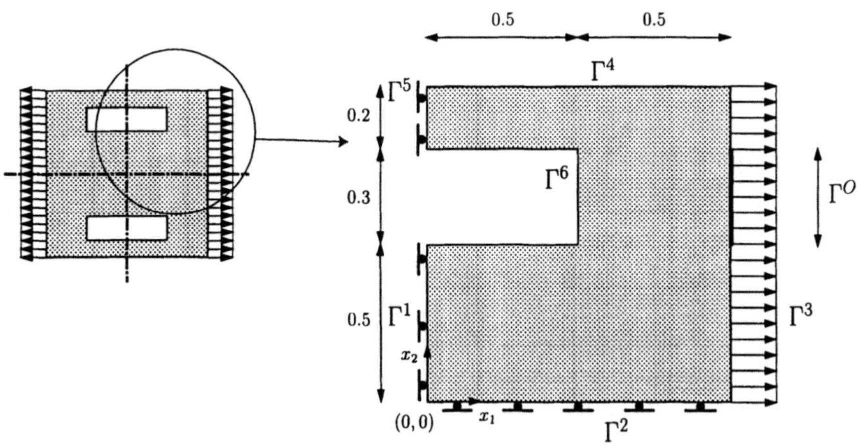

Fig. 8. Geometry for linear elasticity example: one–fourth of a symmetrically loaded plate with two square holes. A normal traction is applied on Γ^3; the outputs of interest are the normal force on Γ^5, and the average normal displacement over the segment $\Gamma^O \subset \Gamma^3$.

The domain boundaries are denoted by Γ^j, $j = 1, \ldots, 6$. A horizontal unit distributed force is applied along Γ^3, and homogeneous essential boundary conditions are enforced on the normal displacement on Γ^1, Γ^2, and Γ^5; stress-free condition apply on all other boundaries and components. The Young's modulus and Poisson ratio are taken to be 1 and 0.3, respectively.

Two outputs are considered: the normal force on the boundary segment Γ^5, denoted by s_1, and the average horizontal displacement over the segment Γ^O, denoted by s_2. These outputs can be expressed as $s_1 = l_1^O(u)$, $s_2 = l_2^O(u)$,

where

$$l_1^O(v) = \int_\Omega \frac{\partial \chi_i}{\partial x_j} \sigma_{ij}(v) d\Omega, \quad l_2^O(v) = \frac{1}{|\Gamma^O|} \int_{\Gamma^O} v_1 ds.$$

We express the force output as a bounded linear functional [16], where summation over repeated indices is assumed $(i, j = 1, 2)$, $\sigma_{ij}(u)$ is the stress tensor, and χ_i is a vector function defined as

$$\chi_1 = \begin{cases} 1 - 2x_1 & : \ (x_1, x_2) \in \Omega' \\ 0 & : \ (x_1, x_2) \in \Omega / \Omega' \end{cases}, \qquad \chi_2 = 0.$$

Here $\Omega' = \{(x_1, x_2) | \ 0 \leq x_1 \leq 0.5, 0.8 \leq x_2 \leq 1.0\}$. Since there are no volume forces, or tractions applied at either Γ^4 or Γ^6, $l_1^O(v)$ is equivalent, for sufficiently smooth functions, to the integral along Γ^5 of the horizontal component of the traction vector.

The same initial triangulation $\mathcal{T}_H^0$ and "truth" mesh $\mathcal{T}_h$ have been considered for both outputs. The initial triangulation consists of 108 triangles. The "truth" mesh has been generated from this initial triangulation by four refinement interactions in which *all* triangles are subdivided; thus, the number of elements in the "truth" mesh is $27,648 = 108 \times 4^4$. In the adaptive process, elements were selected for refinement based on a threshold of $\alpha = 0.3$ for all the iterations. The accuracy goal, r^{obj}, for the relative (half) bound gap is 0.003 (or 0.3%).

For the force output s_1, 16 refinement iterations are required starting from a relative bound gap of 15.64%; only 12 refinement iterations are required for the displacement output s_2, for which the initial relative bound gap is 13.82%. Figures 9 and 10 show the relative bound gap as a function of the number of elements, at each adaptive iteration, for the force and displacement outputs, respectively. We also show the bound gap computed for a sequence of grids obtained by uniform refinement, the finest of which still has four times fewer elements than the "truth" mesh $\mathcal{T}_h$. For both outputs we observe that, the uniformly refined meshes require approximately four times as many elements as the adaptive meshes for comparable levels of accuracy.

We note that, say for the force output, $|(s_{1,pre}^{16} - s_{1,h})/s_{1,pre}^{16}| = 0.0001$, whereas $r_1^{16} = 0.003$, and thus the effectivity of the estimator in the classical sense is not as good as we may hope, even though the convergence *rate* of $\Delta(\mathcal{T}_H)$ is optimal. We emphasize, however, that our bounds provide *certainty* that $|(s_{1,pre}^{16} - s_{1,h})/s_{1,pre}^{16}| \leq 0.003$ — and more importantly, that s_h lies between $s_1^{-,16} = -0.3230$ and $s_1^{+,16} = -0.3211$ — without calculating s_h on the uniformly highly refined (and hence expensive) "truth" mesh $\mathcal{T}_h$. Future work will address techniques by which to reduce the ratio $\Delta/|s_h - s_{pre}|$ as $H \to 0$; this, in turn, will permit to achieve r^{obj} on coarser meshes $\mathcal{T}_H$.

Finally, figures 11 and 12 show the initial, an intermediate, and the final adaptive triangulations obtained for the force and displacement outputs, respectively. Refinement around the singular re-entrant corners is apparent in both cases but, in addition, in the displacement output case, further refinement is observed near the points where l_2^O is "discontinuous".

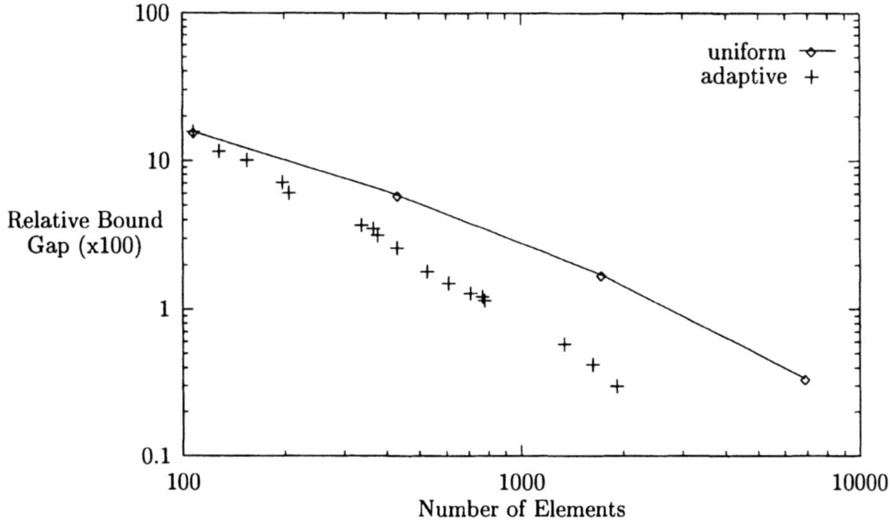

Fig. 9. Bound gap $\Delta(\mathcal{T}_H)$ as a function of the number of elements K_H, for the force output s_1, for meshes generated using uniform and adaptive refinement.

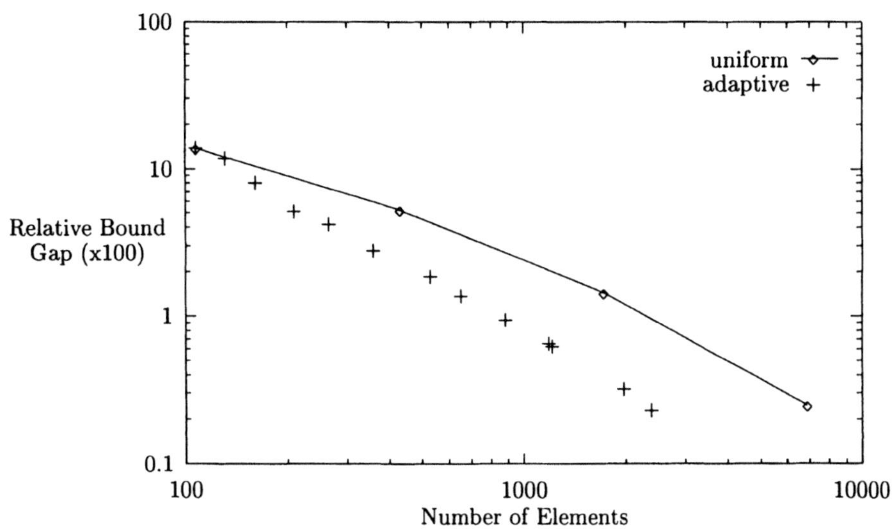

Fig. 10. Bound gap $\Delta(\mathcal{T}_H)$ as a function of the number of elements K_H, for the displacement output s_2, for meshes generated using uniform and adaptive refinement.

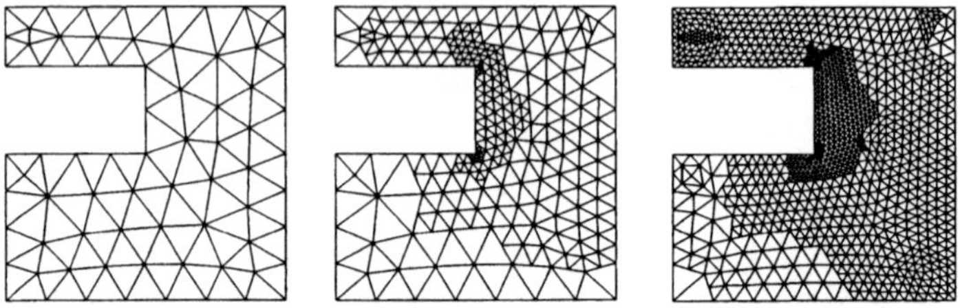

Fig. 11. Adaptively generated meshes for force output s_1. Figure shows $\mathcal{T}_H^0$, $\mathcal{T}_H^8$, and $\mathcal{T}_H^{16}$, with $K_H^0 = 108$, $K_H^8 = 432$ and $K_H^{16} = 1917$ elements; predicted outputs are $s_{1,pre}^0 = -0.3198$, $s_{1,pre}^8 = -0.3215$, and $s_{1,pre}^{16} = -0.3220$; and relative bound gaps are $r_1^8 = 15.64\%$, $r_1^8 = 2.59\%$ and $r_1^{16} = 0.30\%$.

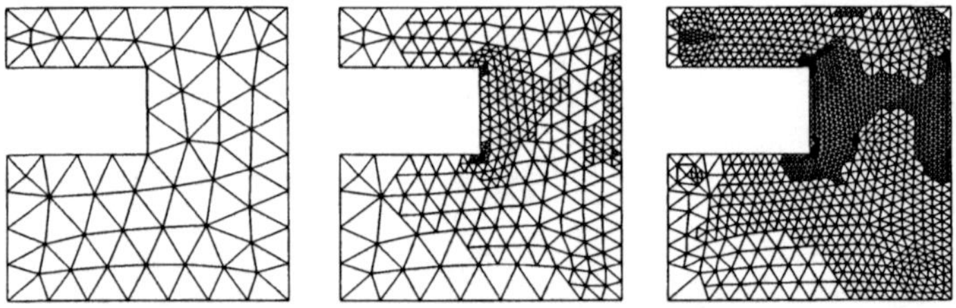

Fig. 12. Adaptively generated meshes for displacement output s^2. Figure shows $\mathcal{T}_H^0$, $\mathcal{T}_H^6$, and $\mathcal{T}_H^{12}$, with $K_H^0 = 108$, $K_H^6 = 531$ and $K_H^{12} = 2382$ elements; predicted outputs are $s_{2,pre}^0 = -0.4056$, $s_{2,pre}^6 = -0.4204$, and $s_{2,pre}^{12} = -0.4226$; and relative bound gaps are $r_2^0 = 13.82\%$, $r_2^6 = 1.83\%$ and $r_2^{12} = 0.23\%$.

8 Stabilized Methods

We now extend out bound procedure to stabilized formulations for convection dominated problems. The most important difference in this case is that due to the presence of the stabilization terms, the corresponding linear and bi-linear forms in the variational statement are a function of the mesh parameter $\delta = (H \text{ or } h)$. As a result standard orthogonality is lost and the bound procedure needs to be modified accordingly.

We consider for illustration purposes the stationary linear convection problem

$$U \cdot \nabla u + \sigma u = f^I \quad \text{in } \Omega, \quad \nabla \cdot U = 0$$

$$u = g^I \quad \text{on } \Gamma_i,$$

where Ω is a bounded domain in $\mathbb{R}^2$ with boundary Γ, σ is a real non–negative parameter, and $U = (U_1, U_2)$ is an incompressible prescribed velocity field. The boundary Γ is decomposed into an inflow boundary $\Gamma_i := \{\mathbf{x} \in \partial\Omega \mid \mathbf{n} \cdot \mathbf{U} < 0\}$ and an outflow boundary $\Gamma_o = \Gamma \backslash \Gamma_i$, where $\mathbf{n}$ is the unit outward normal on Γ. We assume that f^I and g^I are sufficiently regular so that the problem is well posed.

In order to introduce the stabilized weak formulation of this problem we define

$$(w,v) = \int_\Omega wv\,d\mathbf{x}, \qquad \langle w,v\rangle = \int_\Gamma \mathbf{U}\cdot\mathbf{n}\,wv\,ds,$$

$$\langle w,v\rangle_i = \int_{\Gamma_i} \mathbf{U}\cdot\mathbf{n}\,wv\,ds, \quad \langle w,v\rangle_o = \langle w,v\rangle - \langle w,v\rangle_i.$$

The output of interest, $s = l^O(u)$, will be of the form

$$l^O(v) = (f^O, v) + \langle g^O, v\rangle_o.$$

For $\delta = H, h$, we consider the following coarse and fine mesh problems : find $u_\delta \in X_\delta$ such that

$$a_\delta(u_\delta, v) = l_\delta(v), \quad \forall v \in X_\delta.$$

Here,

$$\begin{aligned} a_\delta(w,v) &\equiv (U\cdot\nabla w + \sigma w\ ,\ v + \tau_\delta(U\cdot\nabla w - \sigma w)) - \langle w,v\rangle_i \\ l_\delta(v) &\equiv (f^I\ ,\ v + \tau_\delta(U\cdot\nabla w - \sigma w)) - \langle g^I, v\rangle_i. \end{aligned}$$

Note that, since the boundary conditions are imposed naturally, the functions in X_H and X_h are, in this case, unrestricted on the boundary.

The stabilization parameter τ_δ is taken to be constant over each triangle $T_\delta \in \mathcal{T}_\delta$ and equal to

$$\tau_\delta = [|(\mathbf{l}_{12}\cdot\bar{\mathbf{U}})/(\mathbf{l}_{12}\cdot\mathbf{l}_{12})| + |(\mathbf{l}_{23}\cdot\bar{\mathbf{U}})/(\mathbf{l}_{23}\cdot\mathbf{l}_{23})| + |(\mathbf{l}_{31}\cdot\bar{\mathbf{U}})/(\mathbf{l}_{31}\cdot\mathbf{l}_{31})|]^{-1}.$$

Here $\bar{\mathbf{U}}$ is the average value of $\mathbf{U}$ in the element, and $\mathbf{l}_{ij} = \mathbf{x}_j - \mathbf{x}_i$, where $\mathbf{x}_i$ and $\mathbf{x}_j$ are the coordinates of the element nodes i and j respectively. *A priori* error analysis of the resulting scheme (see [8] for the analysis of a similar algorithm) predicts an optimal convergence rate for the L_2–norm of the error in the $\mathbf{U}$–directional derivative, $||\mathbf{U}\cdot\nabla(u-u_\delta)||_0 \le C\delta||u||_2$, and a near–optimal convergence rate for the L_2–norm of the error in the solution, $||u-u_\delta||_0 \le C\delta^{3/2}||u||_2$. Here $||\cdot||_k$ denotes the usual Sobolev norms.

8.1 Lagrangian Bound Formulation

Repeating the steps of section 6.2, we construct the "energy" equality that will be used to convexify out problem. On mesh h we have

$$a_h^S(e_h, e_h) = R_h(e_h) \equiv l_h(e_h) - a_h(u_H, e_h),$$

where

$$\begin{aligned} a_h^S(w,v) \equiv\ & (U\cdot\nabla w\ ,\ \tau_h\, U\cdot\nabla v) + (\sigma w, v) \\ & - (\sigma w, \tau_h \sigma w) + \frac{1}{2}[\langle w,v\rangle_o - \langle w,v\rangle_i] \end{aligned}$$

which is coercive provided τ_h is sufficiently small — a condition which is not difficult to meet *a priori*. Our Lagrangian $\mathcal{L}_h : \hat{X}_h \times X_h \times Q_h \mapsto \mathbb{R}$ is then,

$$\begin{aligned}\mathcal{L}_h^{\pm}(\hat{v},\theta,q) \equiv \kappa(a_h^S(\hat{v},\hat{v}) - R_h(\hat{v})) \pm l^O(v) \\ + a_h(\hat{v},\theta) - R_h(\theta) + b(\hat{v},q).\end{aligned}$$

Adjoint computation Our approach to computing a suitable adjoint is to derive the continuous adjoint equation and then approximate it on the coarse mesh. The continuous adjoint equation in this case reads,

$$-U\cdot\nabla\psi + \sigma\psi = -f^O \quad \text{in } \Omega, \quad \nabla\cdot U = 0$$

$$\psi = -g^O \quad \text{on } \Gamma_o.$$

The H-mesh approximation of this problem is : find $\psi_H \in X_H$,

$$\begin{aligned}(v - \tau_H\, U\cdot\nabla v\,,\, -U\cdot\nabla\psi_H + \sigma\psi_H) + \langle v,\psi_H\rangle_o \\ = (v - \tau_H\, U\cdot\nabla v\,, -f^O) + \langle v, -g^O\rangle_o, \quad \forall\, v\in X_H\end{aligned}$$

Note that the perturbation of the right hand side is necessary to obtain a high order approximation to the continuous adjoint problem; inattention to this detail yields a much less accurate adjoint ψ_H [12].

Bounds The procedures for equilibration and solution of the local problems are identical to those described in section 6.2. After calculation, the optimized bounds become,

$$\begin{aligned}s^+ &= l^O(u_H) - 2a_h^S(\hat{e}_{0h},\hat{e}_{1h}) - R_h(\psi_H) + 2\,|||\hat{e}_{0h}|||_h\ |||\hat{e}_{1h}|||_h,\\ s^- &= l^O(u_H) - 2a^S(\hat{e}_{0h},\hat{e}_{1h}) - R_h(\psi_H) - 2\,|||\hat{e}_{0h}|||_h\ |||\hat{e}_{1h}|||_h,\end{aligned}$$

where we recall that $|||\cdot|||^2 \equiv a_h^S(\cdot,\cdot)$.

The only difference between these expressions and (21), (22) is the term $R_h(\psi_H)$ which in this case is nonzero due to the lack of orthogonality alluded to earlier.

8.2 Application Example

We consider the numerical solution of the two–dimensional convection equation on a unit square $\Omega =]0,1[^2$. The convection velocity $\mathbf{U}$, is given by $\mathbf{U} = (-x_2, x_1)$; we take $\sigma = 0$ and $f^I = 0$. At the inflow boundary $\partial\Omega_i$, g^I is given by

$$g^I = \begin{cases} 5.0(x_1 - 0.2), & \text{for } 0.2 < x_1 \le 0.4, \quad x_2 = 0\\ 1, & \text{for } 0.4 < x_1 \le 0.6, \quad x_2 = 0\\ 1 - 5(x_1 - 0.6), & \text{for } 0.6 < x_1 \le 0.8, \quad x_2 = 0\\ 0, & \text{otherwise} \end{cases} \quad . \tag{23}$$

The output considered is the weighted integral of the solution on the outflow boundary $\partial\Omega_i$, that is, $f^O = 0$ and

$$g^O = \begin{cases} 10(x_1 - 0.6), & \text{for } 0.6 < x_2 \le 0.7, \quad x_1 = 0 \\ 1 - 10(x_1 - 0.7), & \text{for } 0.7 < x_2 \le 0.8, \quad x_1 = 0 \\ 0, & \text{otherwise} \quad . \end{cases} \tag{24}$$

The computational domain, together with the functions g and g^O, is sketched in figure 13.

A nearly uniform initial mesh consisting of 240 triangles is uniformly refined four times to generate a sequence of meshes with $240 \cdot 4^r$ elements for $r = 1, 2, 3$, and 4. The finest mesh so obtained, consisting of 61,440 triangles, is taken to be the fine mesh $\mathcal{T}_h$. For each of the coarser meshes we calculate the upper and lower bounds; the results are shown in Fig. 14. The observed convergence rate is slightly greater than $O(H^2)$.

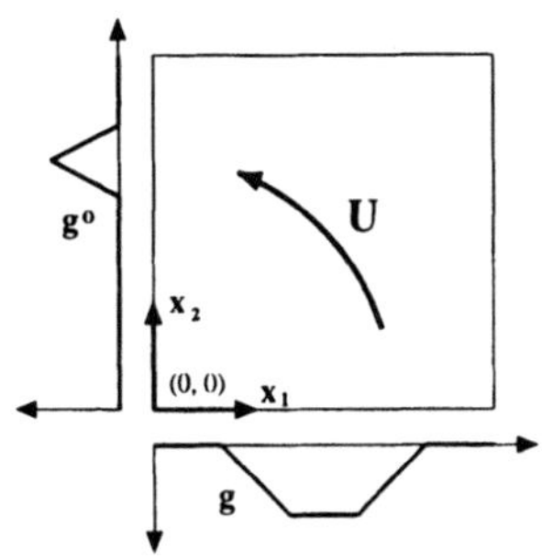

Fig. 13. Two-dimensional linear convection: computational domain with the definition of the functions g and g^O.

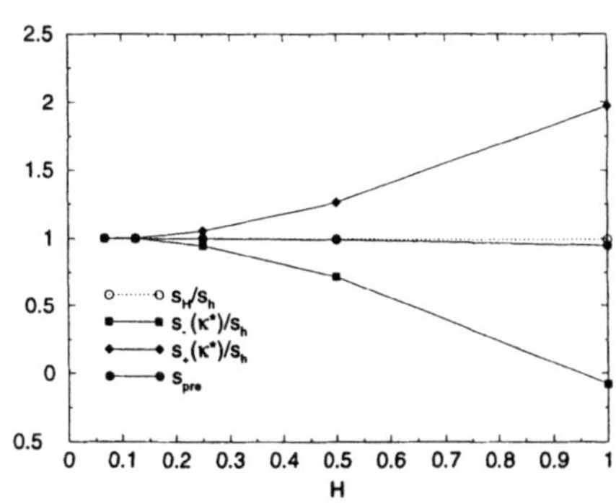

Fig. 14. Two-dimensional linear convection: computed upper and lower bounds as a function of H.

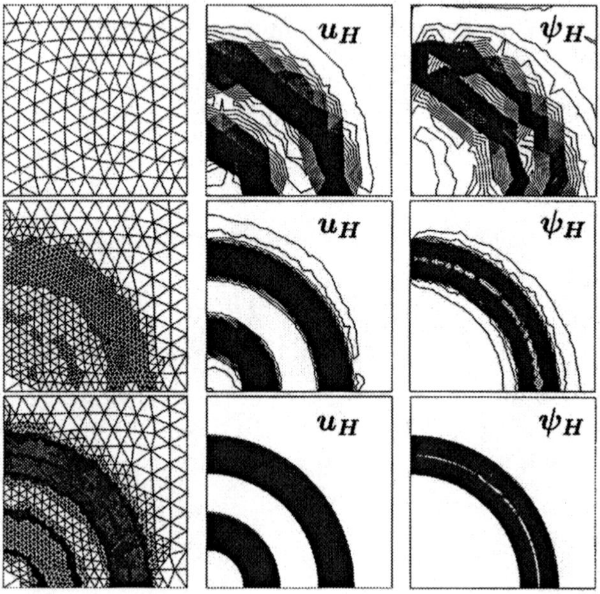

Fig. 15. Two–dimensional linear convection: adaptively generated mesh (left), u_H (middle), and ψ_H (right), for a) the initial mesh of 240 elements and relative bound gap 108%, b) an intermediate mesh, after 5 refinements, of 1,557 elements and relative bound gap 7.10%, and c) the final mesh, after 10 refinements, of 7,791 elements and relative bound gap 0.1%.

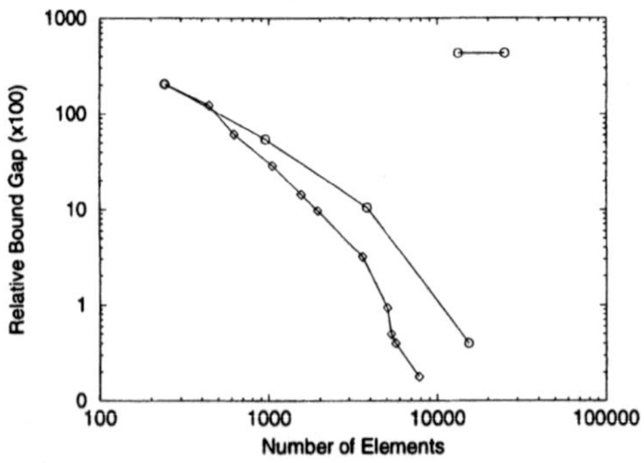

Fig. 16. Two–dimensional linear convection: bound gap $2\Delta/s_{pre}$ as a function of the number of elements in $\mathcal{T}_H$ for meshes generated using uniform and adaptive refinement.

We next implement the adaptive procedure described in section 7. The starting triangulation is the coarsest mesh considered above, which has 240 triangles. On this mesh the error, given by Δ/s_{pre}, is 1.08. In the subsequent adaptive process all those triangles whose contribution to the bound gap is larger than 0.2 that of the largest elemental contribution are refined; the strategy described in [18] is used in order to treat the "hanging nodes" generated during the refinement process. The accuracy goal for Δ/s_{pre} is set to 0.001; a total of 10 meshes are generated before this goal is achieved. The final adapted mesh has 7,791 elements, which compares favorably with the uniform mesh that would be required to obtain a similar level of accuracy. In particular, a three-level uniformly refined mesh has 15,360 triangles, but only produces an accuracy $\Delta/s_{pre} = 0.002$. Fig. 15 shows the mesh and the contours for the computed solution u_H and adjoint ψ_H, initially, after one intermediate refinement step, and at the end of the adaptive process; note the effect of the adjoint (and hence output) structure on the mesh. Finally, Fig. 16 presents $2\Delta/s_{pre}$ as a function of the number of elements in $\mathcal{T}_H$ for both adaptive and uniform refinement

9 Non-coercive problems

The bound procedure developed above is applicable whenever a Lagrangian can be found which has the required properties. When the differential operator considered is non-coercive, and possibly non-linear, the convexification procedure based on the energy equality is not globally valid and therefore our standard approach fails. The basic method can be extended to deal with these more general cases but at the expense of obtaining only asymptotic bounds. That is, bounds are guaranteed once the coarse mesh is such that $H < H_0$. Unfortunately, H_0 is not known *a priori* and as a consequence some certainty is lost. In practice however, this approach has been found to work very well and only, in very rare circumstances and using very inaccurate H-mesh approximations, have we found that the computed bounds are not valid. This procedure is given in detail in [14], and here we only provide a very brief description.

We consider the problem of determining $u \in X$, such that

$$\mathcal{A}(u, v) = 0, \qquad \forall v \in X.$$

Here, $\mathcal{A} : X \times X \mapsto \mathbb{R}$, is linear in the second argument, but possibly non-linear with respect to the first argument. We also consider an output $s = \mathcal{S}(u)$, defined by $\mathcal{S} : X \mapsto \mathbb{R}$. We introduce also the coarse grid solution $u_H \in X_H$, such that

$$\mathcal{A}(u_H, v) = 0, \quad \forall v \in X_H.$$

The forms $\mathcal{A}$ and $\mathcal{S}$ are expanded about u_H,

$$\mathcal{A}(u_H + w, v) - \mathcal{A}(u_H, v) = E(w, v) + F(w, w, v) + G(w, v)$$
$$\mathcal{S}(u_H + w) - \mathcal{A}(u_H) = L(w) + M(w, w) + N(w).$$

Here $L(w)$ is linear, $E(w, v)$ and $M(w, w)$ are bi-linear and $F(w, w, v)$ is tri-linear. The forms $G(w, v)$ and $N(w)$ correspond to the remainder in the expansions and contain the higher order terms.

We consider the symmetric part of $E(\cdot, \cdot)$ defined as

$$E^S(w, v) = \frac{1}{2}(E(w, v) + E(v, w)).$$

Furthermore, we introduce the following splitting of $E^S(w, v)$,

$$E^S(w, v) = E^S_c(w, v) + E^S_a(v, w), \quad E^S_c(w, w) > 0, \ \forall w \in X \ (w \neq 0),$$

that is, we identify the (or a) coercive part of $E^S(w, v)$, with $E^S_c(w, v)$, and $E^S_a(v, w)$ with the remainder.

9.1 Bound Algorithm

The bound algorithm in the non-coercive case comprises the following steps, given below in algorithmic form:

- Solve problem on the coarse mesh: find $u_H \in X_H : \mathcal{A}(u_H, v) = 0, \ \ \forall v \in X_H$,
  and evaluate the primal residual $R^u(v) = -\mathcal{A}(u_H, v), \ \ \forall v \in X_h$.
- Compute the coarse mesh adjoint : find $\psi_H \in X_H : E(v, \psi_H) = -L(v), \ \forall v \in X_H$, and determine the adjoint residual $R^\psi(v) = -L(v) - E(v, \psi_H), \ \ \forall v \in X_h$
- Calculate the hybrid fluxes, $p^u_H \in Q, \ p^\psi_H \in Q$, using the equilibration procedure described in section 5.4.
- Compute the reconstructed error by solving the local problems: find $\hat{e}^u_h \in \hat{X}_h$, and $\hat{e}^\psi_h \in \hat{X}_h$ such that

$$2E^S_c(\hat{e}^u_h, v) = R^u(v) + b(v, p^u_H), \quad \forall v \in \hat{X}_h,$$
$$2E^S_c(\hat{e}^\psi_h, v) = R^\psi(v) + b(v, p^\psi_H), \quad \forall v \in \hat{X}_h.$$

Finally, the bound average is computed as

$$\bar{s} = \mathcal{S}(u_H) - 2E^S_c(\hat{e}^u_h, \hat{e}^\psi_h) - R(\psi_H).$$

Note that, whenever the form $\mathcal{A}$ does not depend on h, the term $R(\psi_H)$ will vanish.

The bound gap is

$$\Delta = 2\sqrt{E^S_c(\hat{e}^u_h, \hat{e}^u_h)\ E^S_c(\hat{e}^\psi_h, \hat{e}^\psi_h)},$$

which, if required, can be written in terms of positive elemental contributions for adaptivity purposes [14]. The upper and lower bounds follow as $s^\pm = \bar{s} \pm \Delta$.

Bound Properties (s^-) Introducing $e = u_h - u_H$ and $\hat{e} = \hat{e}_h^u + \frac{1}{\kappa}\hat{e}_h^\psi$, it is possible to write [14],

$$s^- = \mathcal{S}(u_h) - \kappa E_c^S(e - \hat{e}, e - \hat{e}) + \mathcal{F}(e),$$

with $\kappa = \sqrt{E_c^S(\hat{e}_h^\psi, \hat{e}_h^\psi)/E_c^S(\hat{e}_h^u, \hat{e}_h^u)}$.

For coercive linear problems $\mathcal{F}(e) = 0$ and bounds will obtain for all H. For more general problems, if we can find a coercive splitting (9) such that $E_c^S(e - \hat{e}, e - \hat{e}) \gg \mathcal{F}(e)$ for H sufficiently small then asymmptotic bounds will be obtained. Typically, we will have expect $E_c^S(e-\hat{e}, e-\hat{e}) \sim O(H^p)$ and , $\mathcal{F}(e) \sim O(H^p)$, with $p > q$. This procedure has been applied to a number on non-coercive problems [19,14,21] including the Helmholtz equation, the Burgers equation and the symmetric eigenvalue problem. In all these cases one obtains, using linear elements, $p = 4$, $q = 2$. The approach has also been extended to the Navier-Stokes equations, but in this case further elaboration is necessary and the reader is referred to [13] for further details. Below we give some computational results for the symmetric eigenvalue problem and for the incompressible Navier-Stokes equations.

9.2 Eigenvalue Problem

We consider the following two dimensional eigenvalue problem

$$-\nabla^2 u = \lambda u, \quad \text{in } \Omega, \quad u|_\Gamma = 0.$$

The computational domain is defined in figure 17 and consists of a square domain of unit side length with a rectangular cut out. Here, the outputs

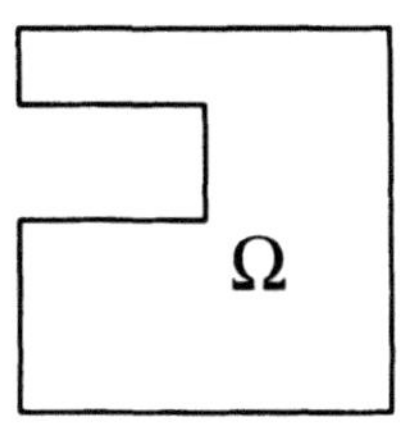

Fig. 17. Computational domain used for the eigenvalue problem.

of interest are the eigenvalues $s = \lambda$, but it is also possible to include the eigenvectors provided they are suitably normalized [14].

By defining

$$a(w, v) \equiv \int_\Omega \nabla w \cdot \nabla v \, d\Omega, \quad m(w, v) \equiv \int_\Omega w \nabla v \, d\Omega,$$

the variational problem becomes: find $(u, \lambda) \in X \times \mathbb{R}$ such that

$$a(u, v) - \lambda m(u, v) - \sigma(m(u, u) - 1) = 0, \quad \forall (v, \sigma) \in X \times \mathbb{R}.$$

In terms of the terminology introduced above we write : find $(u,\lambda) \in X \times \mathbb{R}$ such $\mathcal{A}((u,\lambda),(v,\sigma)) = 0$, $\forall (v,\sigma) \in X \times \mathbb{R}$. Here, $\mathcal{A}((w,\mu),(v,\sigma)) \equiv a(w,v) - \mu m(w,v) - \sigma(m(w,w) - 1)$. The corresponding bi-linear form becomes $E((w,\mu),(v,\sigma)) \equiv a(w,v) - \lambda_H m(w,v) - \mu m(u_H,v) + 2\sigma m(u_H,w)$ and we simply choose $E_c^S((w,\mu),(v,\sigma)) = a(w,v)$ as the coercive component of the symmetric part of $E(\cdot,\cdot)$. The results are summarized in figures 18 and 19. Figure 18 shows the contours of the first three eigenmodes and corresponding eigenvalues computed on a very fine mesh $\mathcal{T}_h$ consisting of 27,648 elements. This mesh is obtained by subdividing each element of the coarse mesh $\mathcal{T}_H$, shown in figure 19 into $4^4 = 256$ elements. Figure 19 shows the results computed for the first three eigenmodes on mesh $\mathcal{T}_H$ of 108 elements and mesh $\mathcal{T}_{H/4}$ of 1728 elements. In all cases bounds are obtained and the bound gap is found to converge at an approximate rate of $\mathcal{O}(H^2)$. For this problem, it can be shown that the eigenvalue computed on the coarse mesh, λ_H, is an upper bound for λ_h. Our bound procedure does in fact predict that $s^+ = \lambda_H$.

$\lambda_h^1 = 38.546$ $\lambda_h^2 = 60.473$ $\lambda_h^3 = 78.458$

Fig. 18. Contours of the first three eigenmodes and corresponding eigenvalues computed on the reference mesh $\mathcal{T}_h$.

9.3 Navier-Stokes Equations

We consider now the problem of natural convection in a two dimensional domain. This is the problem for which results were presented earlier in section 1.1. The incompressible Navier-Stokes equations are solved together with the energy equation. Boussinesq's approximation is used to model buoyancy effects. Thus, the set of partial equations to be solved are

$$-\frac{\partial}{\partial x_j}\left(\frac{\partial u_i}{\partial x_j} + \frac{\partial u_j}{\partial x_i}\right) + \frac{\partial (u_j u_i)}{\partial x_j} + \frac{\partial p}{\partial x_i} = -Gr\,\theta \hat{g}_i$$

$$\frac{\partial u_i}{\partial u_i} = 0$$

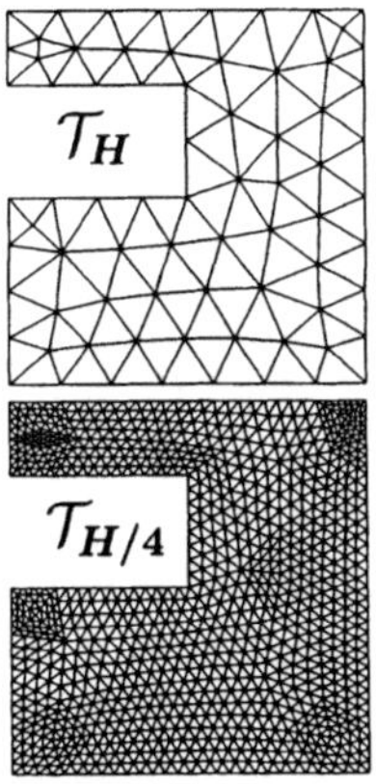

$\lambda_h^1 = 38.546$	$\lambda_h^2 = 60.473$	$\lambda_h^3 = 78.458$
$s^+ = 42.023$	$s^+ = 65.216$	$s^+ = 87.346$
$s^- = 34.234$	$s^- = 56.448$	$s^- = 70.315$
$s^+ - s^- = 7.789$	$s^+ - s^- = 14.504$	$s^+ - s^- = 17.030$
$s^+ = 38.845$	$s^+ = 60.779$	$s^+ = 78.995$
$s^- = 38.164$	$s^- = 60.232$	$s^- = 78.033$
$s^+ - s^- = 0.681$	$s^+ - s^- = 0.546$	$s^+ - s^- = 0.961$

Fig. 19. Computed results for eigenvalue problem. Table shows the upper bound the lower bound and the bound gap for the first three eigenvalues and for each of the two meshes shown on the left consisting of 108 and 1728 linear triangular elements.

$$-\frac{1}{Pr}\frac{\partial^2\theta}{\partial x_j \partial x_j} + \frac{\partial(u_j\theta)}{\partial x_j} = 0,$$

where the summation convention over repeated indices applies, $\mathbf{u} = (u_1, u_2)$ is the velocity field, p is the pressure field, and θ is the temperature field, and $\hat{\mathbf{g}} = (\hat{g}_1, \hat{g}_2)$ is the unit vector in the direction of gravity. The flow is governed by two non-dimensional parameters: the Prandtl number, Pr, and the Grashof number, Gr.

The example considered is that used in the introduction of these notes. The computational boundary is shown in figure 20. The boundary conditions are $u = 0$ on all boundaries, and $\theta = 0$ on the right vertical boundary, a prescribed heat flux $\frac{1}{Pr}\frac{\partial\theta}{\partial n} = Q(= 1)$ on the boundaries shown, and adiabatic conditions on all other boundaries. For the simulations we have assumed that $Pr = 1$ and $Gr = 50,000$

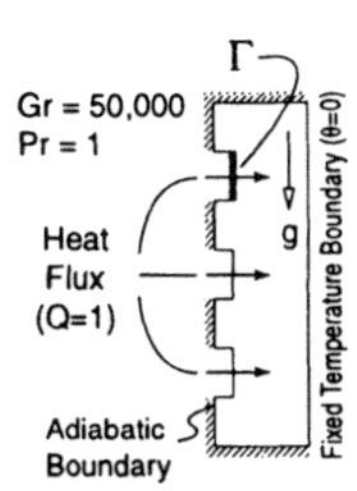

Fig. 20. Computational domain for the natural convection problem.

The output of interest is the average temperature over Γ, as shown in figure 20,

$$s(u,p,\theta) = \frac{1}{|\Gamma|}\int_\Gamma \theta\, ds.$$

The discretization of the Navier-Stokes equations is performed using the Crouzeix-Raviart element with linear discontinuos pressures and quadratic plus cubic bubble representation of the velocity components. The detail application of the bound procedure, including adaptivity, to this problem can be found in [13]. A prescribed accuracy of 1% is specified and the adaptive algorithm is initiated. Figure 3 shows the initial mesh $\mathcal{T}_H^0$ of 264 and the final mesh $\mathcal{T}_H^2$ obtained after two adaptations, of 680 elements, together with the computed isotherms and streamlines.

	$\mathcal{T}_H^0$	$\mathcal{T}_H^1$	$\mathcal{T}_H^2$	$\mathcal{T}_{H/2}$
# of elements	264	416	628	1056
s^-	0.253	0.270	0.273	0.272
s^+	0.296	0.282	0.279	0.279
Δ	0.0212	0.0057	0.0026	0.0036

Finally, the above table shows the number of elements and bounds computed on the initial mesh and two successively refined meshes. A reduction of the relative bound gap from $\approx$ 8% for the initial mesh, to less than 1% for the final mesh, is achieved by optimal refinement with a final mesh which contains only slightly more than twice the number of elements of the initial mesh. These results contrast with the last column which shows the figures corresponding to a uniformly subdivided mesh. In the calculations reported here, the truth mesh is obtained by subdividing each coarse mesh element into 6^2 elements.

10 Conclusions

We have presented a rigorous procedure for computing bounds for functional outputs of solutions of partial differential equations. The procedure allows us to bound the find mesh solution at the cost of only solving coarse mesh global problems. In addition, the local problem solutions are symmetric and linear. For coercive problems the bounds are uniform whereas for general non-coercive and/or non-linear problems the bounds are only asymmmptotic. In practice however, we have found that asymmmptotic bounds are very reliable and only on very rare occasions, solving on very coarse meshes, have we not obtained bounds. The effectivities computed are of order $\mathcal{O}(1)$ (typically between 1 and 6-7) which we consider "good" if one takes into account the high level of certainty delivered. Finally, the procedure presented lends itself naturally to adaptive mesh refinement by allowing the bound gap to be decomposed into the sum of positive elemental contributions which can be used as refinement indicators.

Acknowledgments

These notes present a summary of research results undertaken by the authors which span a several years. We would like to acknowledge the collaboration with many colleagues, in particular, M. Paraschivoiu, Y. Maday, and L. Machiels. Over the years this work has been sponsored by several funding sources including NASA, DARPA, and the Singapore-MIT Alliance. Finally, we would like to thank H. Ciria for the careful reading of this manuscript.

References

1. M. Ainsworth and T.J. Oden, *A unified approach to a posteriori error estimation based on element residual methods*, Numer. Math., 65:23-50, 1993.
2. M. Ainsworth and J.T. Oden, *A posteriori Error Estimation in Finite ELement Analysis*, Wiley-Interscience, 2000,
3. I. Babuška and W.C. Rheinboltd, *A posteriori error estimates for the finite element method*, Int. J. Num. Meth. in Egngr., 12:1597-1615, 1978.
4. R. Bank and A. Weiser, *Some a posteriori error estimators for elliptic partial differential equations*, Math. Comp., 44:283-301, 1985.
5. R. Becker and R. Rannacher, *Weighted a posteriori error control in finite element methods*, IWR Preprint 96-1 (SFB359), Heildelberg, 1996.
6. D. Bertsekas, *Nonlinear Programming*, Athena Scientific, 1995.
7. F. Brezzi and M. Fortin, *Mixed and hybrid finite element methods*, Springer-Verlag, New York, 1991.
8. C. Johnson, U. Nävert and J. Pitkäranta, *Finite Element Methods for Linear Hyperbolic Problems*, Comp. Meth. in Appl. Mech. and Engrg., 45 (1984), pp. 285–312.
9. P. Ladevèze and D. Leguillon, *Error estimate procedure in the finite element method and applications*, SIAM J. Numer. Anal., 20:485-509, 1983.
10. L.D. Landau and E.M. Lifshitz, *Fluid Mechanics*, Course of Theoretical Physiscs, vol. 6, Pergamon Press 1982.
11. L. Machiels, Y. Maday and A.T. Patera, *A "Flux-Free" nodal Neumann subproblem approach to output bounds for partial differential equations*, to appear in C.. Acad.Sci. Paris, 1999.
12. L. Machiels, A.T. Patera, J. Peraire and Y. Maday, *A general framework for finitie element a posteriori error control: application to linear and nonlinear convection-dominated problems*, ICFD Conference on Numerical Methods for Fluid Dynamics, Oxford, 1998.
13. L. Machiels, J. Peraire and A.T. Patera, *A posteriori finite element output bounds for the incompressible Navier-Stokes equations; Application to a natural convection problem*, J. Comp. Phys., 172, 401-425, 2001.
14. Y. Maday, A.T. Patera and J. Peraire, *A general formulation for a posteriori bounds for output functionals of partial differential equations; application to the eigenvalue problem*, C.. Acad.Sci. paris, t. 328, Serie I, p. 823-828, 1999.
15. M. Paraschivoiu and A.T. Patera, *A hierarchy duality approach to bounds for the outputs of partial differential equations*, Comput. Methods Appl. Mech. Engrg., 15 Ladevèze and J.T. Oden editors. Elsevier, 1998.

16. M. Paraschivoiu, J. Peraire and A.T. Pàtera, *A posteriori finite element bounds for linear-functional outputs of elliptic partial differential equations*, Comput. Methods Appl. Mech. Engrg., 150:289-312 (1997).
17. A.T. Patera and E. Ronquist, *A general output bound result: application to discretization and iteration error estimation and control*, Mathematical Models and Methods in Applied Science, to appear.
18. J. Peraire and A.T. Patera, *Bounds for linear-functional outputs of coercive partial differential equations: local indicators and adpative refinement*, in Advances in Adaptive Computational Methods in Mechanics, P. Ladevèze and J.T. Oden editors. Elsevier, 1998.
19. J. Peraire and A.T. Patera, *Asymmptotic a posteriori finite element bounds for the outputs of non-coercive problems: the Helmholtz and Burgers equations*, Comp. Meth. Appl. Mech. Engrg., 171, 77-86, 1999.
20. C. Prud'homme, D. Rovas, K. Veroy, L. Machiels, Y. Maday, A.T. Patera and G. Turinici, *Reliable real-time solution of parametrized partial differential equations: reduced-basis output bond methods*, Journal of Fluids Engineering, 2002 (to appear).
21. J. Sarrate, J. Peraire and A.T. Patera, *A posteriori bounds for non-linear outputs of solutions of the Helmholtz equation*, Int. J. Num. Meth. Fluids, 31, 17-36, 1999.
22. G. Strang, *Introduction to applied mathematics*, Wellesley-Cambridge Press, 1986.

Computable Error Estimators and Adaptive Techniques for Fluid Flow Problems

Serge Prudhomme and J. Tinsley Oden

Texas Institute for Computational and Applied Mathematics,
The University of Texas at Austin, Austin, Texas 78712, U.S.A.

Abstract. The present paper reviews several methods for *a posteriori* error estimation of finite element approximations. The first part (Section 1) is concerned with global error estimators which provide for estimates and bounds in global norms (so-called energy norms). In particular, explicit and implicit residual methods are described in detail. The second part of the paper (Section 2) deals with a more recent technique, in which the numerical error in finite element approximations of general problems are estimated in terms of quantities of interest. The concept of goal-oriented adaptivity which embodies mesh adaptation procedures designed to control error in specific quantities is also described. Both global and goal-oriented error estimation methods are illustrated on model problems based on the Poisson, convection-diffusion, Stokes and steady-state Navier-Stokes equations. The third part of the paper (Section 3) presents preliminary work on *a posteriori* error estimation for time-dependent problems, namely the parabolic heat equation and the incompressible Navier-Stokes equations. The methods deal with estimating the approximation errors due to the discretization in space using either explicit or implicit error estimators.

Contents

1 Global Error Estimation by Residual Methods

1.1 Introduction

Theories and methods of *a posteriori* error estimation provide a means for assessing numerical error in finite element approximations of partial differential equations and for developing adaptive meshing algorithms for optimizing finite element meshes. Progress in computer performance has allowed analysts to investigate increasingly more complex problems depicting physical events, for which it becomes more difficult to evaluate the reliability of the solutions. *A posteriori* error estimation proves to be an unavoidable adjunct to the computations in order to automatically verify that the model equations are solved accurately. The other attractive feature of error estimation is to provide with refinement indicators to be used in adaptive methods in order to control or reduce numerical errors. Although other criteria could be used for adaptation, error estimates in specific measures are highly desirable to indicate where and when to refine a solution procedure. A vast literature on *a posteriori* error estimation and adaptive methods in finite element and other discretization methods is now available. Several books have recently been published (cf. [60,4,11,37]). It is noteworthy, however, that since the pioneering work of Babuška and Rheinboldt in the late seventies (see in particular [9]), most of the works have been devoted to elliptic and parabolic linear problems in the framework of finite element methods.

In Computational Fluid Dynamics, and to the best knowledge of the authors, very few papers have been published on the subject in the case of finite volume methods. Preliminary works are in progress in the case of discontinuous Galerkin methods. In addition, most of the effort has until recently focused on estimating the error in global norms such as the so-called energy norm or the L^2 norm. Although these methods are still very efficient in controlling errors, they provide on the other hand limited information to the analyst about the accuracy of the solution with respect to specific physical quantities.

New methods have recently been developed to estimate errors in so-called quantities of interest. One important feature of these methods is that they often involve global error estimates in the energy norm. We therefore review in this paper various approaches for deriving *a posteriori* error estimates in global norms in the case of steady-state problems of interest in Computational Fluid Dynamics.

Following this brief introduction, we present in Section 1.2 the model problems and some useful notations. In Section 1.3, we introduce the residual functional and show how it relates to the error. We briefly describe in Section 1.4 recovery-based error estimators and then give a detailed account in Section 1.5 on residual methods. We then propose an adaptive strategy based on these global error estimators in Section 1.6.

1.2 Preliminaries

Notations. Let Ω denote a bounded open domain in $\mathbb{R}^d$, $d = 1, 2$, or 3, with Lipschitz continuous boundary $\partial\Omega$, and let Γ_D and Γ_N denote portions of the boundary such that $\bar{\Gamma}_D \cup \bar{\Gamma}_N = \partial\Omega$, $\Gamma_D \cap \Gamma_N = \varnothing$.

Let s be a positive integer and S an open region in $\mathbb{R}^d$ (S may stand for the whole domain Ω, for a subdomain of Ω or for a portion of the boundary). Denote by $H^s(S)$ the Sobolev space of functions in $L^2(S)$ with distributional derivatives of order $\leq s$ in $L^2(S)$ as well. The L^2 norm is denoted by $\|\cdot\|_0$ and the H^s norm by $\|\cdot\|_s$. We will also use H^s seminorms, we shall denote by $|\cdot|_s$.

For scalar-valued functions, we will in general use V to refer to the space of admissible solutions:

$$V = \{v \in H^1(\Omega);\ v = 0 \text{ on } \Gamma_D\},$$

the boundary values being understood in the sense of traces of H^1-functions. In the case of vector-valued functions, we will use boldfaced letters; for example, the space of admissible solutions will read:

$$\boldsymbol{V} = (V)^d = V \times V \times \ldots \times V.$$

We will also consider the subspace Q of $L^2(\Omega)$ functions defined as:

$$Q = \{q \in L^2(\Omega); \textstyle\int_\Omega q \, \mathrm{d}x = 0\}.$$

Let $\mathcal{P}_h$ be a partition of Ω, in other words, $\mathcal{P}_h$ is a finite collection of N_e open elements K_i, $i = 1, 2, \ldots, N_e$. The partition is assumed to satisfy $\bar{\Omega} = \bigcup_{K_i \in \mathcal{P}_h} \bar{K}_i$, and $K_i \cap K_j = \varnothing$, whenever i is different from j. The size h_K of an element K is measured in terms of its diameter. We then associate with the partition $\mathcal{P}_h$ the parameter $h = \max_{K \in \mathcal{P}_h} h_K$. In addition, the boundary of an element K is denoted by ∂K, and the unit normal vector outward from K by $\mathbf{n}$.

We consider here finite element spaces of hierarchical shape functions:

$$V^h = \{v \in V; v_{|K} = \hat{v} \circ F_K^{-1}, \hat{v} \in P_{p_K}(\widehat{K}),\ \forall\ K \in \mathcal{P}_h\} \tag{1}$$

where F_K is an affine mapping from the master element $\widehat{K}$ to the element K in physical space, and $P_{p_K}(\widehat{K})$ is the space of polynomial functions defined on $\widehat{K}$ of degree at most p_K. We employ here *hierarchical shape functions* based on the integrated Legendre polynomials $\phi^{(k)}(\xi)$ of degree k (see [58]) which are defined as:

$$\phi^{(k)}(\xi) = \frac{(P_k(\xi) - P_{k-2}(\xi))}{\sqrt{2(2k-1)}}, \qquad \xi \in [-1,1], \quad k = 2,3,\ldots$$

Here the P_k's are the so-called Legendre polynomials and we note that all $\phi^{(k)}$ vanish at $\xi = \pm 1$. The shape functions are usually classified into the following types: *nodal*, *edge* and *interior bubble* functions in two dimensions and *nodal*, *edge*, *face* and *interior bubble* functions in three dimensions. As an example, we list below the shape functions for quadrilateral elements:

Nodal shape functions: We use the classical Lagrangian functions associated with each vertex of the quadrilateral element:

$$\nu(\xi,\eta) = \frac{1}{4}(1 \pm \xi)(1 \pm \eta), \qquad \xi,\eta \in [-1,1].$$

Edge shape functions: With each edge of an element, we associate $p_K - 1$ edge functions. In the case of the first edge, between nodes 1 and 2, they are of the form:

$$\varphi^{(k)}(\xi,\eta) = \frac{1}{2}(1-\eta)\phi^{(k)}(\xi), \qquad \xi,\eta \in [-1,1], \quad k = 2,\ldots,p_K.$$

Interior bubble shape functions: Finally, we define the $(p_K - 1)^2$ interior bubble functions as:

$$\theta^{(k,j)}(\xi,\eta) = \phi^{(k)}(\xi)\phi^{(j)}(\eta), \qquad \xi,\eta \in [-1,1], \quad k,j = 2,\ldots,p_K.$$

Note that the support of any interior bubble function exactly coincides with just one element, in other words, interior bubble functions vanish everywhere except in one element.

Edge and interior bubble functions of degree two are shown in Fig. 1 and Fig. 2, respectively, in the case of quadrilateral elements.

The particular choice of hierarchical shape functions allows us to easily parametrize the approximation process and increase or decrease the orders of the approximations. In particular, it allows us to conveniently set up and solve the auxiliary problems for the error estimates, and, for incompressible fluid flow simulations, to satisfy the discrete *inf-sup* condition (8) by simply using different polynomial degrees for the approximation of the velocity and pressure variables.

Model problems. We shall consider in this paper the following steady-state model problems relevant to computational fluid mechanics:

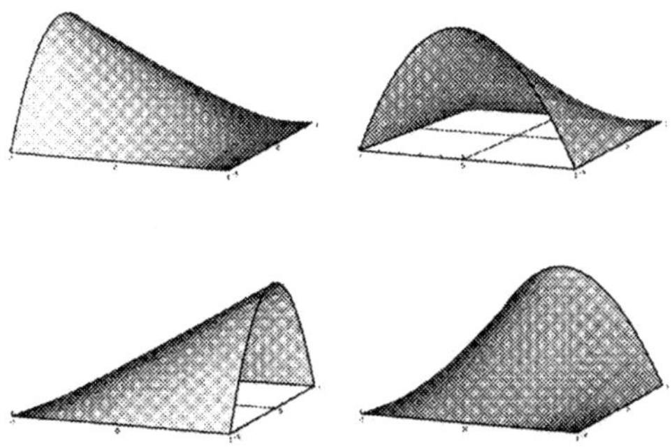

Fig. 1. Shape functions φ of degree 2 associated with each edge of a quadrilateral element.

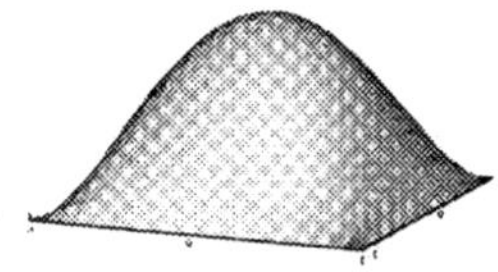

Fig. 2. Interior bubble function θ of degree 2 in a quadrilateral element.

A) **Poisson Problem:** Find the scalar-valued function u defined on $\bar{\Omega}$ such that:

$$\left.\begin{aligned} -\Delta u &= f, \quad && \text{in } \Omega \\ u &= 0, && \text{on } \Gamma_D \\ \mathbf{n} \cdot \nabla u &= g, && \text{on } \Gamma_N \end{aligned}\right\} \tag{2}$$

where $f \in L^2(\Omega)$, $g \in L^2(\Gamma_N)$.

B) **Convection-Diffusion Problem:** Find the scalar-valued function u defined on $\bar{\Omega}$ such that:

$$\left.\begin{aligned} -\Delta u + \alpha \cdot \nabla u &= f, \quad && \text{in } \Omega \\ u &= 0, && \text{on } \Gamma_D \\ \mathbf{n} \cdot \nabla u &= g, && \text{on } \Gamma_N \end{aligned}\right\} \tag{3}$$

where f and g are as above, and where α is a smooth vector-valued function such that $\nabla \cdot \alpha = 0$ (solenidal function).

C) **Stokes Problem:** Find the velocity vector field $\boldsymbol{u} = \boldsymbol{u}(\boldsymbol{x})$ and the pressure field $p = p(\boldsymbol{x})$ such that:

$$\left.\begin{aligned} -\Delta \boldsymbol{u} + \nabla p &= \boldsymbol{f}, \quad && \text{in } \Omega \\ \nabla \cdot \boldsymbol{u} &= 0, && \text{in } \Omega \\ \boldsymbol{u} &= 0, && \text{on } \Gamma_D \\ \mathbf{n} \cdot \nabla \boldsymbol{u} &= \boldsymbol{g}, && \text{on } \Gamma_N \end{aligned}\right\} \tag{4}$$

where $\boldsymbol{f}$ and $\boldsymbol{g}$ are vector-valued data functions defined on $(L^2(\Omega))^d$ and $(L^2(\Gamma_N))^d$ respectively.

D) Navier-Stokes Problem: Find the velocity vector field $\boldsymbol{u} = \boldsymbol{u}(\boldsymbol{x})$ and the pressure field $p = p(\boldsymbol{x})$ such that:

$$\left.\begin{aligned} -\Delta \boldsymbol{u} + \boldsymbol{u} \cdot \nabla \boldsymbol{u} + \nabla p = \boldsymbol{f}, &\quad \text{in } \Omega \\ \nabla \cdot \boldsymbol{u} = 0, &\quad \text{in } \Omega \\ \boldsymbol{u} = 0, &\quad \text{on } \Gamma_D \\ \mathbf{n} \cdot \nabla \boldsymbol{u} = \boldsymbol{g}, &\quad \text{on } \Gamma_N \end{aligned}\right\} \tag{5}$$

where the data are chosen as in the Stokes problem.

Weak formulations. We proceed by presenting the functional setting in order to introduce the finite element formulations of the above problems. Weak forms of partial differential equations are obtained by multiplying the equations by suitable test functions, by integrating over the whole domain, and integrating by parts the high-order terms.

We show as an example how to derive the weak form of the Poisson equation. Multiplying the equation by v and integrating over Ω, we obtain:

$$\int_\Omega (-\Delta u)\, v \,\mathrm{d}x = \int_\Omega f v \,\mathrm{d}x.$$

Then, integrating by part the first integral, we arrive at:

$$\int_\Omega \nabla u \cdot \nabla v \,\mathrm{d}x = \int_\Omega f v \,\mathrm{d}x + \int_{\partial\Omega} (\mathbf{n} \cdot \nabla u)\, v \,\mathrm{d}s.$$

Finally, applying the Neumann boundary condition and choosing v such that $v = 0$ on Γ_D yield the weak form of the Poisson equation, i.e.

$$\int_\Omega \nabla u \cdot \nabla v \,\mathrm{d}x = \int_\Omega f v \,\mathrm{d}x + \int_{\Gamma_N} g v \,\mathrm{d}s.$$

We remark that all the integrals in the above equation are well defined whenever u and v belong to the space V. Moreover, if u is a solution of (2), it also solves the associated weak form; vice-versa, if u satisfies the weak form and is smooth enough, it also satisfies the Poisson problem.

For the sake of simplicity in notation, we now introduce linear, bilinear and trilinear forms in order to set up the weak formulations of the above problems. For u, v in V, we have:

$$\begin{aligned} A(u, v) &= \int_\Omega \nabla u \cdot \nabla v \,\mathrm{d}x, \\ C(\boldsymbol{\alpha}, u, v) &= \int_\Omega (\boldsymbol{\alpha} \cdot \nabla) v \,\mathrm{d}x, \\ F(v) &= \int_\Omega f v \,\mathrm{d}x + \int_{\Gamma_N} g v \,\mathrm{d}s. \end{aligned}$$

Similarly, for $\boldsymbol{u}$, $\boldsymbol{v}$, $\boldsymbol{w}$ in $\boldsymbol{V}$ and $q \in Q$, we have:

$$\begin{aligned} A(\boldsymbol{u},\boldsymbol{v}) &= \int_\Omega \nabla\boldsymbol{u} : \nabla\boldsymbol{v}\,\mathrm{d}x, \\ B(q,\boldsymbol{v}) &= -\int_\Omega q\nabla\cdot\boldsymbol{v}\,\mathrm{d}x, \\ C(\boldsymbol{w},\boldsymbol{u},\boldsymbol{v}) &= \int_\Omega (\boldsymbol{w}\cdot\nabla\boldsymbol{u})\boldsymbol{v}\,\mathrm{d}x, \\ F(\boldsymbol{v}) &= \int_\Omega \boldsymbol{f}\cdot\boldsymbol{v}\,\mathrm{d}x + \int_{\Gamma_N} \boldsymbol{g}\cdot\boldsymbol{v}\,\mathrm{d}s. \end{aligned}$$

We now recast all the problems introduced in the previous section in the following general weak form. Find $U \in \mathcal{V}$ such that:

$$\mathcal{B}(U,Z) = \mathcal{F}(Z), \qquad \forall\, Z \in \mathcal{V}. \tag{6}$$

Note that $\mathcal{B}$ is nonlinear in the case of the Navier-Stokes equations.

A) Poisson Problem: Choose $\mathcal{V} = V$, $U = u$, $Z = v \in V$, and

$$\begin{aligned} \mathcal{B}(u,v) &= A(u,v), \\ \mathcal{F}(v) &= F(v). \end{aligned}$$

B) Convection-Diffusion Problem: Choose $\mathcal{V} = V$, $U = u$, $Z = v \in V$, and

$$\begin{aligned} \mathcal{B}(u,v) &= A(u,v) + C(\boldsymbol{\alpha},u,v), \\ \mathcal{F}(v) &= F(v). \end{aligned}$$

C) Stokes Problem: Choose $\mathcal{V} = \boldsymbol{V}\times Q$, $U = (\boldsymbol{u},p)$, $Z = (\boldsymbol{v},q) \in \boldsymbol{V}\times Q$, and

$$\begin{aligned} \mathcal{B}((\boldsymbol{u},p),(\boldsymbol{v},q)) &= A(\boldsymbol{u},\boldsymbol{v}) + B(p,\boldsymbol{v}) + B(q,\boldsymbol{u}), \\ \mathcal{F}((\boldsymbol{v},q)) &= F(\boldsymbol{v}). \end{aligned}$$

The weak form of the Stokes problem can alternatively be rewritten as follows: Find $\boldsymbol{u} \in \boldsymbol{V}$, $p \in Q$ such that

$$\begin{aligned} A(\boldsymbol{u},\boldsymbol{v}) + B(p,\boldsymbol{v}) &= F(\boldsymbol{v}), &\quad& \forall\, \boldsymbol{v} \in \boldsymbol{V}, \\ B(q,\boldsymbol{u}) &= 0, && \forall\, q \in Q. \end{aligned}$$

The first equation is the momentum equation, the second the continuity equation.

D) Navier-Stokes Problem: As before, $\mathcal{V} = V \times Q$, $U = (u, p)$, $Z = (v, q) \in V \times Q$, and

$$\mathcal{B}((\boldsymbol{u},p),(\boldsymbol{v},q)) = A(\boldsymbol{u},\boldsymbol{v}) + C(\boldsymbol{u},\boldsymbol{u},\boldsymbol{v}) + B(p,\boldsymbol{v}) + B(\boldsymbol{u},q),$$
$$\mathcal{F}((\boldsymbol{v},q)) = F(\boldsymbol{v}).$$

Alternatively, we also have:

$$\begin{aligned} A(\boldsymbol{u},\boldsymbol{v}) + C(\boldsymbol{u},\boldsymbol{u},\boldsymbol{v}) + B(p,\boldsymbol{v}) &= F(\boldsymbol{v}), \qquad \forall\, \boldsymbol{v} \in \boldsymbol{V}, \\ B(q,\boldsymbol{u}) &= 0, \qquad \forall\, q \in Q. \end{aligned}$$

The weak formulations for the model problems are summarized in Table 1.

Table 1. Weak formulations for the model problems.

	$\mathcal{V}$	U	Z	$\mathcal{B}(U,Z)$	$\mathcal{F}(Z)$
(A)	V	u	v	$A(u,v)$	$F(v)$
(B)	V	u	v	$A(u,v) + C(\boldsymbol{\alpha},u,v)$	$F(v)$
(C)	$\boldsymbol{V} \times Q$	$(\boldsymbol{u},p)$	$(\boldsymbol{v},q)$	$A(\boldsymbol{u},\boldsymbol{v}) + B(p,\boldsymbol{v}) + B(q,\boldsymbol{u})$	$F(\boldsymbol{v})$
(D)	$\boldsymbol{V} \times Q$	$(\boldsymbol{u},p)$	$(\boldsymbol{v},q)$	$A(\boldsymbol{u},\boldsymbol{v}) + C(\boldsymbol{u},\boldsymbol{u},\boldsymbol{v}) + B(p,\boldsymbol{v}) + B(q,\boldsymbol{u})$	$F(\boldsymbol{v})$

Finite element approximations. Finite element approximations of the above problems will be obtained by using the standard Galerkin approach, for which the trial functions belong to the same finite element spaces $\mathcal{V}^h$ as the test functions. A finite element version of Problem (6) consists in finding a function $U_h \in \mathcal{V}^h \subset \mathcal{V}$ such that

$$\mathcal{B}(U_h, Z) = \mathcal{F}(Z), \qquad \forall\, Z \in \mathcal{V}^h. \tag{7}$$

Here, the finite element space $\mathcal{V}^h$ is taken as V^h as given in (1) for the Poisson and convection-diffusion equations. In the case of the Stokes and Navier-Stokes equations, we shall choose $\mathcal{V}^h$ as $\boldsymbol{V}^h \times Q^h$, where $\boldsymbol{V}^h = (V^h)^d \subset \boldsymbol{V}$ and $Q^h \subset Q$. However, it is well known that $\boldsymbol{V}^h$ and Q^h should be constructed such that the discrete ***inf-sup*** condition is satisfied (see [30,31] for more details). In other words, the discrete formulation is stable if there exists a constant $\beta_h > 0$ such that:

$$\sup_{\boldsymbol{v}_h \in \boldsymbol{V}^h, \boldsymbol{v}_h \neq 0} \frac{|B(q_h, \boldsymbol{v}_h)|}{|\boldsymbol{v}_h|_1} \geq \beta_h \|q_h\|_0, \qquad \forall\, q_h \in Q^h. \tag{8}$$

In the case of hierarchical shape functions, we shall use the following rule inspired by the work of Suri *et al.* [55,21] on locking-free *h*-*p* elements: given an element K and a spectral order p_K on K for the velocity variable, the pressure is chosen of degree at most $p_K - 1$ for the edge functions and at most $p_K - 2$ for the interior bubble functions. Note that the Taylor-Hood element [31] (biquadratic in $\boldsymbol{v}$, bilinear in p) is consistent with this rule.

1.3 Numerical Error and Residual

No matter how sophisticated finite element methods used to solve mathematical models are, all results from these involve numerical errors. The errors actually originate from various sources; there are modeling errors, geometrical errors, integration errors, round-off errors, algorithmic errors, discretization errors, *etc.* Modeling errors are introduced when, a mathematical abstraction is put forth as a characterization of an actual physical event. Other modeling errors arise when a simplified more tractable model is used in place of a sophisticated mathematical model that cannot be used to deliver useful information because of its complexity. For example, turbulence phenomena are generally simulated on computers using Reynolds Averaged Navier-Stokes (RANS) or Large-Eddy Simulation (LES) models rather than Direct Numerical Simulations (DNS) of the Navier-Stokes equations. Geometrical errors occur when the boundary curves are approximated by piecewise polynomial functions, i.e. the representation of continuous functions by discrete approximations to render the analysis to a form acceptable by a digital computer. For example, "Kinks" appear along the boundary which perturb the behavior of the solution in their vicinity. Integration errors result from estimating the various integrals using approximate integration rules, while algorithmic errors are due to the algorithm which is used to solve a given set of equations (e.g. Newton method for nonlinear equations, iterative solvers). Finally, discretization errors are due to the discretization scheme, either in space or in time.

We shall concentrate in this paper in estimating the errors due to finite element discretizations in the spatial dimension. In order to do so, we assume that all the other types of errors (modeling, geometrical, integration, round-off, algorithmic) are kept negligible during the computations as compared to the approximation errors.

In the general case, the numerical error due to the spatial discretization is defined as:

$$E = U - U_h \tag{9}$$

where $U \in \mathcal{V}$, $U_h \in \mathcal{V}^h \subset \mathcal{V}$ (we envisage here conforming finite element methods only). It immediately follows that the error E belongs to $\mathcal{V}$ as well. Replacing U by $U_h + E$ in (6) and subtracting $\mathcal{B}(U_h, Z)$ from both sides of the equation, the numerical error is then governed by:

$$\mathcal{B}(U_h + E, Z) - \mathcal{B}(U_h, Z) = \mathcal{R}_h(Z), \qquad \forall\ Z \in \mathcal{V}, \tag{10}$$

where $\mathcal{R}_h$ is a linear functional defined on $\mathcal{V}$, called the residual and given by:

$$\mathcal{R}_h(Z) = \mathcal{F}(Z) - \mathcal{B}(U_h, Z). \tag{11}$$

One property we shall use very often in error estimation is the so-called orthogonality property. In terms of the residual, the property states that the residual vanishes on $\mathcal{V}^h$; i.e.

$$\mathcal{R}_h(Z) = 0, \qquad \forall\, Z \in \mathcal{V}^h. \tag{12}$$

This directly follows from the definition of the approximation U_h obtained by the Galerkin method.

Remark 1. If $\mathcal{B}(\cdot,\cdot)$ is a bilinear form on $\mathcal{V} \times \mathcal{V}$, the error equation simplifies to:

$$\mathcal{B}(E, Z) = \mathcal{R}_h(Z), \qquad \forall\, Z \in \mathcal{V}.$$

We then observe that this problem is as complex to solve as the one for U. Therefore the main objective of ***a posteriori*** error estimation is not to get an approximation of the function E, but rather to estimate a measure of the error which is either easily computable, or of interest to the user. The residual $\mathcal{R}_h$, being the source of the error, can obviously be postprocessed, in some way to yield information on the error in some appropriate norms. This is the subject of what follows. Other type of measures, so-called quantities of interest, will be studied in the second section of this paper. □

We list in Table 2 the residual functionals for Problems (A–D) and in Table 3 the corresponding forms which govern the approximation errors as established in (10). We remark that for the Stokes and Navier-Stokes problems, the error equation can be decoupled into two equations, one related to the momentum equation and the other to the continuity equation. The error equations for the Stokes problem then read:

$$\begin{aligned} A(\boldsymbol{e}_u, \boldsymbol{v}) + B(e_p, \boldsymbol{v}) &= \mathcal{R}_h^m(\boldsymbol{v}), \qquad && \forall\, \boldsymbol{v} \in \boldsymbol{V}, \\ B(q, \boldsymbol{e}_u) &= \mathcal{R}_h^c(q), && \forall\, q \in Q, \end{aligned}$$

where $\boldsymbol{e}_u = \boldsymbol{u} - \boldsymbol{u}_h$ and $e_p = p - p_h$. Here $\mathcal{R}_h^m$ and $\mathcal{R}_h^c$ respectively denote the residuals in the momentum and continuity equations:

$$\begin{aligned} \mathcal{R}_h^m(\boldsymbol{v}) &= F(\boldsymbol{v}) - A(\boldsymbol{u}_h, \boldsymbol{v}) - B(p_h, \boldsymbol{v}), \\ \mathcal{R}_h^c(q) &= -B(q, \boldsymbol{u}_h). \end{aligned}$$

Similarly, the error equations for the Navier-Stokes equations are:

$$\begin{aligned} A(\boldsymbol{e}_u, \boldsymbol{v}) + C(\boldsymbol{e}_u, \boldsymbol{u}_h, \boldsymbol{v}) &+ C(\boldsymbol{u}_h, \boldsymbol{e}_u, \boldsymbol{v}) \\ +C(\boldsymbol{e}_u, \boldsymbol{e}_u, \boldsymbol{v}) + B(e_p, \boldsymbol{v}) &= \mathcal{R}_h^m(\boldsymbol{v}), \qquad && \forall\, \boldsymbol{v} \in \boldsymbol{V}, \\ B(q, \boldsymbol{e}_u) &= \mathcal{R}_h^c(q), && \forall\, q \in Q, \end{aligned}$$

and the residuals $\mathcal{R}_h^m$ and $\mathcal{R}_h^c$ are given here by:

$$\mathcal{R}_h^m(\boldsymbol{v}) = F(\boldsymbol{v}) - A(\boldsymbol{u}_h, \boldsymbol{v}) - C(\boldsymbol{u}_h, \boldsymbol{u}_h, \boldsymbol{v}) - B(p_h, \boldsymbol{v}),$$
$$\mathcal{R}_h^c(q) = -B(q, \boldsymbol{u}_h).$$

Table 2. Residual functionals for each model problem (A–D).

	E	$\mathcal{R}_h(Z)$
(A)	e_u	$F(v) - A(u_h, v)$
(B)	e_u	$F(v) - A(u_h, v) - C(\boldsymbol{\alpha}, u_h, v)$
(C)	$(\boldsymbol{e}_u, e_p)$	$F(\boldsymbol{v}) - A(\boldsymbol{u}_h, \boldsymbol{v}) - B(p_h, \boldsymbol{v}) - B(q, \boldsymbol{u}_h)$
(D)	$(\boldsymbol{e}_u, e_p)$	$F(\boldsymbol{v}) - A(\boldsymbol{u}_h, \boldsymbol{v}) - C(\boldsymbol{u}_h, \boldsymbol{u}_h, \boldsymbol{v}) - B(p_h, \boldsymbol{v}) - B(q, \boldsymbol{u}_h)$

Table 3. Forms governing the error for each model problem (A–D).

	E	$\mathcal{B}(U_h + E, Z) - \mathcal{B}(U_h, Z)$
(A)	e_u	$A(e_u, v)$
(B)	e_u	$A(e_u, v) + C(\boldsymbol{\alpha}, e_u, v)$
(C)	$(\boldsymbol{e}_u, e_p)$	$A(\boldsymbol{e}_u, \boldsymbol{v}) + B(e_p, \boldsymbol{v}) + B(q, \boldsymbol{e}_u)$
(D)	$(\boldsymbol{e}_u, e_p)$	$A(\boldsymbol{e}_u, \boldsymbol{v}) + C(\boldsymbol{u}_h, \boldsymbol{e}_u, \boldsymbol{v}) + C(\boldsymbol{e}_u, \boldsymbol{u}_h, \boldsymbol{v}) + C(\boldsymbol{e}_u, \boldsymbol{e}_u, \boldsymbol{v})$
		$+B(e_p, \boldsymbol{v}) + B(q, \boldsymbol{e}_u)$

There essentially exist two approaches to estimate the error in energy-like norms: recovery methods and residual approaches. These are described in the following sections.

1.4 Recovery-based Methods

Recovery-based error estimators were suggested by Zienkiewicz and Zhu [63] as early as 1987. Improved techniques were proposed later by the same authors [64,65]. The main motivation underlying these methods follows the simple observation that piecewise continuous finite element solutions U_h generally exhibit discontinuous gradients at the interface of the elements. If the exact solution U to be sought is smooth enough, such jumps in the gradients of U_h provide an indication that the numerical solutions are erroneous. The approach consists then in recovering piecewise continuous gradients $\boldsymbol{G}(U_h)$

from the finite lement approximation U_h and to calculate the "difference" as the error measure $\mathcal{E}$, i.e.

$$\mathcal{E} \approx \eta = \sqrt{\int_{\Omega} |\boldsymbol{G}(U_h) - \nabla U_h|^2 \, \mathrm{d}x}. \tag{13}$$

Various methods have been proposed in the literature to compute $\boldsymbol{G}(U_h)$ and the reader is referred to [64,65] for an account on the Zienkiewicz-Zhu patch recovery technique. Some of the approaches rely on *superconvergence* properties of the finite element method obtained at specific points in the element (see also [4]).

Although the method proves to be accurate and robust in many situations (see for example the numerical investigation by Babuška *et al* [14]), Ainsworth and Oden [4] write about the *cautionary tale*: "Recovery-based estimators possess a number of attractive features that have led to their popularity. In particular, their ease of implementation, generality, and ability to produce quite accurate estimators have led to their widespread adoption, especially in the engineering community. However, the estimators also have drawbacks." They proceed by presenting a one-dimensional example in which "the recovery-based estimator produces an estimated error of zero, while the actual error can be arbitrarily large."

We have also experienced some misfortune with the method while using it in an adaptive scheme for the simulation of shock waves in compressible flows. Indeed, using classical finite element or finite volume methods, the shock is generally smeared across a few elements, and thus is captured as a linear approximation with very steep gradient rather than a discontinuity. It follows that the gap in the gradient across element boundaries is very small and the recovery-based estimator predicts near zero errors at the location of the shock. Any adaptive scheme has then the tendency to refine elements around the shock, but not at the core, as desired. Such an adapted mesh is shown in Fig. 3 from [49]. While remedies have been developed to overcome such problems, one may question the reliability of such an error estimator.

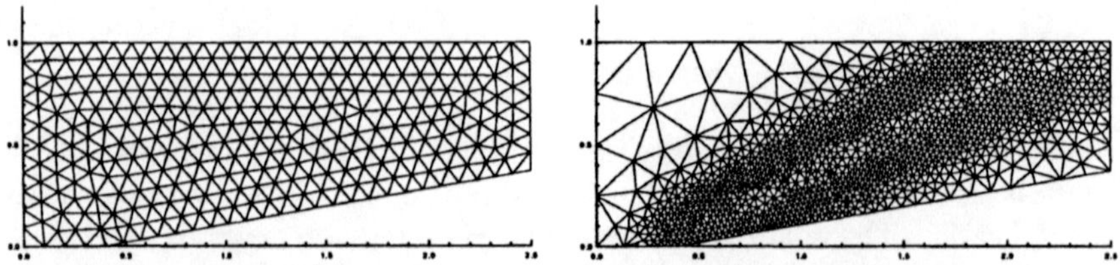

Fig. 3. Initial and adapted meshes for the simulation of a supersonic flow over a ramp. The adaptive scheme is based on a recovery-based error estimator.

1.5 Residual Methods

As suggested earlier, the residual contains necessary and sufficient information to determine the numerical error. The motivation behind residual methods is then to postprocess, in an effective manner, the residual in order to derive a computable error estimator. In this section, we show how the residual relates to the error in some appropriate norms and describe various techniques to evaluate the norm of the residual.

Relationships between residuals and errors. In many cases, it is possible to show that the norm of the residual is equivalent to the norm of the error, *i.e.* there exist strictly positive constants C_l and C_u such that:

$$C_l \|E\|_{\mathcal{V}} \leq \|\mathcal{R}_h\|_* \leq C_u \|E\|_{\mathcal{V}},$$

where $\|E\|_{\mathcal{V}}$ denotes a norm of E in $\mathcal{V}$ and $\|\mathcal{R}_h\|_*$ the norm of the residual with respect to $\|\cdot\|_{\mathcal{V}}$. Such a result can be established in an abstract setting for the case of bilinear forms as stated in the following theorem.

Theorem 1. *Let $E = U - U_h$ be the approximation error in U_h where U and U_h are respectively the solution of (6) and (7). Assume that $\mathcal{B}(\cdot,\cdot)$ is a continuous bilinear form on $\mathcal{V} \times \mathcal{V}$, i.e. there exists a constant $M > 0$ such that:*

$$|\mathcal{B}(U,Z)| \leq M \|U\|_{\mathcal{V}} \|Z\|_{\mathcal{V}}, \qquad \forall\, U, Z \in \mathcal{V}. \tag{14}$$

in addition, assume that $\mathcal{B}(\cdot,\cdot)$ satisfies the inf-sup condition; there exists a constant $\gamma > 0$ such that for all $U \in \mathcal{V}$:

$$\sup_{Z \in \mathcal{V}, Z \neq 0} \frac{|\mathcal{B}(U,Z)|}{\|Z\|_{\mathcal{V}}} \geq \gamma \|U\|_{\mathcal{V}}. \tag{15}$$

Then, there exist C_l, $C_u > 0$, independent of the mesh size h, such that

$$C_l \|E\|_{\mathcal{V}} \leq \|\mathcal{R}_h\|_* \leq C_u \|E\|_{\mathcal{V}}, \tag{16}$$

where the norm of the residual is defined as

$$\|\mathcal{R}_h\|_* = \sup_{Z \in \mathcal{V}, Z \neq 0} \frac{|\mathcal{R}_h(Z)|}{\|Z\|_{\mathcal{V}}}. \tag{17}$$

Proof. We first establish the lower bound. Starting from the definition of the norm of the residual, using (6) and the continuity of the bilinear form, we have:

$$\begin{aligned} \|\mathcal{R}_h\|_* &= \sup_{Z \in \mathcal{V}, Z \neq 0} \frac{|\mathcal{R}_h(Z)|}{\|Z\|_{\mathcal{V}}} = \sup_{Z \in \mathcal{V}, Z \neq 0} \frac{|\mathcal{B}(E,Z)|}{\|Z\|_{\mathcal{V}}} \\ &\leq \sup_{Z \in \mathcal{V}, Z \neq 0} \frac{M \|E\|_{\mathcal{V}} \|Z\|_{\mathcal{V}}}{\|Z\|_{\mathcal{V}}} \leq M \|E\|_{\mathcal{V}}. \end{aligned}$$

Now, for the upper bound, starting from the *inf-sup* condition and using (6), we obtain:

$$\gamma\|E\|_{\mathcal{V}} \leq \sup_{Z\in\mathcal{V},Z\neq 0} \frac{|\mathcal{B}(E,Z)|}{\|Z\|_{\mathcal{V}}} = \sup_{Z\in\mathcal{V},Z\neq 0} \frac{|\mathcal{R}_h(Z)|}{\|Z\|_{\mathcal{V}}} = \|\mathcal{R}_h\|_*.$$

In brief, we have shown that:

$$C_l\|E\|_{\mathcal{V}} \leq \|\mathcal{R}_h\|_* \leq C_u\|E\|_{\mathcal{V}},$$

with $C_l = \gamma$ and $C_u = M$. □

We emphasize here that the norm of the residual provides an upper bound on the error depending on the stability or *inf-sup* constant γ. We now apply the result of the previous theorem to our model problems:

A) Poisson Problem: In this case, the norm of the error is chosen as $\|e\|_V = \sqrt{A(e,e)}$ since A is positive definite and symmetric (inner product). Then, it is clear that:

$$|\mathcal{B}(u,v)| \leq \|u\|_V\|v\|_V, \qquad \forall\, u,v \in V,$$

and that

$$\sup_{v\in V,v\neq 0} \frac{|\mathcal{B}(u,v)|}{\|v\|_V} \geq \frac{|\mathcal{B}(u,u)|}{\|u\|_V} \geq \|u\|_V, \qquad \forall\, u \in V.$$

Therefore $C_l = C_u = 1$ and $\|\mathcal{R}_h\|_* = \|e_u\|_V$. In other words the norm of the residual is exactly equal to the norm of the residual, which makes this particular case ideal.

B) Convection-Diffusion Problem: We also choose the norm for the error as $\|e\|_V = \sqrt{A(e,e)}$. It can be shown that the bilinear form $C(\boldsymbol{\alpha},\cdot,\cdot)$ is constinuous on $V \times V$, i.e. there exists a positive constant M_0, dependent on $\boldsymbol{\alpha}$ such that:

$$|C(\boldsymbol{\alpha},u,v)| \leq M_0\|u\|_V\|v\|_V, \qquad \forall\, u,v \in V,$$

so that:

$$|\mathcal{B}(u,v)| \leq (1+M_0)\|u\|_V\|v\|_V, \qquad \forall\, u,v \in V.$$

Moreover, since $\nabla\cdot\boldsymbol{\alpha} = 0$, we have the fundamental result: $C(\boldsymbol{\alpha},u,u) = 0$, $\forall\, u \in V$ (see Girault and Raviart [30]). Therefore, for all $u \in V$, we have:

$$\sup_{v\in V,v\neq 0} \frac{|\mathcal{B}(u,v)|}{\|v\|_V} \geq \frac{|\mathcal{B}(u,u)|}{\|u\|_V} = \frac{|A(u,u) + C(\boldsymbol{\alpha},u,u)|}{\|u\|_V} = \|u\|_V.$$

It follows that $C_l = 1$ and $C_u = 1 + M_0$ so that:

$$\|e_u\|_V \leq \|\mathcal{R}_h\|_* \leq (1+M_0)\|e_u\|_V.$$

C) Stokes Problem: We introduce here the norm:

$$\|(\boldsymbol{e}_u, e_p)\|_{\boldsymbol{V}\times Q} = \|\boldsymbol{e}_u\|_{\boldsymbol{V}} + \|e_p\|_0,$$

where $\|\boldsymbol{e}_u\|_V = \sqrt{A(\boldsymbol{e}_u, \boldsymbol{e}_u)}$ and $\|\cdot\|_0$ is the $L^2(\Omega)$-norm. The corresponding norms for the residuals $\mathcal{R}_h^m$ and $\mathcal{R}_h^c$ are given as:

$$\|\mathcal{R}_h^m\|_* = \sup_{\boldsymbol{v}\in\boldsymbol{V},\boldsymbol{v}\neq 0} \frac{|\mathcal{R}_h^m(\boldsymbol{v})|}{\|\boldsymbol{v}\|_V}, \qquad \|\mathcal{R}_h^c\|_* = \sup_{q\in Q,q\neq 0} \frac{|\mathcal{R}_h^c(q)|}{\|q\|_0},$$

so that:

$$\|\mathcal{R}_h\|_* = \sup_{\substack{(\boldsymbol{v},q)\in\boldsymbol{V}\times Q\\(\boldsymbol{v},q)\neq 0}} \frac{|\mathcal{R}_h(\boldsymbol{v},q)|}{\|(\boldsymbol{v},q)\|_{V\times Q}} = \|\mathcal{R}_h^m\|_* + \|\mathcal{R}_h^c\|_*.$$

Referring to [42,43,50], it can be shown that

$$C_l\|(\boldsymbol{e}_u, e_p)\|_{V\times Q} \leq \|\mathcal{R}_h\|_* \leq C_u\|(\boldsymbol{e}_u, e_p)\|_{V\times Q}$$

where C_l and C_u are positive constants which depend on the *inf-sup* and continuity constants for the bilinear form $B(\cdot,\cdot)$.

D) Navier-Stokes Problem: In this case, the form $\mathcal{B}(\cdot,\cdot)$ is nonlinear and Theorem 1 can not be applied as is. However, using the same norms as above, it can be shown that:

$$C_l\|(\boldsymbol{e}_u, e_p)\|_{V\times Q} \leq \|\mathcal{R}_h\|_* \leq C_u\|(\boldsymbol{e}_u, e_p)\|_{V\times Q},$$

where C_u depends on the continuity constants for $C(\cdot,\cdot,\cdot)$ and $B(\cdot,\cdot)$, and where C_l, for small values of $\|\boldsymbol{e}_u\|_V$, depends on the *inf-sup* constant for $B(\cdot,\cdot)$ and the continuity constant for $C(\cdot,\cdot,\cdot)$. This result is valid for small data for which we know that there exists a unique bounded (in $(H^1(\Omega)^d)$) solution of the steady-state Navier-Stokes equations (for more details, see [3,33,4] and references therein).

We conclude from these theoretical results that the norm of the residual should in general provides reasonable estimate of the error (in global norms usually suggested by the forms in the error equation) under the condition that the constants C_l and C_u do not take values far from unity. The values of these constants depend on the problem at hand. At best, they are both equal to one when the bilinear form $\mathcal{B}$ is symmetric and positive definite as it is the case for the Poisson problem and error is measured in the so-called energy norm, $\|e_u\| = \sqrt{\mathcal{B}(e_u, e_u)}$. For the convection-diffusion problem, only one of the constants is equal to one. However, in more general cases, the constants C_l and C_u depend on other constants whose value are not known or difficult to evaluate. It is then necessary to rely on numerical experiments to analyze the relationship between the residuals and the errors.

The next task in *a posteriori* error estimation is to evaluate the norm of residuals. The norm of the residual may sometimes be straightforward to compute. For instance, the norm of $\mathcal{R}_h^c$ can be evaluated exactly, i.e.

$$\|\mathcal{R}_h^c\|_* = \|\nabla \cdot \boldsymbol{u}_h\|_0, \tag{18}$$

where $\boldsymbol{u}_h$ is the discrete velocity field. However, norms of residual which are defined in terms of $H^1(\Omega)$-norm are more complicated to compute. In the case of $\mathcal{R}_h^m$, for example, using the Riesz Representation theorem, we know that there exists a unique function $\boldsymbol{\phi} \in \boldsymbol{V}$ such that

$$\|\boldsymbol{\phi}\|_V = \|\mathcal{R}_h^m\|_* \tag{19}$$

and

$$A(\boldsymbol{\phi}, \boldsymbol{v}) = \mathcal{R}_h^m(\boldsymbol{v}), \qquad \forall\, \boldsymbol{v} \in \boldsymbol{V}. \tag{20}$$

Note that $A(\cdot,\cdot)$ is actually the inner product associated with the norm $\|\cdot\|_V$. Evaluating the norm of $\mathcal{R}_h^m$ is then equivalent to solving a global problem on the whole domain Ω.

A technical aspect of *a posteriori* error estimation is then to design methods which circumvent solving such global problems. We present in what follows several approaches. We have classified the methods into explicit or implicit methods. Explicit methods are those which do not require solving any auxiliary problems. On the other hand, implicit methods involve the solution of local or global problems, set up in such a way that the solving cost is a fraction of the computational cost spent to obtain the finite element solution.

Explicit methods. For the sake of simplicity, we describe explicit methods on the Poisson model problem. We recall that the residual in this special case reads $\mathcal{R}_h(v) = F(v) - A(u_h, v)$, $v \in V$. Moreover, the norm of the residual is given as

$$\|\mathcal{R}_h\|_* = \sup_{v \in V, v \neq 0} \frac{|\mathcal{R}_h(v)|}{\|v\|_V}.$$

The starting point in explicit methods is to decompose the residual functional into elementwise contributions. We have:

$$\begin{aligned}
\mathcal{R}_h(v) &= \int_\Omega f v \,\mathrm{d}x + \int_{\Gamma_N} g v \,\mathrm{d}s - \int_\Omega \nabla u_h \cdot \nabla v \,\mathrm{d}x \\
&= \sum_{K \in \mathcal{P}_h} \left\{ \int_K f v \,\mathrm{d}x + \int_{\partial K \cap \Gamma_N} g v \,\mathrm{d}s - \int_K \nabla u_h \cdot \nabla v \,\mathrm{d}x \right\} \\
&= \sum_{K \in \mathcal{P}_h} \left\{ \int_K (f v + \Delta u_h v) \,\mathrm{d}x + \int_{\partial K \cap \Gamma_N} g v \,\mathrm{d}s - \int_{\partial K} (\mathbf{n} \cdot \nabla u_h)\, v \,\mathrm{d}s \right\}.
\end{aligned}$$

Introducing the interior residual r_K and boundary residual $j_{\gamma,K}$, defined for each element K of the partition as:

$$r_K = f + \Delta u_h, \qquad \forall\, \boldsymbol{x} \in K, \tag{21}$$

and

$$j_{\gamma,K} = \begin{cases} \mathbf{n} \cdot \nabla u_h - g, & \text{if } \gamma \in \partial K \cap \Gamma_N \\ \mathbf{n} \cdot \nabla u_h, & \text{if } \gamma \in \partial K \cap \Gamma_D \\ \frac{1}{2}[(\mathbf{n} \cdot \nabla u_h)_{|K} + (\mathbf{n} \cdot \nabla u_h)_{|L}], & \text{if } \gamma \in \partial K \backslash \partial \Omega \end{cases} \qquad (22)$$

a new expression for the residual is obtained:

$$\mathcal{R}_h(v) = \sum_{K \in \mathcal{P}_h} \int_K r_K \, v \, \mathrm{d}x - \int_{\partial K} j_{\gamma,K} \, v \, \mathrm{d}s.$$

Let v_h be an arbitrary function in V^h. Since $\mathcal{R}_h$ vanishes on V^h (orthogonality property), we then obtain:

$$\mathcal{R}_h(v) = \sum_{K \in \mathcal{P}_h} \int_K r_K \, (v - v_h) \, \mathrm{d}x - \int_{\partial K} j_{\gamma,K} \, (v - v_h) \, \mathrm{d}s.$$

Using Cauchy-Schwartz, we have:

$$|\mathcal{R}_h(v)| \leq \sum_{K \in \mathcal{P}_h} \left\{ \|r_K\|_{0,K} \, \|v - v_h\|_{0,K} + \|j_{\gamma,K}\|_{0,\partial K} \, \|v - v_h\|_{0,\partial K} \right\}.$$

In order to proceed, we need to estimate $\|v - v_h\|_{0,K}$ and $\|v - v_h\|_{0,\partial K}$. For that purpose, let $v_h = \mathcal{I}_h v$, where $\mathcal{I}_h : V \longrightarrow V^h$ is a linear bounded operator such that (see Bernardi and Girault [18] and Ainsworth and Oden [4]):

$$\|v - \mathcal{I}_h v\|_{0,K} \leq C h_K \, |v|_{1,\tilde{K}} \, ,$$
$$\|v - \mathcal{I}_h v\|_{0,\gamma} \leq C h_K^{1/2} \|v\|_{1,\tilde{K}},$$

where C is a positive constant which depends only on the regularity of the partition. Here γ represents any edge of K and $\tilde{K}$ denotes the patch of elements associated with K. We therefore have:

$$\begin{aligned} |\mathcal{R}_h(v)| &\leq C \sum_{K \in \mathcal{P}_h} \|r_K\|_{0,K} \, h_K \, |v|_{1,\tilde{K}} + \|j_{\gamma,K}\|_{0,\partial K} \, h_K^{1/2} \|v\|_{1,\tilde{K}} \\ &\leq C \sum_{K \in \mathcal{P}_h} \sqrt{h_K^2 \|r_K\|_{0,K}^2 + h_K \|j_{\gamma,K}\|_{0,\partial K}^2} \times \sqrt{|v|_{1,\tilde{K}}^2 + \|v\|_{1,\tilde{K}}^2} \\ &\leq C \sqrt{\sum_{K \in \mathcal{P}_h} h_K^2 \|r_K\|_{0,K}^2 + h_K \|j_{\gamma,K}\|_{0,\partial K}^2} \times \sqrt{\sum_{K \in \mathcal{P}_h} |v|_{1,\tilde{K}}^2 + \|v\|_{1,\tilde{K}}^2} \\ &\leq C \|v\|_{1,\Omega} \sqrt{\sum_{K \in \mathcal{P}_h} h_K^2 \|r_K\|_{0,K}^2 + h_K \|j_{\gamma,K}\|_{0,\partial K}^2}. \end{aligned}$$

Finally, using Poincaré's Inequality (see Schwab [54]), i.e. $\|v\|_{1,\Omega} \leq C \|v\|_V$, we then obtain:

$$\frac{|\mathcal{R}_h(v)|}{\|v\|_V} \leq C \sqrt{\sum_{K \in \mathcal{P}_h} h_K^2 \|r_K\|_{0,K}^2 + h_K \|j_{\gamma,K}\|_{0,\partial K}^2},$$

which becomes,

$$\|\mathcal{R}_h\|_* \le C \sqrt{\sum_{K\in\mathcal{P}_h} h_K^2 \|r_K\|_{0,K}^2 + h_K \|j_{\gamma,K}\|_{0,\partial K}^2}.$$

We remark that the constant C takes on different values at various places in the above development and is basically unknown. Nevertheless, the quantities η_K defined as:

$$\eta_K^2 = h_K^2 \|r_K\|_{0,K}^2 + h_K \|j_{\gamma,K}\|_{0,\partial K}^2 \tag{23}$$

can be combined together to provide an explicit *a posteriori* error estimator η of the norm of the residual (and *a fortiori* of the error) according to:

$$\eta = \sqrt{\sum_{K\in\mathcal{P}_h} \eta_K^2}. \tag{24}$$

Because the constant C is unknown, this estimator cannot be considered as a guaranteed upper bound on the norm of the residual.

One purpose of calculating implicit residual error estimates (see next subsection) is actually to remove such constants in the bound on the norm of the residual. Finally, we remark that the local quantities η_K can be used to construct refinement indicators for the design of an adaptative strategy.

Implicit methods. Implicit methods have been thus called as they all involve solving auxiliary problems. The methods can be further classified into three types, whether the auxiliary problems are set up on each element, on subdomains (patches of elements), or on the whole domain (whole partition). These methods are respectively called element, subdomain and global residual methods and are described below.

Element residual methods. We describe in the following how to derive upper and lower bounds on the norm of the residual using element residual methods. From the definition of the norm of the residual, we observe that if it is possible to construct two functions, $\chi \in V_-$ and $\phi \in V_+$ ($\phi \notin V$), where V_- and V_+ are vector spaces such that $V_- \subseteq V \subseteq V_+$, the residual would then be bounded as:

$$\frac{|\mathcal{R}_h(\chi)|}{\|\chi\|_V} \le \|\mathcal{R}_h\|_* \le \frac{|\widetilde{\mathcal{R}}_h(\phi)|}{\|\phi\|_V}$$

provided that one can find a proper extension $\widetilde{\mathcal{R}}_h$ of $\mathcal{R}_h$ to the space V_+. We show below how to obtain such functions χ and ϕ.

Once again, we find convenient to describe the method in the case of the Poisson problem. Note that it can be easily extended to other model problems, with only a few minor changes. Several works have been devoted to the element residual method (see, for instance, [15,2–4] and references therein). A natural approach consists in considering a vector space V_+, which contains V, and to construct an extension of $\mathcal{R}_h$ on V_+.

One such space is the so-called *broken space.* We first introduce local spaces V_K associated with each element $K \in \mathcal{P}_h$ such as:

$$V_K = \{v \in H^1(K)/\mathbb{R};\ v = 0 \text{ on } \Gamma_D \cap \partial K\},$$

where $H^1(K)/\mathbb{R}$ is the quotient space of equivalence classes of functions in $H^1(K)$ differing by constants. This choice ensures that the auxiliary problems (see below) are solvable in the sense that they admit a unique solution. The broken space $H^1(\mathcal{P}_h) \supset V$ is then defined as:

$$H^1(\mathcal{P}_h) = \{v \in L^2(\Omega); v_K = v|_K \in V_K, \forall\ K \in \mathcal{P}_h\}, \tag{25}$$

with associated norm

$$||v||_{H^1(\mathcal{P}_h)} = \sqrt{\sum_{K \in \mathcal{P}_h} \int_K \nabla v \cdot \nabla v \, \mathrm{d}x}.$$

For $v \in V$, note that $||v||_{H^1(\mathcal{P}_h)} = ||v||_V$. We also introduce the bilinear forms $A_K(\cdot,\cdot)$ and the restrictions of $\mathcal{R}_h(\cdot)$ on each element K of the partition as:

$$\begin{aligned} A_K(u,v) &= \int_K \nabla u \cdot \nabla v \, \mathrm{d}x, \\ \mathcal{R}_{h,K}(v) &= \int_K fv \, \mathrm{d}x + \int_{\partial K \cap \Gamma_N} gv \, \mathrm{d}s - \int_K \nabla u_h \cdot \nabla v \, \mathrm{d}x. \end{aligned}$$

As the norm $||\cdot||_V$ is induced by the bilinear form $A(\cdot,\cdot)$ on V, the norm $||\cdot||_{\mathcal{P}_h}$ is given by the inner product $\sum_K A_K(\cdot,\cdot)$ on $H^1(\mathcal{P}_h)$. We will then use $A_K(\cdot,\cdot)$ to set up the auxiliary problems on each element. The form $A_K(\cdot,\cdot)$ can also be viewed, in the particular case of the Poisson problem, as the restriction of $A(\cdot,\cdot)$ on element K. Note that this is not necessarily true for other model problems. Finally, we can extend the residual $\mathcal{R}_h$ to the whole space $H^1(\mathcal{P}_h)$ by simply decomposing $\mathcal{R}_h$ into the elementwise contributions $\mathcal{R}_{h,K}(\cdot)$. However, it is convenient at this stage to add another contribution which involves fluxes on the interior element boundaries, such that the extension $\widetilde{\mathcal{R}}_h$ reads:

$$\widetilde{\mathcal{R}}_h(v) = \sum_K \mathcal{R}_{hK}(v_{|K}) + \sum_K \int_{\partial K} g_K v_{|K} \, \mathrm{d}s.$$

The flux functions g_K need only satisfy the condition

$$\sum_K \int_{\partial K} g_K v_{|K} \, \mathrm{d}s = 0, \qquad \forall\ v \in V, \tag{26}$$

to ensure that $\widetilde{\mathcal{R}}_h$ is equal to $\mathcal{R}_h$ on V.

Theorem 2. *For each element K in the partition, let $\phi_K \in V_K$ denote the solution of the local problem:*

$$A_K(\phi_K, v) = \mathcal{R}_{h,K}(v) + \int_{\partial K} g_K v \, \mathrm{d}s, \qquad \forall\, v \in V_K, \tag{27}$$

with the g_K's such that (26) holds, and define $\phi \in H^1(\mathcal{P}_h)$ such that $\phi_{|K} = \phi_K$. Then, the function ϕ provides an upper bound on the residual:

$$\|\mathcal{R}_h\|_* \leq \|\phi\|_{H^1(\mathcal{P}_h)} = \sqrt{\sum_K A_K(\phi_K, \phi_K)}. \tag{28}$$

Proof. Starting from the definition of the residual, using condition (26), applying the fact that $V \subset H^1(\mathcal{P}_h)$, and, finally, using (27) on each element, we have

$$\begin{aligned}
\|\mathcal{R}_h\|_* &= \sup_{v \in V\setminus\{0\}} \frac{|\mathcal{R}_h(v)|}{\|v\|_V} = \sup_{v \in V\setminus\{0\}} \frac{|\sum_K \mathcal{R}_{h,K}(v_{|K})|}{\|v\|_V} \\
&= \sup_{v \in V\setminus\{0\}} \frac{|\sum_K \mathcal{R}_{h,K}(v_{|K}) + \sum_K \int_{\partial K} g_K v_{|K} \, \mathrm{d}s|}{\|v\|_{H^1(\mathcal{P}_h)}} \\
&\leq \sup_{v \in H^1(\mathcal{P}_h)\setminus\{0\}} \frac{|\sum_K (\mathcal{R}_{h,K}(v_{|K}) + \int_{\partial K} g_K v_{|K} \, \mathrm{d}s)|}{\|v\|_{H^1(\mathcal{P}_h)}} \\
&\leq \sup_{v \in H^1(\mathcal{P}_h)\setminus\{0\}} \frac{\sum_K |A_K(\phi_K, v_{|K})|}{\|v\|_{H^1(\mathcal{P}_h)}} \\
&\leq \sup_{v \in H^1(\mathcal{P}_h)\setminus\{0\}} \frac{\sum_K \sqrt{A_K(\phi_K, \phi_K)} \sqrt{A_K(v_{|K}, v_{|K})}}{\|v\|_{H^1(\mathcal{P}_h)}} \\
&\leq \sup_{v \in H^1(\mathcal{P}_h)\setminus\{0\}} \frac{\|\phi\|_{H^1(\mathcal{P}_h)} \|v\|_{H^1(\mathcal{P}_h)}}{\|v\|_{H^1(\mathcal{P}_h)}} \leq \|\phi\|_{H^1(\mathcal{P}_h)}
\end{aligned}$$

which is the stated result. □

Note that the auxiliary problems are well-posed on the spaces V_K deprived of the constant modes. We also emphasize here that the functions g_K have not yet been specified. The upper bound actually holds for arbitrary functions g_K as long as the condition (26) is satisfied. For example, these functions could all be set to zero. However, this choice does not provide optimal estimates in the sense that they generally overestimate the norm of the residual.

We now describe methods which have been developed to construct specific functions g_K. The choice is obviously not unique. First, we observe that (26) necessarily holds if:

$$\begin{aligned}
g_K &= 0, \qquad \forall\, \gamma \subset \partial\Omega, \\
g_K + g_L &= 0, \qquad \forall\, \gamma = \bar{K} \cap \bar{L}, \;\; K \neq L.
\end{aligned}$$

We present two approaches to further determine the fluxes g_K:

1) **Flux-splitting method:** In the first approach, proposed in [10,15] and also investigated in [41], the fluxes g_K are computed on each edge of the partition as:

$$g_K = \begin{cases} \frac{1}{2}(\mathbf{n} \cdot \nabla u_h)_K - \frac{1}{2}(\mathbf{n} \cdot \nabla u_h)_L, & \text{if } \gamma \not\subset \partial\Omega, \\ 0, & \text{if } \gamma \subset \partial\Omega. \end{cases} \tag{29}$$

This method is attractive since the computation of the fluxes is relatively simple to implement. On the other hand, the robustness of the error estimator is not guaranteed and numerical experiments show that the upper bound may take on in some cases large values.

2) **Equilibrated flux-splitting method:** An alternative method is to compute the fluxes g_K such that they satisfy:

$$\mathcal{R}_{h,K}(v) + \int_{\partial K} g_K v \,\mathrm{d}s = 0, \qquad \forall\, v \in V_K^h,$$

where V_K^h is the finite element space defined on each element K such as:

$$V_K^h = \{v \in H^1(K); v = \hat{v} \circ F_K^{-1}, \hat{v} \in P_{p_K}(\widehat{K}), v = 0 \text{ on } \partial K \cap \Gamma_D\}.$$

In other words, V_K^h is the space of polynomial shape functions up to the degree p_K. The above condition is suggested by the fact that the residual vanishes on V^h. We also observe that the auxiliary problems now satisfy the compatibility condition for constant modes, i.e.

$$\mathcal{R}_{h,K}(1) + \int_{\partial K} g_K \,\mathrm{d}s = 0.$$

Extensive details of the method can be found in [3,4,35,36]. The error estimator based on this approach generally yields very accurate estimates for two dimensional problems; however it proves to be extremely difficult to implement in three dimensions in the case of irregular meshes (meshes with hanging nodes).

So far, the auxiliary problems (27) have been defined on the infinite dimensional spaces V_K. These problems can at best be approximated in local finite element spaces in order to derive computable error estimators. In the following, we describe via a two-dimensional example how to construct these local finite element spaces.

Suppose that the mesh is made of quadrilateral elements and that the solution u_h is approximated by piecewise linear polynomials constructed from the nodal shape functions ν_i, $i = 1, \ldots, 4$, i.e. $p_K = 1$. Let W_K denote the finite element space consisting of *edge* and *interior bubble* functions of degree strictly greater than one and up to $p_K + q_K$, $q_K \geq 1$.

Hence, solving for an approximation $\psi_K \in V_K^h \cup W_K$ (resp. W_K) of ϕ_K which satisfies:

$$A_K(\psi_K, v) = \mathcal{R}_{h,K}(v) + \int_{\partial K} g_K v \, \mathrm{ds}, \quad \forall \, v \in V_K^h \cup W_K \text{ (resp. } W_K), \tag{30}$$

the "upper bound" estimate η_{upp} is computed as:

$$\eta_{upp} = \sqrt{\sum_K A_K(\psi_K, \psi_K)}. \tag{31}$$

We remark that the quantity η_{upp} is not guaranteed to provide an upper bound on the norm of the residual. Indeed, introducing the local norm $\|v\|_K = \sqrt{A_K(v,v)}$ it is straightforward to show that, for each element,

$$\begin{aligned}
\|\psi_K\|_K &\le \sup_{v \in V_K^h \cup W_K \setminus \{0\}} \frac{|\mathcal{R}_{h,K}(v) + \int_{\partial K} g_K v \, \mathrm{ds}|}{\|v\|_K} \\
&\le \sup_{v \in V_K \setminus \{0\}} \frac{|\mathcal{R}_{h,K}(v) + \int_{\partial K} g_K v \, \mathrm{ds}|}{\|v\|_K} \\
&\le \sup_{v \in V_K \setminus \{0\}} \frac{A_K(\phi_K, v)}{\|v\|_K} \\
&\le \sup_{v \in V_K \setminus \{0\}} \frac{\|\phi_K\|_K \|v\|_K}{\|v\|_K} \le \|\phi_K\|_K
\end{aligned}$$

so that $\eta_{upp} \le \|\phi\|_{H^1(\mathcal{P}_h)}$. However, based on our numerical experiments, η_{upp} generally provides an upper bound when h becomes sufficiently small.

Remark 2. Since the local problems (27) are independently solved on each element, the global error function ψ, defined such that $\psi_{|K} = \psi_K, \forall \, K \in \mathcal{P}_h$, is generally discontinuous across the elements. □

We proceed by describing how a simple postprocessing of the above estimator yields a lower bound on the norm of the residual. To our best knowledge, only a few works on lower bound estimators have been published to date. In [42,44], the authors proposed to solve a global problem on the space of continuous perturbations consisting of edge and interior bubble functions (see global residual method below). Huerta *et al.* [32] proposed to solve local problems using interior bubble functions only. However, the method yields poor lower bounds on the norm of the residual and some postprocessing is needed in order to improve their accuracy. In [12,57], Babuška and Strouboulis proposed a method in which the interelement gaps in the discontinuous error function ψ are postprocessed. In the present approach, inspired from the latter but different in its implementation, we construct a continuous function χ derived from ψ and compute the lower bound η_{low} such as:

$$\eta_{low} = \frac{|\mathcal{R}_h(\chi)|}{\|\chi\|_V} \le \|\mathcal{R}_h\|_*.$$

This idea was also suggested in [12] but disregarded (we remark here that the authors wrote $\mathcal{R}_h(\chi)$ instead of $|\mathcal{R}_h(\chi)|$ – the absolute value is actually needed to ensure that the lower bound η_{low} takes on positive values).

For the sake of clarity, we show next how to construct such a function χ in the case of quadrilateral elements. We first suppose that the solution $\psi_K \in W_K$ of (30), where W_K is the set of polynomial functions of degree two, can be extended by zero outside element K such that:

$$\psi = \sum_{K\in\mathcal{P}_h} \psi_K = \sum_{K\in\mathcal{P}_h} \left\{ \left(\sum_{\gamma=1}^{4} a_{\gamma,K}\, \varphi_{\gamma,K} \right) + b_K\, \theta_K \right\}$$

where $a_{\gamma,K}$ and b_K are known coefficients and:

$$\varphi_{\gamma,K} = \begin{cases} \phi_\gamma^{(2)} \circ F_K^{-1}, & \text{in } K, \\ 0, & \text{in } \Omega\backslash K, \end{cases} \qquad \theta_K = \begin{cases} \theta^{(2,2)} \circ F_K^{-1}, & \text{in } K, \\ 0, & \text{in } \Omega\backslash K. \end{cases}$$

We remark that since the interior bubble functions θ_K all vanish on the element boundaries, only the edge coefficients makes ψ discontinuous. We naturally postprocess the edge functions first.

Step 1: Edge functions. Let γ be an edge shared by elements K and L and let $\varphi_{\gamma,K}$ and $\varphi_{\gamma,L}$ be the edge functions associated with γ in K and L respectively. We consider, for instance, the discontinuous function ψ_d:

$$\psi_d(\mathbf{x}) = \begin{cases} a_{\gamma,K}\, \varphi_{\gamma,K}(\mathbf{x}), & \text{if } \mathbf{x} \in K, \\ a_{\gamma,L}\, \varphi_{\gamma,L}(\mathbf{x}), & \text{if } \mathbf{x} \in L, \\ 0, & \text{if } \mathbf{x} \notin K, L, \end{cases}$$

which is displayed in Fig. 4 for arbitrary values of $a_{\gamma,K}$ and $a_{\gamma,L}$ ($a_{\gamma,K}$ is supposed to be different from $a_{\gamma,L}$). We construct a continuous function χ_c from ψ_d simply by taking the average of $a_{\gamma,K}$ and $a_{\gamma,L}$. Hence, as illustrated in Fig. 4, χ_c reads:

$$\chi_c(\mathbf{x}) = \begin{cases} \frac{1}{2}(a_{\gamma,K} + a_{\gamma,L})\, \varphi_{\gamma,K}(\mathbf{x}), & \text{if } \mathbf{x} \in K, \\ \frac{1}{2}(a_{\gamma,K} + a_{\gamma,L})\, \varphi_{\gamma,L}(\mathbf{x}), & \text{if } \mathbf{x} \in L, \\ 0, & \text{if } \mathbf{x} \notin K, L. \end{cases}$$

Applying this procedure to all interior interfaces of the mesh, we eventually obtain a continuous function χ_c from ψ as:

$$\chi_c = \sum_{K\in\mathcal{P}_h} \left\{ \left(\sum_{\gamma=1}^{4} \underline{a}_{\gamma,K}\, \varphi_{\gamma,K} \right) + b_K\, \theta_K \right\}, \tag{32}$$

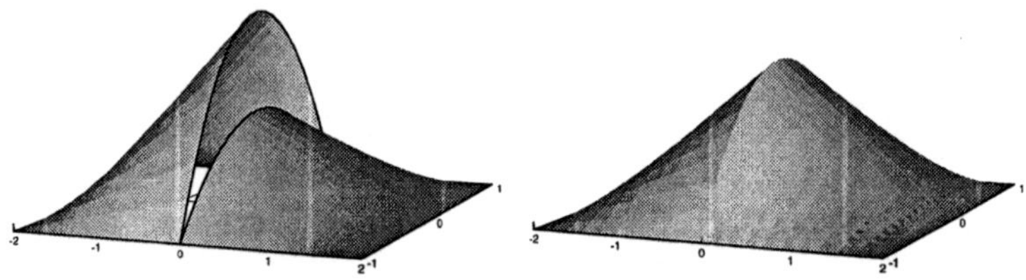

Fig. 4. Recovery of a continuous function χ_c (right) from a discontinuous function ψ_d (left).

where the coefficients $\underline{a}_{\gamma,K}$ stand for averaged values of the coefficients $a_{\gamma,K}$, i.e.

$$\underline{a}_{\gamma,K} = \begin{cases} \frac{1}{2}(a_{\gamma,K} + a_{\gamma,L}) & \text{if } \gamma = \bar{K} \cap \bar{L}, K \cap L = \varnothing, \\ a_{\gamma,K} & \text{if } \gamma \subset \partial\Omega. \end{cases} \tag{33}$$

Since the function χ_c thus defined is continuous, it becomes a possible candidate for a computable lower bound. However, preliminary numerical experiments revealed that the resulting bound was sometimes inaccurate. This is actually explained, following a careful analysis, by the fact that the functions ψ and χ_c could be very different in the interior of some elements of the mesh. We show in the next example how the averaging procedure described above can dramatically modify the shape of the initial function ψ.

Example 1. Let K be an element in the mesh and let us consider the following discontinuous function as displayed in Fig. 5:

$$\begin{cases} 2\left(\sum_{\gamma=1}^{4} \varphi_{\gamma,K}(\mathbf{x})\right) - \dfrac{3}{2}\theta_K(\mathbf{x}), & \text{if } \mathbf{x} \in K, \\ 0, & \text{if } \mathbf{x} \notin K. \end{cases}$$

This function, after postprocessing the coefficients associated with each edge, yields a continuous function, whose restriction on K is given by:

$$\left(\sum_{\gamma=1}^{4} \varphi_{\gamma,K}(\mathbf{x})\right) - \frac{3}{2}\theta_K(\mathbf{x}).$$

This continuous function, also shown in Fig. 5, is quite deformed when compared to the original function. Such a deformation during postprocessing is held responsible for getting lower bounds of poor accuracy.

This example clearly indicates that it is also necessary to postprocess the coefficients b_K associated with the interior bubble functions.

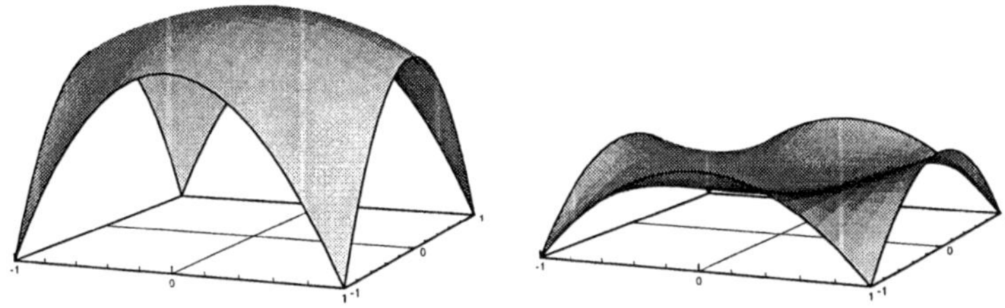

Fig. 5. Derfomation of a bubble function after averaging of the edge functions only (see Example 1).

Step 2: Bubble functions. We propose to compute new coefficients $\underline{b}_K$ such that, for a given element K,

$$A_K(\underline{b}_K\theta_K, \theta_K) = \mathcal{R}_{h,K}(\theta_K) - \sum_{\gamma\subset\partial K} A_K(\underline{a}_{\gamma,K}\,\varphi_{\gamma,K}, \theta_K). \tag{34}$$

In other words, this is equivalent to solving the local problems (27) supplemented by Dirichlet boundary conditions:

$$\psi_K(\mathbf{x}) = \underline{a}_{\gamma,K}\,\varphi_{\gamma,K}(\mathbf{x}), \qquad \forall\,\mathbf{x}\in\gamma,\ \forall\,\gamma\subset\partial K.$$

Step 3: Lower bound. The continuous function χ to be used for the computation of the lower bound is given as:

$$\chi = \sum_{K\in\mathcal{P}_h}\left\{\left(\sum_{\gamma=1}^{4}\underline{a}_{\gamma,K}\,\varphi_{\gamma,K}\right) + \underline{b}_K\,\theta_K\right\}$$

where the coefficients $\underline{a}_{\gamma,K}$ and $\underline{b}_K$ are given by (33) and (34) respectively. The lower bound η_{low} is then obtained as:

$$\eta_{low} = \frac{|\mathcal{R}_h(\chi)|}{\|\chi\|_V}. \tag{35}$$

We observe that η_{low} is a guaranteed lower bound on the norm of the residual. The attractive features of this postprocessing are: 1) η_{low} is relatively cheap to obtain once the upper bound has been computed, 2) the difference between η_{low} and η_{upp} can provide us with some information on the reliability of η_{upp}.

Subdomain residual methods. In subdomain residual methods, the local problems are set up on patches of elements rather than on one element. The method was initially designed by Babuška and Rheinbolt [9], extended to meshes with hanging nodes in [8], and applied to steady-state Navier-Stokes problems in [33]. We observe that the local residual quantities defined on each patch can be resolved either explicitly or implicitly. This is shown in [4]. We also mention the recent work of Carstensen and Funkel [20] and Morin *et al.* [38] on subdomain residual methods which seems to be promising.

Global residual method. We recall here that the goal is to derive an estimate of the norm of the residual. In the case of the Poisson and Convection-Diffusion problems, using the Riesz Representation theorem, it is known that there exists a unique function $\phi \in V$ such that $\|\phi\|_V = \|\mathcal{R}_h\|_*$ and

$$A(\phi, v) = \mathcal{R}_h(v), \qquad \forall\, v \in V.$$

Note that the governing equation for ϕ is identical to the error equation associated with the Poisson problem, hence $\phi = e_u$ in this case. For the Convection-Diffusion problem, the problems governing the e_u and ϕ are different; therefore, ϕ is no longer equal to e_u.

The approach here consists in estimating the function ϕ by a finite element approximation ψ defined on a finite element space of bubble functions of higher degree than the polynomial degree used to compute the solution u_h. Let W denote such a space such that $W \neq \{0\}$, $W \cap V^h = \{0\}$ and $W \cup V^h \subset V$. The choice for W is not unique; nevertheless, it may be conveniently constructed within a *h-p* finite element data structure from layers of piecewise polynomial basis functions involving the shape functions (edge and interior bubbles in two dimensions) between $p_K + 1$ and $p_K + q_K$, $q_K \geq 1$, where p_K is the order of the solution u_h for each element K of the mesh. We then seek $\psi \in W$ such that:

$$A(\psi, v) = \mathcal{R}_h(v), \qquad \forall\, v \in W.$$

Although global, the above problem is efficiently solved using the Conjugate-Gradient method performing only a few iterations (we take advantage of the fact that all edge and interior bubble functions vanish at the nodes). In doing so, we can show that (cf. [50,42,43])

$$\sqrt{(1-\sigma^2)(1-\gamma^2)}\|\mathcal{R}_h\|_* \leq \|\psi\|_V \leq \|\mathcal{R}_h\|_*$$

where σ is inversely related to the dimension of W (σ should tend to zero as more and more degrees of freedom are added to W) and γ represents a measure of the angle between V^h and W (in the optimal case, if W were orthogonal to V^h with respect to the $A(\cdot,\cdot)$ inner product, then the constant γ would be equal to zero. The other conclusion is that the estimator η,

$$\eta = \|\psi\|_V, \tag{36}$$

which can be decomposed into elementwise contributions η_K by taking the restriction of $\|\psi\|_V$ over each element, provides a guaranteed lower bound on the norm of the residual. The method is easily implemented in an existing finite element code since it reuses most of the resources already required to solve for the solution u_h.

1.6 Adaptation Strategy

The simplest strategy to control and reduce approximation errors consists of an iterative procedure whose steps are described below. The strategy is rather general as it does not require any specific information about the type of problem.

We have studied in the preceding sections various methods to estimate errors with respect to a global measure norm. The relative error reads in the general case:

$$e_{\text{rel}} = \frac{\|E\|_{\mathcal{V}}}{\|U\|_{\mathcal{V}}}.$$

Since both the exact solution U and the error E are unknown, we use the available approximations instead. Then, the mesh needs to be adapted whenever

$$\frac{\eta}{\|U_h\|_{\mathcal{V}}} \geq C_{\text{tol}} \tag{37}$$

where η is the global estimate of the error E with respect to the norm $\|\cdot\|_{\mathcal{V}}$.

The main objective in mesh adaptation is to locally refine the elements which exhibit large sources of error. In other words, it means refining the elements whose elementwise contributions of the residual, the source of error, are the largest. We have seen that η, the error estimate, is above all an estimate of the norm of the residual. Therefore, the elementwise contributions η_K defined for each element K such that

$$\eta^2 = \sum_{K \in \mathcal{P}_h} \eta_K^2$$

can serve as refinement indicator. Indeed, performing a simple normalization of the contributions by the largest one, the strategy consists in refining an element K of the mesh if

$$\frac{\eta_K}{\max_J \{\eta_J\}} \geq C_{\text{adp}} \tag{38}$$

where C_{adp} is a user-prescribed parameter ranging from zero to one. The refinement procedure consists in dividing any tagged element into subelements, possibly allowing for "hanging nodes".

We note that the quantities η_K characterize the contribution to the global error estimate η due to residuals in element K, they do not represent the error in element K. The error in K depends upon the residuals in K plus those in elements outside and remote from K. This latter error is called *pollution error* and exists because our methods, up to this point, are designed to estimate global errors in norms of functions defined over the entire domain Ω. Nevertheless, the indicators η_K can be used to drive an effective adaptive process that controls global error by selective mesh refinement.

2 Goal-Oriented Error Estimation and Adaptation

2.1 Introduction

Solution adaptive meshing implies the selection of adequate error measures so as to properly assess the quality of numerical approximations. A satisfactory measure is essential for constructing new meshes that will better meet the goals of the analyst. Traditionally, *a posteriori* error estimates have been global in nature, providing upper bounds of approximation error in energy or energy-type norms defined over the entire domain of solutions. Very recently, new methods based on duality techniques (see [16,47,45,46,17,51,44] for example) have been developed which provide for the calculation of error bounds of local quantities of interest, such as pointwise values of solutions or their gradients. We also mention the early work of Gartland [28] who proposed computable error bounds on pointwise quantities using the dual problem. Such estimates make possible significant generalizations of existing adaptive strategies, providing for so-called *goal-oriented* adaptive methods which adapt meshes to yield good approximations of local quantities of interest.

We review in this section of the paper the various methods allowing for error estimation with respect to quantities of interest. The review is organized as follows. In Section 2.2, we provide a few examples of quantities of interest. We present in Section 2.3 the detailed approach which allows to derive the desired error estimates using a general model problem. The results are then briefly described in the case of the Poisson equation, the convection-diffusion equations and the Stokes and Navier-Stokes problems. In Section 2.4, we introduce the general methodology. We then suggest an adaptive strategy in Section 2.5 followed by numerical examples in Section 2.6.

2.2 Quantity of Interest

In goal-oriented error estimation, analysts specify the goal of their calculations by identifying a quantity of interest that can be computed using the solution. The quantity of interest is represented by a functional defined on the space of admissible solutions, the functional being either linear or nonlinear.

The quantity of interest is a linear functional. This is the desired case for which the theory is well defined and rigorous. Note that many quantities of interest fall into this category. Here are a few examples in terms of the velocity field $\boldsymbol{u}$.

Example 2 (Averaged values). The average of the velocity field $\boldsymbol{u}$ in a prescribed direction $\boldsymbol{\nu}$, where $\boldsymbol{\nu}$ is a uniform unit vector, is represented by the linear functional

$$L(\boldsymbol{u}) = \frac{1}{|\omega|} \int_{\omega} \boldsymbol{u} \cdot \boldsymbol{\nu} \, \mathrm{d}x. \tag{39}$$

The error in $L(\boldsymbol{u}_h)$ is then defined by:

$$\mathcal{E} = L(\boldsymbol{u}) - L(\boldsymbol{u}_h) = L(\boldsymbol{e}) = \frac{1}{|\omega|}\int_\omega \boldsymbol{e}\cdot\boldsymbol{\nu}\,\mathrm{d}x.$$

Example 3 (Pointwise values). We compare two approaches for the error estimation of pointwise quantities. In the first approach, the quantity of interest is given as:

$$L(\boldsymbol{u}) = u(\boldsymbol{x}_0),$$

which defines the value of the velocity component in the x-direction u at a given point $\boldsymbol{x}_0$ in the domain. However, when u is a function of $H^1(\Omega)$, the quantity is not necessarily defined for $d \geq 2$. For that reason, we prefer to consider for the quantity of interest the local average or weighted value of the solution around the point $\boldsymbol{x}_0$:

$$L_\varepsilon(\boldsymbol{u}) = \int_\Omega u(\boldsymbol{x})\,k_\varepsilon(\boldsymbol{x}-\boldsymbol{x}_0)\,\mathrm{d}x, \tag{40}$$

where k_ε is a mollifying kernel defined as:

$$k_\varepsilon(\boldsymbol{x}) = \begin{cases} C\exp\left(|\boldsymbol{x}|^2/\epsilon^2 - 1\right)^{-1} & \text{if } |\boldsymbol{x}| < \varepsilon \\ 0 & \text{if } |\boldsymbol{x}| \geq \varepsilon \end{cases} \tag{41}$$

with C, which depends on d, ε and $\boldsymbol{x}_0$, chosen so as to satisfy:

$$\int_\Omega k_\varepsilon(\boldsymbol{x}-\boldsymbol{x}_0)\,\mathrm{d}x = 1.$$

We remark that $L_\varepsilon(\boldsymbol{u})$ converges to $u(\boldsymbol{x}_0)$ as ε goes to zero when u is continuous in a neighborhood of $\boldsymbol{x}_0$.

The quantity of interest is a nonlinear functional. In this case, it is necessary to linearize the quantity of interest functional. We give one example below to show how to proceed.

Example 4 (Kinetic energy). The functional associated with the kinetic energy of a flow reads:

$$N(\boldsymbol{u}) = \frac{1}{2}\int_\Omega \boldsymbol{u}^2\,\mathrm{d}x$$

where $\boldsymbol{u}$ is the velocity field. The error in $N(\boldsymbol{u}_h)$, i.e. $\mathcal{E} = N(\boldsymbol{u}) - N(\boldsymbol{u}_h)$, is then given as:

$$\mathcal{E} = \frac{1}{2}\int_\Omega (\boldsymbol{u}_h+\boldsymbol{e})^2 - \boldsymbol{u}_h^2\,\mathrm{d}x = \int_\Omega \boldsymbol{u}_h\cdot\boldsymbol{e}\,\mathrm{d}x + \frac{1}{2}\int_\Omega \boldsymbol{e}^2\,\mathrm{d}x,$$

that is,

$$\mathcal{E} = N'(\boldsymbol{u}_h;\boldsymbol{e}) + \frac{1}{2}N''(\boldsymbol{u}_h;\boldsymbol{e},\boldsymbol{e}),$$

where $N'(\boldsymbol{u}_h;\cdot)$ and $N''(\boldsymbol{u}_h;\cdot,\cdot)$ denote the first and second derivatives of N with respect to the first variable $\boldsymbol{u}_h$, respectively. Neglecting the second-order term involving the error, assuming the error is small, we then consider the linear functional:

$$L(\boldsymbol{e}) = \int_\Omega \boldsymbol{u}_h \cdot \boldsymbol{e}\, \mathrm{d}x \approx \mathcal{E}.$$

Many other quantities of interest can be defined. In particular, in computational fluid dynamics, we may be interested in the lift and drag coefficients, in the circulation around a closed subregion, in the wall shear stress in specific areas of a wing profile, *etc.* All these quantities can naturally be defined in terms of linear or nonlinear functionals of the solution.

2.3 The Step-by-Step Approach

Theoretical analysis. We assume that we are interested in solving a boundary value problem of the type: Find $U \in \mathcal{V}$ such that:

$$\mathcal{B}(U, Z) = \mathcal{F}(Z), \qquad \forall\, Z \in \mathcal{V}, \tag{42}$$

where $\mathcal{V}$ is a Hilbert space, $\mathcal{B}(\cdot,\cdot)$ a bilinear form on $\mathcal{V}\times\mathcal{V}$ and $\mathcal{F}(\cdot)$ a bounded linear form on $\mathcal{V}$. We have encountered several types of problems which may be cast as above in Section 1 of this paper. Let $\mathcal{V}^h$ denote a finite element space associated with $\mathcal{V}$ and $U_h \in \mathcal{V}^h$ an approximation of U which satisfies:

$$\mathcal{B}(U_h, Z) = \mathcal{F}(Z), \qquad \forall\, Z \in \mathcal{V}^h.$$

The error in the approximation U_h is then defined as $E = U - U_h$, $E \in \mathcal{V}$.

Finally, let L denote a bounded linear functional in $\mathcal{V}'$ and suppose that the goal of the computational analysis is to evaluate as accurately as possible the quantity $L(U)$. The accuracy of the finite element approximation is then assessed with respect to the quantity L, that is, the error measure $\mathcal{E} \in \mathbb{R}$ is given by:

$$\mathcal{E} = L(U) - L(U_h).$$

Since L is linear, this can be written as:

$$\mathcal{E} = L(U) - L(U_h) = L(U - U_h) = L(E).$$

Remark 3. Recall that L is a linear functional. It follows that the error quantity $L(E)$ can take on negative values, and may be equal to zero even though the error E is nonzero. □

We could just compute the quantity $L(E)$ knowing the error E. Unfortunately, it is, in general, as expensive to solve for E as to solve for U since the error is the solution of:

$$\mathcal{B}(E, Z) = \mathcal{F}(Z) - \mathcal{B}(U_h, Z) = \mathcal{R}_h(Z), \qquad \forall\, Z \in \mathcal{V}^h. \tag{43}$$

where $\mathcal{R}_h(\cdot)$ is the residual functional. Alternatively, one can try to relate $L(E)$ to the residual $\mathcal{R}_h$ without having to compute E. Such an approach is justified because the residual, as the source of error, contains the information which drives the error.

The starting point in our derivation is to find the relationship between $L(E)$ and $\mathcal{R}_h$; namely, one aims at finding a bounded linear functional W, if it exists, such that:

$$L(E) = W(\mathcal{R}_h).$$

At this point, the functional W should belong to the bidual of $\mathcal{V}$, as it is a functional defined on the dual space $\mathcal{V}'$ ($\mathcal{R}_h \in \mathcal{V}'$). However, because $\mathcal{V}$ was assumed a Hilbert space, and *a fortiori* a reflexive space, every element of the bidual can be associated with an element of the space itself. In particular, identifying W with an element in $\mathcal{V}$, the above relation becomes:

$$L(E) = \mathcal{R}_h(W).$$

We refer to W as the influence function with respect to L, as it indicates the influence of the residual on the quantity $L(E)$. Note that the influence function is dependent on the quantity L in which we are interested.

We proceed by showing how to derive the influence function. Using the error equation (43), it is clear that, since $W \in \mathcal{V}$:

$$L(E) = \mathcal{B}(E, W),$$

and such a relation is necessarily satisfied whenever $W \in \mathcal{V}$ is the solution of:

$$\mathcal{B}(Z, W) = L(Z), \qquad \forall\, Z \in \mathcal{V}.$$

This auxiliary problem is referred to as the adjoint or dual problem of the primal problem (42). We will alternatively refer to the influence function as the dual solution or adjoint variable. In general, existence and uniqueness of the solution W are proved in the same way as for the primal problem.

Remark 4. The influence function is the solution of a infinite dimensional problem which cannot in general be calculated exactly. The idea is then to approximate the influence function using finite elements. Note however that when W is approximated by $W^h \in \mathcal{V}^h$, we then have

$$\mathcal{R}_h(Z_h) = 0,$$

since the residual vanishes on $\mathcal{V}^h$ ("orthogonality property"). In conclusion, the dual problem needs to be approximated in a space larger than $\mathcal{V}^h$. □

We now describe different approaches that have been proposed to obtain the solution of the dual problem:

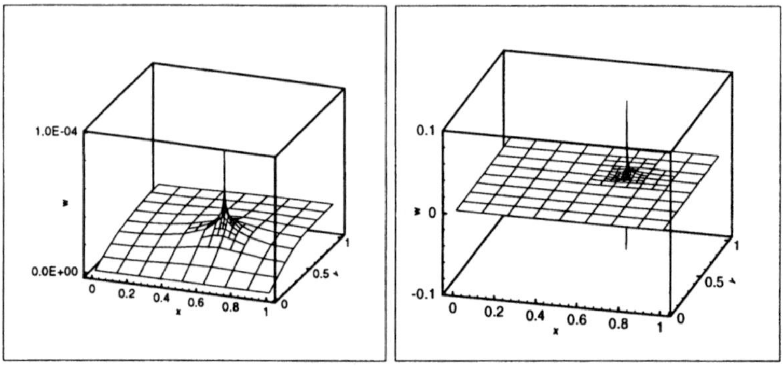

Fig. 6. Examples of computed influence functions: the plot on the left corresponds to the pointwise value of a function u, the one on the right to its pointwise partial derivative in the x-direction.

1. In some rare cases, the influence function can be determined analytically, at least partially. Indeed, if one is interested in the error at a point in the domain, the influence function is simply the Green function. For certain classes of problems, for instance the Poisson problem, the Green function can be given in terms of a linear combination of the fundamental solution of the Laplace operator with respect to the point of interest and a harmonic function in Ω chosen such that the Green function vanishes on the Dirichlet boundary Γ_D. Such a method was originally used to extract accurate pointwise stresses or derivatives from finite element solutions. For more details, we refer the reader to [1,5–7,19,39,40,61,62].
2. The second approach consists in computing a global finite element approximation of the dual problem. However, in view of the previous remark, the finite element space should have a larger dimension than the solution finite element space $\mathcal{V}^h$, using, for example, higher-degree polynomial functions on the same mesh, or a mesh with more elements. This approach is, for example, considered in Becker and Rannacher [17]. The major drawback in this method is that the cost in solving the dual problem is greater than for the primal problem in the case of linear problems. Taking advantage of the fact that the influence function often presents local features, especially when the quantity of interest is of local nature, we suggest here that the approximation of the influence function be computed on an adapted mesh based on error estimates in the energy norm. The error in the quantity of interest would then be estimated by:

$$\eta = \mathcal{R}_h(W_H) \approx L(E),$$

where the parameter H refers to the adapted mesh for the dual problem. The main difficulty is then to handle two different meshes in the finite element code.

3. In the last approach, the influence function is approximated in the space $\mathcal{V}^h$. In other words, we seek the approximation $W_h \in \mathcal{V}^h$ such that

$$\mathcal{B}(Z, W_h) = L(Z), \qquad \forall\, Z \in \mathcal{V}^h.$$

Since $\mathcal{R}_h(W_h) = 0$, we then have:

$$L(E) = \mathcal{R}_h(W) = \mathcal{R}_h(W) - \mathcal{R}_h(W_h) = \mathcal{R}_h(W - W_h) = \mathcal{R}_h(\Sigma)$$

where $\Sigma = W - W_h \in \mathcal{V}$ denotes the error in the approximate influence function. In other words, the quantity $L(E)$ is given in terms of the residual $\mathcal{R}_h$, which is known, and the error Σ, which we know how to estimate in global norms. We show in the next subsection how to derive estimates and bounds on $L(E)$.

Error estimates using the error in influence function.

General case: Starting with $L(E) = \mathcal{R}_h(\Sigma)$ and using the Cauchy-Schwartz inequality, we obtain:

$$|L(E)| \leq ||\mathcal{R}_h||_* ||\Sigma||_{\mathcal{V}}.$$

We have actually shown in Section 1 how to estimate the norm of the residual. Moreover, we may obtain an estimate of $||\Sigma||_{\mathcal{V}}$ by a residual method, in which case we may show that the error in the approximate influence function is actually equivalent to the residual associated with the dual problem:

$$\mathcal{R}_h^W(Z) = L(Z) - \mathcal{B}(Z, W_h), \qquad \forall\, Z \in \mathcal{V}.$$

The main drawback in the present approach is that the global residual and error norms, and thus their product, may be several orders of magnitude larger than the quantity $L(E)$.

Let $(\cdot,\cdot)_{\mathcal{V}}$ denote the inner product associated with the norm $||\cdot||_{\mathcal{V}}$ (recall that $\mathcal{V}$ is assumed to be a Hilbert space). Then, recall from the Riesz Representation theorem that there exists a unique function $\Phi \in \mathcal{V}$ such that:

$$(\Phi, Z)_{\mathcal{V}} = \mathcal{R}_h(Z), \qquad \forall\, Z \in \mathcal{V}.$$

It follows that the quantity $L(E)$ may be rewritten:

$$|L(E)| = |(\Phi, \Sigma)_{\mathcal{V}}| \leq ||\Phi||_{\mathcal{V}} ||\Sigma||_{\mathcal{V}}.$$

Remark 5. Let $\boldsymbol{a}$ and $\boldsymbol{b}$ be two vectors in $\mathbb{R}^2$. The scalar product of two vectors is defined as:

$$|\boldsymbol{a} \cdot \boldsymbol{b}| = |\cos(\theta)|\; ||\boldsymbol{a}||\; ||\boldsymbol{b}|| \leq ||\boldsymbol{a}||\; ||\boldsymbol{b}||$$

where θ denotes the angle between $\boldsymbol{a}$ and $\boldsymbol{b}$ and $||\cdot||$ the Euclidian norm in $\mathbb{R}^2$. In other words, if $\cos(\theta)$ takes on small values, i.e. when $\boldsymbol{a}$ and $\boldsymbol{b}$ are nearly orthogonal, then $|\boldsymbol{a} \cdot \boldsymbol{b}|$ would be much smaller than the product $\|\boldsymbol{a}\|\,\|\boldsymbol{b}\|$. By analogy, let us suppose now that Φ and Σ are two functions defined on $[0,1]$ in a one dimensional case such as shown in Fig. 7. The product of these two functions is qualitatively shown in Fig. 7 as well. It is clear in this case that $|(\Phi,\Sigma)_\mathcal{V}|$ will be much smaller than the quantity $\|\Phi\|_\mathcal{V}\|E\|_\mathcal{V}$. Such a situation does often occur when the solutions U and W exhibit marked local variations; for example, when U is singular at a corner of the domain, and when the quantity of interest consists in finding the value of the solution at a point for W. □

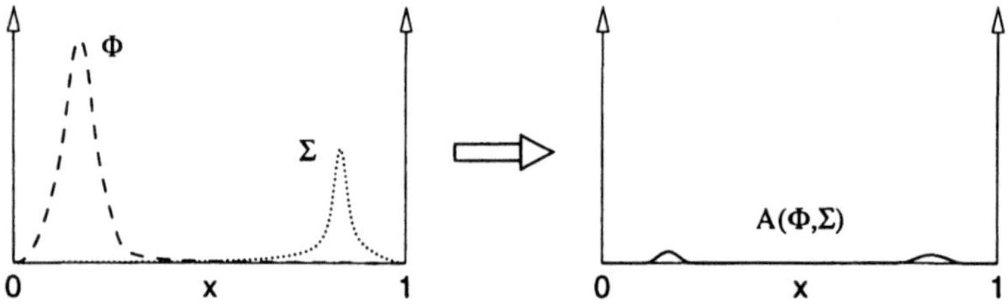

Fig. 7. Qualitative example of functions Φ and Σ for which $|(\Phi,\Sigma)_\mathcal{V}|$ will be much smaller than the quantity $\|\Phi\|_\mathcal{V}\|E\|_\mathcal{V}$.

An alternative approach is to decompose the global inner-product $(\cdot,\cdot)_\mathcal{V}$ into elementwise inner-products $(\cdot,\cdot)_{K,\mathcal{V}}$ and to use the Cauchy-Schwartz inequality on each elementwise contributions. Hence

$$L(E) = (\Phi,\Sigma)_\mathcal{V} = \sum_{K\in\mathcal{P}_h} (\Phi,\Sigma)_{K,\mathcal{V}}$$

so that:

$$|L(E)| \leq \sum_{K\in\mathcal{P}_h} |(\Phi,\Sigma)_{K,\mathcal{V}}| \leq \sum_{K\in\mathcal{P}_h} \|\Phi\|_{K,\mathcal{V}}\|\Sigma\|_{K,\mathcal{V}}.$$

Suppose that implicit estimates η^Φ and η^Σ of $\|\Phi\|_\mathcal{V}$ and $\|\Sigma\|_\mathcal{V}$, respectively, have been computed. We have seen that it is always possible to compute elementwise contributions η_K^Φ and η_K^Σ to η^Φ and η^Σ, i.e.

$$\eta^\Phi = \sqrt{\sum_{K\in\mathcal{P}_h} (\eta_K^\Phi)^2},$$

$$\eta^\Sigma = \sqrt{\sum_{K\in\mathcal{P}_h} (\eta_K^\Sigma)^2}.$$

An estimate η^L of $L(E)$ is then obtained as:

$$\eta^L = \sum_{K\in\mathcal{P}_h} (\eta_K^\Phi) \times (\eta_K^\Sigma),$$

in which case the elementwise contributions can naturally be determined as:

$$\eta_K^L = (\eta_K^\Phi) \times (\eta_K^\Sigma).$$

Remark 6. Such an approach has to be used with care. Indeed, η^Σ is a global error estimate of $\|\Sigma\|_\mathcal{V}$. When the estimate is decomposed into elementwise contributions, it does not necessarily imply that η_K^Σ produces a reliable estimate of $\|\Sigma\|_{K,\mathcal{V}}$ at the element level. This is due to "pollution effects" as observed by Babuška *et al.* [13] (recall our comments at the end of Section 1.6). □

In a similar manner, we can exploit the ideas used in deriving explicit estimators in order to derive upper bounds on the quantity $L(E)$. Recall the the residual functional can be defined in terms of element interior residuals r_K and boundary residuals $j_{\gamma,K}$ such that:

$$\mathcal{R}_h(Z) = \sum_{K\in\mathcal{P}_h} \int_K r_K Z \,\mathrm{d}x - \int_{\partial K} j_{\gamma,K} Z \,\mathrm{d}s.$$

Therefore, replacing Z by $(W - \mathcal{I}_h W)$, where $\mathcal{I}_h$ defines an arbitrary operator from V to V^h, we would get:

$$\begin{aligned} L(E) &= \sum_{K\in\mathcal{P}_h} \int_K r_K (W - \mathcal{I}_h W)\,\mathrm{d}x - \int_{\partial K} j_{\gamma,K}(W - \mathcal{I}_h W)\,\mathrm{d}s \\ &\le \sum_{K\in\mathcal{P}_h} \|r_K\|_{0,K}\,\|W - \mathcal{I}_h W\|_{0,K} + \|j_{\gamma,K}\|_{0,\partial K}\,\|W - \mathcal{I}_h W\|_{0,\partial K}. \end{aligned}$$

This method, called the Dual-Weighted Residual method (DWR), is followed by Becker and Rannacher (see [17] and earlier contributions). The difficulty here is to estimate the quantities $\|W - \mathcal{I}_h W\|_{0,K}$ and $\|W - \mathcal{I}_h W\|_{0,\partial K}$. The authors in [17] propose to use local higher-order approximations or approximations by difference quotients. They do not report however satisfactory results using either of those two methods (see Table 5.1 in [17]).

Case where $\mathcal{B}(\cdot,\cdot)$ defines an inner-product: In this particular case, from the equation for the error E, we have:

$$L(E) = \mathcal{R}_h(\Sigma) = \mathcal{B}(E, \Sigma).$$

Since $\mathcal{B}(\cdot,\cdot)$ is symmetric positive-definite, it also defines an inner product with associated norm $\|Z\|_\mathcal{V} = \sqrt{\mathcal{B}(Z,Z)}$. As suggested in [5–7], we may apply the *parallelogram identity* for scalar product to obtain:

$$L(E) = \mathcal{B}(E, \Sigma) = \frac{1}{4}\|E + \Sigma\|_\mathcal{V}^2 - \frac{1}{4}\|E - \Sigma\|_\mathcal{V}^2. \tag{44}$$

Since E and Σ can vary by several orders of magnitude, it is useful to scale the errors by a scaling factor $s \in \mathbb{R}$ (see [45,48]) such that:

$$L(E) = \mathcal{B}(sE, s^{-1}\Sigma) = \frac{1}{4}\|sE + s^{-1}\Sigma\|_{\mathcal{V}}^2 - \frac{1}{4}\|sE - s^{-1}\Sigma\|_{\mathcal{V}}^2. \tag{45}$$

The scaling factor s is chosen so that the quantities $\|sE\|_{\mathcal{V}}$ and $\|s^{-1}\Sigma\|_{\mathcal{V}}$ have same amplitudes, i.e. $\|sE\|_{\mathcal{V}} = \|s^{-1}\Sigma\|_{\mathcal{V}}$, which implies that:

$$s = \sqrt{\frac{\|\Sigma\|_{\mathcal{V}}}{\|E\|_{\mathcal{V}}}}. \tag{46}$$

Such a choice of the scaling factor s is justified because it minimizes the quantities $\|sE + s^{-1}\Sigma\|_{\mathcal{V}}^2$ and $\|sE - s^{-1}\Sigma\|_{\mathcal{V}}^2$. Moreover, the scaler s ensures the *scalability* of the relationship above. If we multiply the load $\mathcal{F}(\cdot)$ in the original problem by a factor a, the term on the right-hand side of the above equation just needs to be multiplied by a to obtain the new error $L(E)$. This result is not true if we do not include the scaling factor.

The identity (45) establishes the relation between the error quantity $L(E)$ and energy norms of linear combinations of E and Σ. We have shown in Section 1 how to compute global errors in energy norm.

Bounds for linear functional outputs of partial differential equations
Paraschivoiu, Patera and Peraire [45,48] suggested to use upper bounds on the global errors to derive bounds on the error in the quantity of interest. Let η_{upp}^+ and η_{upp}^- denote global error estimates such that:

$$0 \le \|sE + s^{-1}\Sigma\|_{\mathcal{V}} \le \eta_{upp}^+,$$
$$0 \le \|sE - s^{-1}\Sigma\|_{\mathcal{V}} \le \eta_{upp}^-.$$

Then, they obtain bounds on $L(E)$:

$$-\frac{1}{4}(\eta_{upp}^-)^2 \le -\|sE - s^{-1}\Sigma\|_{\mathcal{V}} \le L(E) \le \|sE + s^{-1}\Sigma\|_{\mathcal{V}} \le \frac{1}{4}(\eta_{upp}^+)^2. \tag{47}$$

Remark 7. We believe that such bounds may not be reliable since the global norms $\|sE - s^{-1}\Sigma\|_{\mathcal{V}}$ and $\|sE + s^{-1}\Sigma\|_{\mathcal{V}}$ can be in some applications much larger than the quantity $L(E)$ itself. To illustrate this, let us consider the vectors $\boldsymbol{a} = (1,0)$ and $\boldsymbol{b} = (\varepsilon, \sqrt{1-\varepsilon^2})$, $\varepsilon \le 1$, in $\mathbb{R}^2$. Note that $\|\boldsymbol{a}\| = \|\boldsymbol{b}\| = 1$ so that $s = 1$. Then:

$$\frac{1}{4}\|\boldsymbol{a} + \boldsymbol{b}\|^2 = \frac{1}{2}(1+\varepsilon)$$
$$\frac{1}{4}\|\boldsymbol{a} - \boldsymbol{b}\|^2 = \frac{1}{2}(1-\varepsilon)$$

whereas:

$$\boldsymbol{a} \cdot \boldsymbol{b} = \varepsilon,$$

and the conclusion is straightforward when we take ε arbitrarily small. □

Goal-oriented error estimation
Alternatively, Strouboulis *et al.* [56] and Oden and Prudhomme [51,44] have suggested to compute lower and upper bounds η^+_{low}, η^+_{upp}, η^-_{low} and η^-_{upp} such that:

$$\eta^+_{low} \leq \|sE + s^{-1}\Sigma\|_{\mathcal{V}} \leq \eta^+_{upp},$$
$$\eta^-_{low} \leq \|sE - s^{-1}\Sigma\|_{\mathcal{V}} \leq \eta^-_{upp}.$$

Then, the quantities

$$\eta^L_{low} = \frac{1}{4}(\eta^+_{low})^2 - \frac{1}{4}(\eta^-_{upp})^2, \tag{48}$$

$$\eta^L_{upp} = \frac{1}{4}(\eta^+_{upp})^2 - \frac{1}{4}(\eta^-_{low})^2, \tag{49}$$

provide a lower and upper bounds on $L(E)$, i.e.

$$\eta^L_{low} \leq L(E) \leq \eta^L_{upp}. \tag{50}$$

Moreover, taking the average of η^L_{low} and η^L_{upp} we obtain the error estimate η^L:

$$\eta^L = \frac{1}{2}\left(\eta^L_{low} + \eta^L_{upp}\right) \approx L(E). \tag{51}$$

Remark 8 (Quality of bounds and estimates). Quality of error estimates is usually quantified in terms of the effectivity index, which corresponds to the ratio between the error estimate and the exact error (measured with respect to the global norm or the quantity of interest). Here, we evaluate the quality of the bounds η^L_{low} and η^L_{upp} in terms of the effectivity indices of η^+_{low}, η^+_{upp}, η^-_{low} and η^-_{upp}. Let the effectivity indices be:

$$\lambda^-_{low} = \frac{\eta^-_{low}}{\|sE - \Sigma/s\|_{\mathcal{V}}}, \qquad \lambda^-_{upp} = \frac{\eta^-_{upp}}{\|sE - \Sigma/s\|_{\mathcal{V}}},$$
$$\lambda^+_{low} = \frac{\eta^+_{low}}{\|sE + \Sigma/s\|_{\mathcal{V}}}, \qquad \lambda^+_{upp} = \frac{\eta^+_{upp}}{\|sE + \Sigma/s\|_{\mathcal{V}}}.$$

Then, we have:

$$\begin{aligned}
\eta^L_{low} &= \frac{1}{4}(\eta^+_{low})^2 - \frac{1}{4}(\eta^-_{upp})^2 \\
&= \frac{1}{4}(\lambda^+_{low})^2\|sE + \Sigma/s\|^2_{\mathcal{V}} - \frac{1}{4}(\lambda^-_{upp})^2\|sE - \Sigma/s\|^2_{\mathcal{V}} \\
&= \frac{1}{4}(\lambda^+_{low})^2\|sE + \Sigma/s\|^2_{\mathcal{V}} - \frac{1}{4}(\lambda^-_{upp})^2\|sE + \Sigma/s\|^2_{\mathcal{V}} + (\lambda^-_{upp})^2 L(E)
\end{aligned}$$

and, assuming $L(E)$ to be different from zero, we obtain the following effectivity index for the lower bound on $L(E)$:

$$\lambda^L_{low} = \frac{\eta^L_{low}}{L(E)} = (\lambda^-_{upp})^2 + \frac{1}{4}\left((\lambda^+_{low})^2 - (\lambda^-_{upp})^2\right)\frac{\|sE + \Sigma/s\|^2_{\mathcal{V}}}{L(E)}.$$

In the same manner, we have for the upper bound

$$\lambda_{upp}^{L} = \frac{\eta_{upp}^{L}}{L(E)} = (\lambda_{low}^{-})^2 + \frac{1}{4}\left((\lambda_{upp}^{+})^2 - (\lambda_{low}^{-})^2\right)\frac{\|sE + \Sigma/s\|_{\mathcal{V}}^2}{L(E)}.$$

Therefore, the quality of the bounds directly depends on the ratio given by $\|sE + \Sigma/s\|_{\mathcal{V}}^2/L(E)$. This ratio can take on large values depending on the quantity of interest. Also, when adaptivity aims at controlling $L(E)$ instead of $\|E\|_{\mathcal{V}}$, this ratio may have the tendency to increase. In consequence, in order to obtain effectivity indices λ_{low}^{L} and λ_{upp}^{L} close to one, the effectivity indices with respect to the quantities in the energy norm should be excellent, i.e. very close to one, so that $(\lambda_{low}^{+})^2 - (\lambda_{upp}^{-})^2$ and $(\lambda_{upp}^{+})^2 - (\lambda_{low}^{-})^2$ are close to zero. Otherwise we may expect λ_{low}^{L} and λ_{upp}^{L} to deteriorate when the ratio becomes very large. On the other hand, the effectivity index of the estimate η^L is given by:

$$\begin{aligned}\lambda^{L} = \frac{\eta^{L}}{L(E)} &= \frac{1}{2}(\lambda_{low}^{+})^2 + \frac{1}{8}\left((\lambda_{low}^{-})^2 - (\lambda_{low}^{-})^2\right)\frac{\|sE + \Sigma/s\|_{\mathcal{V}}^2}{L(E)} \\ &= \frac{1}{2}(\lambda_{upp}^{+})^2 + \frac{1}{8}\left((\lambda_{upp}^{-})^2 - (\lambda_{upp}^{-})^2\right)\frac{\|sE + \Sigma/s\|_{\mathcal{V}}^2}{L(E)}.\end{aligned}$$

This time, the difference $(\lambda_{low}^{+})^2 - (\lambda_{low}^{-})^2$ and $(\lambda_{upp}^{+})^2 - (\lambda_{upp}^{-})^2$ are very close to zero since we expect the global error estimator to provide effectivity indices $\lambda_{low}^{+} \approx \lambda_{low}^{-}$ and $\lambda_{upp}^{+} \approx \lambda_{upp}^{-}$. It follows that the quality of η^L should not depend on the ratio $\|sE + \Sigma/s\|_{\mathcal{V}}^2/L(E)$, however large it may be. This type of behavior for the estimate and bounds has frequently been observed in numerical experiments. □

Application to the model problems:

A) Poisson Problem: The bilinear form resulting from the Poisson equation is symmetric positive-definite and thus defines an inner product. In this case, one can apply the parallelogram identity. Moreover, it has been shown in Section 1 that lower and upper bounds on the error could be derived.

B) Convection-Diffusion Problem: The dual problem associated with the convection-diffusion problem consists in finding the influence function $\omega \in V$ such that:

$$A(v,\omega) + C(\alpha, v, \omega) = L(v), \qquad \forall v \in V.$$

In this case, the error in the quantity of interest is determined as:

$$|L(e_u)| = |\mathcal{R}_h(\varepsilon)| \leq \sum_{K\in\mathcal{P}_h} |A_K(\phi,\varepsilon)|$$

where ε in the error in the influence function and ϕ the Riesz representer of $\mathcal{R}_h$ in V. We have shown in Section 1 that the error of the convection-diffusion could be bounded above by the residual. Therefore, an estimate of the quantity of interest can be given as:

$$|L(e_u)| \approx \eta^L = \sum_{K \in \mathcal{P}_h} \eta_K^u \cdot \eta_K^\omega,$$

where η_K^u and η_K^ω are the elementwise contributions to the global norm of the residuals associated with the primal solution and the influence function respectively.

C) **Stokes Problem:** The dual problem of the Stokes problem governs the influence functions $(\boldsymbol{\omega}, \rho) \in \boldsymbol{V} \times Q$ such that:

$$\begin{aligned} A(\boldsymbol{v}, \boldsymbol{\omega}) + B(\rho, \boldsymbol{v}) &= L(\boldsymbol{v}, 0), \qquad \forall\, \boldsymbol{v} \in \boldsymbol{V}, \\ B(q, \boldsymbol{\omega}) &= L(0, q), \qquad \forall\, q \in Q. \end{aligned}$$

Defining the error in the influence functions as $\boldsymbol{\varepsilon} = \boldsymbol{\omega} - \boldsymbol{\omega}_h$ and $\xi = \rho - \rho_h$, we then have:

$$|L(\boldsymbol{e}_u, e_p)| = |\mathcal{R}_h^m(\boldsymbol{\varepsilon}) + \mathcal{R}_h^c(\xi)|.$$

Let $\boldsymbol{\phi}$ be the representer of $\mathcal{R}_h^m$ in V, therefore:

$$\begin{aligned} |L(\boldsymbol{e}_u, e_p)| &\le \sum_{K \in \mathcal{P}_h} |A(\boldsymbol{\phi}, \boldsymbol{\varepsilon})| + |B(\xi, \boldsymbol{u}_h)| \\ &\le \sum_{K \in \mathcal{P}_h} \sqrt{\|\boldsymbol{\phi}\|_V^2 + \|\nabla \cdot \boldsymbol{u}_h\|_0^2} \times \sqrt{\|\boldsymbol{\varepsilon}\|_V^2 + \|\xi\|_0^2}. \end{aligned}$$

The reader is referred to [42,43,50] for more details.

D) **Navier-Stokes Problem:** The Navier-Stokes equations are, of course, nonlinear. Nevertheless the dual problem can still be found following the same steps as for the linear case after linearization of the error equation. This is also explained in the next section.

2.4 Generalization to Nonlinear Problems

The general approach for goal-oriented error estimation can be presented in terms of optimal control theory. Let $\mathcal{B}(\cdot;\cdot)$ be a differentiable semilinear form and $\mathcal{F}(\cdot)$ a linear functional defined on $\mathcal{V}$. Moreover, let $N(\cdot)$ denote a functional defined on $\mathcal{V}$, which may be nonlinear.

We then suppose that our computational goal is to estimate the quantity $N(U)$, where $U \in \mathcal{V}$ is constrained to be the solution of the differential equation:

$$\mathcal{B}(U; Z) = \mathcal{F}(Z), \qquad \forall\, Z \in \mathcal{V}. \tag{52}$$

In [17], an "optimal control" approach to this problem is proposed based on the constrained minimization problem,

$$\boxed{\begin{array}{l} \text{Find } U \in \mathcal{V} \text{ such that} \\ \qquad\qquad N(U) = \inf_{V \in M} Q(V) \\ \text{where} \\ \quad M = \{V \in \mathcal{V};\ B(V;Z) = F(Z),\ \forall\ Z \in \mathcal{V}\} \end{array}} \tag{53}$$

The minima U correspond to a saddle point $(U, W) \in \mathcal{V} \times \mathcal{V}$ of the Lagrangian,

$$\mathcal{L}(V, Z) = N(V) + \mathcal{F}(Z) - \mathcal{B}(V; Z).$$

Assuming that (52) admits a unique solution, the minimization problem is then trivial. The stationary points $\{U, W\} \in \mathcal{V} \times \mathcal{V}$ are then solutions of the Euler-Lagrange system:

$$\mathcal{B}(U; Z) = \mathcal{F}(Z), \qquad \forall\ Z \in \mathcal{V}, \tag{54}$$

$$\mathcal{B}'(U; Z, W) = N'(U; Z), \qquad \forall\ Z \in \mathcal{V}. \tag{55}$$

where $\mathcal{B}'$ and N' are the derivatives of $\mathcal{B}$ and N respectively. Equation (54) is the primal problem for U; (55) is the adjoint or dual problem for W, and, with U specified, is a linear variational equation for the influence function W corresponding to the choice N of the quantity of interest.

Remark 9. If $\mathcal{B}(\cdot, \cdot)$ is a bilinear form on $\mathcal{V} \times \mathcal{V}$, then $\mathcal{B}' = \mathcal{B}$. Likewise if N is a linear functional, $N' = N$. The Euler-Lagrange system would then read:

$$\begin{aligned} \mathcal{B}(U, Z) &= \mathcal{F}(Z), \qquad \forall\ Z \in \mathcal{V}, \\ \mathcal{B}(Z, W) &= N(Z), \qquad \forall\ Z \in \mathcal{V}. \end{aligned}$$

□

Finite element approximations of the Euler-Lagrange system are given as $(U_h, W_h) \in \mathcal{V}^h \times \mathcal{V}^h$ such that:

$$\begin{aligned} \mathcal{B}(U_h; Z) &= \mathcal{F}(Z), \qquad \forall\ Z \in \mathcal{V}^h, \\ \mathcal{B}'(U_h; Z, W_h) &= N'(U_h; Z), \qquad \forall\ Z \in \mathcal{V}^h. \end{aligned} \tag{56}$$

Let $\mathcal{R}_h(\cdot)$ be the residual associated with the primal problem:

$$\mathcal{R}_h(Z) = \mathcal{F}(Z) - \mathcal{B}(U_h; Z), \qquad Z \in \mathcal{V}.$$

Then, after a lengthy and tedious computation, it is possible to show that the error in the quantity of interest is given by:

$$N(U) - N(U_h) = \min_{Z_h \in \mathcal{V}^h} \mathcal{R}_h(W - Z_h) + \Delta$$

where Δ denotes the remainder in terms of higher-order derivatives of $\mathcal{B}$ and N and higher-order terms in the errors themselves. In other words, neglecting Δ, we have:

$$|N(U) - N(U_h)| \leq |\mathcal{R}_h(W - Z_h)|$$

where Z_h is any function in $\mathcal{V}^h$ since $\mathcal{R}_h(Z_h) = 0$. We refer the reader to [17] for more details on error estimation for nonlinear problems.

2.5 Adaptive Strategy

The strategy to control or reduce the error with respect to the quantity of interest follows the same steps as for the adaptation procedure based on global error estimates. The relative error reads in the present case:

$$e_{\text{rel}} = \frac{|L(E)|}{|L(U)|}.$$

Note that this relation should be used with care as $L(U)$ may become very small in certain cases. Also we use absolute values of the quantities as they may take on negative values. As before, the mesh needs to be adapted whenever

$$\frac{|\eta^L|}{|L(U_h)|} \geq C_{\text{tol}}$$

where η^L is an estimate of the error in the quantity of interest L. We have seen earlier that the estimate can be decomposed into elementwise contributions η_K such that

$$\eta^L = \sum_{K \in \mathcal{P}_h} \eta_K^L.$$

Therefore, the quantities η_K^L can be used to derive refinement indicators. Indeed, after normalization of the contributions by the largest one, an element K is a candidate for refinement if

$$\frac{|\eta_K^L|}{\max_J\{|\eta_J^L|\}} \geq C_{\text{adp}}$$

where C_{adp} is a user-prescribed parameter ranging from zero to one.

2.6 Numerical Examples

The driven cavity. We consider here the standard driven cavity problem in the unit square $\Omega = (0,1) \times (0,1)$. The load $\boldsymbol{f}$ on the right-hand side of the Stokes equations is set to zero. The velocity is constrained to be zero everywhere on $\partial\Omega$ except on the top boundary $\Gamma = \{x = (x,y);\ 0 \leq x \leq 1,\ y = 1\}$ where $\boldsymbol{u}$ satisfies:

$$\begin{aligned} u_1(x,y) &= 4x(1-x), \\ u_2(x,y) &= 0. \end{aligned}$$

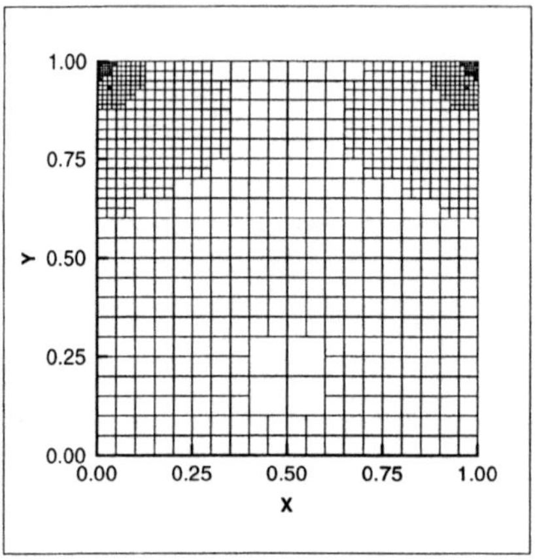

Fig. 8. Globally adapted mesh (N_e = 940) for the driven cavity.

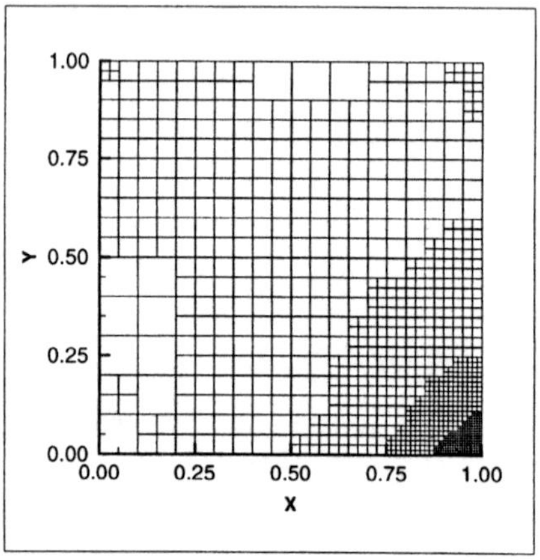

Fig. 9. Goal-oriented adapted mesh ($N_e = 1030$) w.r.t. u_1 in lower-right corner of the cavity.

We choose to prescribe a parabolic profile for u_1 in order to regularize the solution in the two upper corners.

The driven cavity flow exhibits very weak local features in the lower corners of the cavity. It follows that when we use an adaptive strategy based on the global error estimate, we essentially refine around the upper corners, where the sources of error are predominant due to the singularity in the pressure variable. Such a fine adapted mesh is shown in Fig. 8. As a consequence, the approximated velocity is poorly resolved in the lower corners, which is clearly observed in Fig. 10 where the contour lines of the component u_1 have been plotted.

We then appeal to the goal-oriented adaptive strategy in order to construct an optimal mesh designed to capture the local feature in the lower-right corner. Let Ω_s be a truncated circle of center $\boldsymbol{x}_0$ and radius r as shown in Fig. 11. We define a linear functional based on the weighted average of u_1 over the subdomain Ω_s:

$$L(\boldsymbol{u}, p) = \int_{\Omega} k_\varepsilon(\boldsymbol{x} - \boldsymbol{x}_0)\, u_1(\boldsymbol{x})\, \mathrm{d}x$$

where k_ε is the mollifying function (41). Using the adaptive strategy with respect to that goal, we obtain the mesh shown in Fig. 9, which is refined in the region of interest, but also mildly in the upper corners. The new contour lines for u_1 are then greatly improved, as illustrated in Fig. 11.

Channel flow past an obstacle. We have selected this example to highlight again the advantages of the goal-oriented adaptive strategy. The computational domain, on which the initial mesh is superimposed, is shown in Fig. 12,

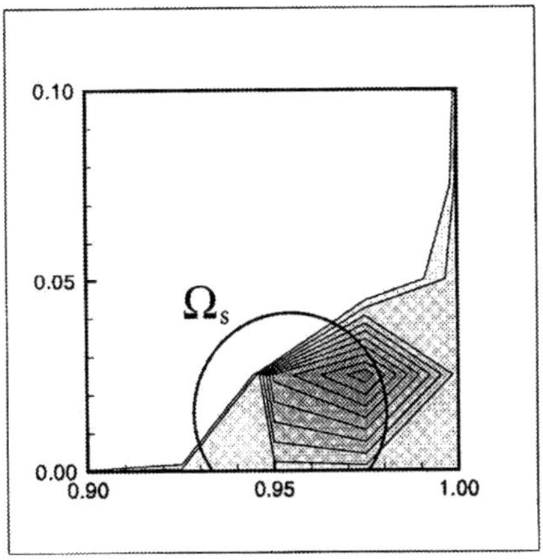

Fig. 10. Contour lines of u_1 on the globally adapted mesh in the lower-right corner of the cavity.

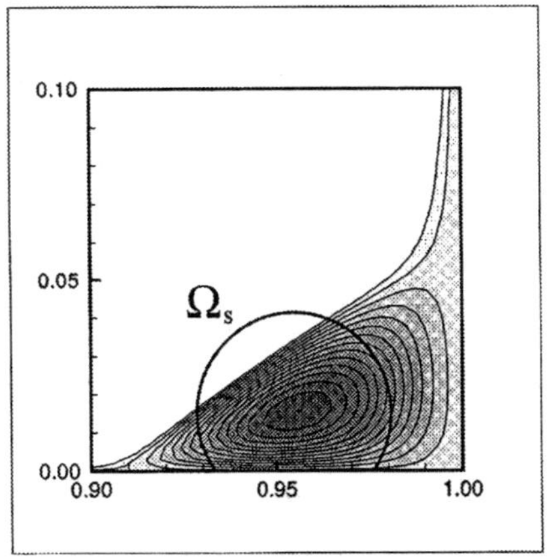

Fig. 11. Contour lines of u_1 on the goal-oriented adapted mesh in the lower-right corner of the cavity.

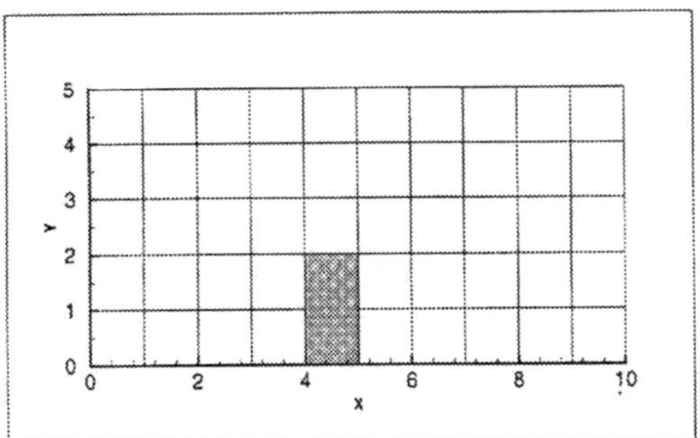

Fig. 12. Computational domain and initial mesh for the channel flow past an obstacle.

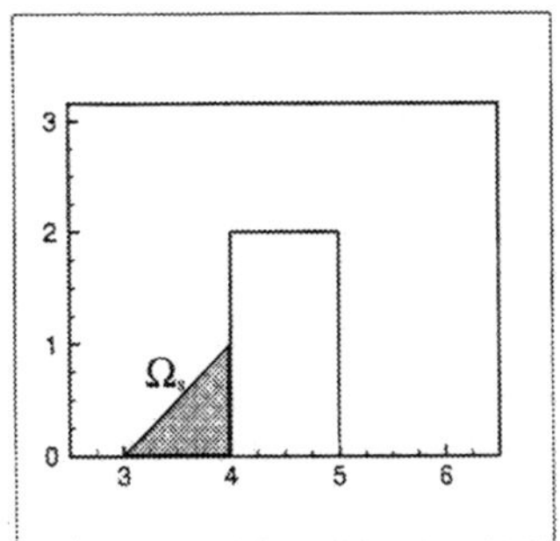

Fig. 13. Subdomain Ω_s.

where the obstacle is represented by the grey area. Along the lateral walls and obstacle, the velocity vector is set to zero. A Poiseuille-type velocity profile is prescribed at the inflow located on the left side of the domain. At the outflow, we use the artificial boundary condition $-p\mathbf{n} + \partial \boldsymbol{u}/\partial n = \mathbf{0}$ where $\mathbf{n}$ denotes the outward normal to the boundary. The velocity and pressure fields are approximated by piecewise biquadratic and bilinear polynomials respectively.

We suppose that our goal is to study the recirculation region upstream of the obstacle. The quantity of interest is chosen to be the *circulation* $\mathcal{C}$ around the closed curve $\partial\Omega_s$ enclosing the subregion Ω_s shown in Fig. 13. It follows that:

$$L(\boldsymbol{u},p) = \mathcal{C} = \int_{\partial\Omega_s} \boldsymbol{u}\cdot\boldsymbol{t}\,\mathrm{d}s = \int_{\Omega_s} \zeta(\boldsymbol{u})\,\mathrm{d}x$$

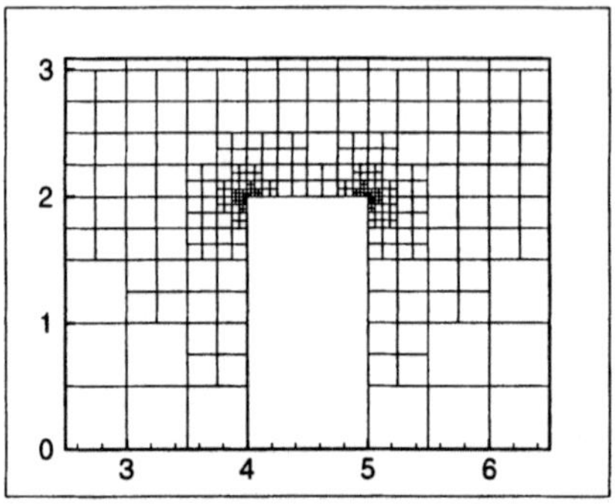

Fig. 14. Adapted mesh based on the global adaptive strategy.

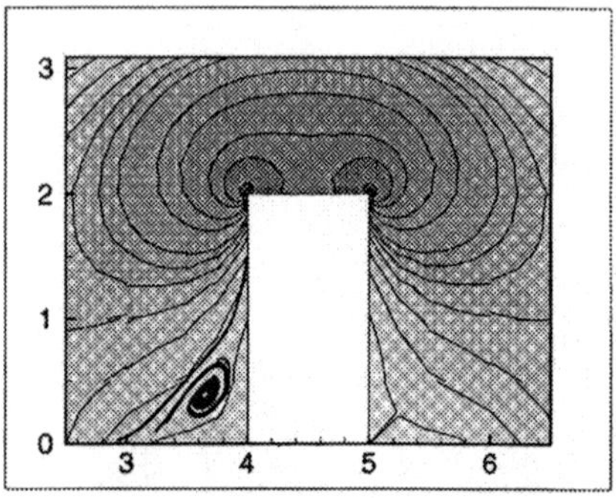

Fig. 15. Streamlines and vorticity contours on the globally adapted mesh.

where $\boldsymbol{t}$ denotes the unit tangent along the path of integration $\partial\Omega_s$ and $\zeta(\boldsymbol{u})$ the vorticity of the flow:

$$\zeta(\boldsymbol{u}) = \frac{\partial u_2}{\partial y} - \frac{\partial u_1}{\partial x}.$$

The choice for this quantity of interest is justified as the recirculation is generally visualized by tracing the streamlines, which, in two-dimensional flows, are the isolines of the streamfunction Ψ. Furthermore, the streamfunction is related to the vorticity by the relation $\Delta\Psi = \zeta$.

We now compare the performance of the global and goal-oriented adaptive strategies. In the first test, the mesh is adapted based on the global error estimator. We show in Fig. 14 the final mesh obtained after 12 adaptation cycles. It contains 561 elements and, as expected, exhibits highly refined regions around the top corners of the obstacle due to the large errors produced by the discontinuity in the pressure. In Fig. 15, we display the vorticity contours and trace streamlines (actually pathlines, since these are identical for steady flows). We observe that the streamlines are very poorly resolved. In the second test, the mesh is adapted according to the goal-oriented strategy. We show in Fig. 16 the final mesh obtained after 11 adaptation cycles. This time, the mesh contains 477 elements and is essentially refined in the neighborhood of Ω_s, as expected, but also around the upper corners of the obstacle. This clearly indicates that the flow upstream of the obstacle is influenced by the large sources of error around the upper corners. Moreover, the vorticity and streamlines are well resolved as shown in Fig. 17 and compare very well to the results found in [27], at least qualitatively.

In Fig. 18, we show the evolution of the relative error $L(\boldsymbol{e}_u, e_p)$ associated with the quantity of interest. In order to do so, we have computed an overkilled solution $(\hat{\boldsymbol{u}}, \hat{p})$ on a very fine mesh, from which we have deduced an accurate approximation of the error in the circulation:

$$L(\boldsymbol{e}_u, e_p) \approx L(\hat{\boldsymbol{u}}, \hat{p}) - L(\boldsymbol{u}_h, p_h).$$

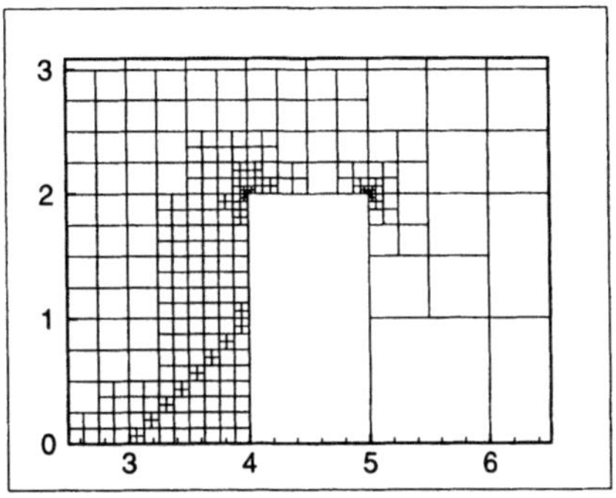

Fig. 16. Adapted mesh based on the goal-oriented adaptive strategy.

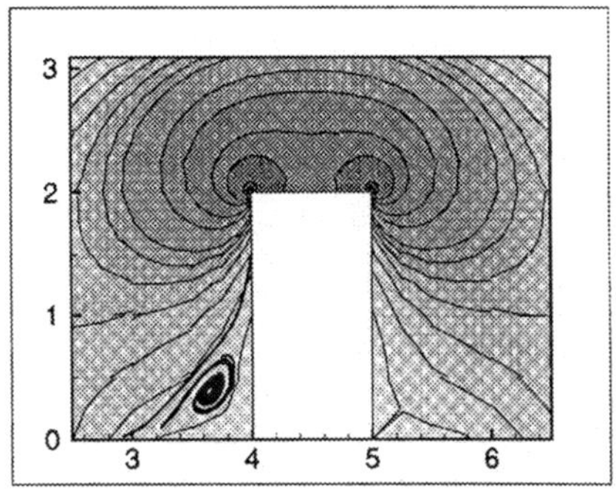

Fig. 17. Streamlines and vorticity contours of the goal-oriented adapted mesh.

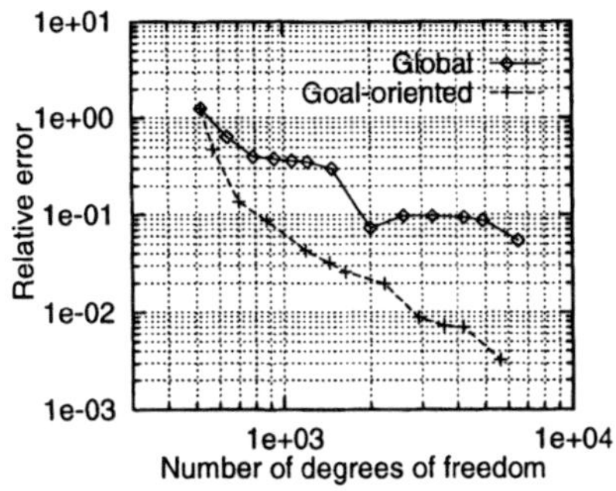

Fig. 18. Relative error w.r.t. the *circulation* using the global and goal-oriented adaptive strategies.

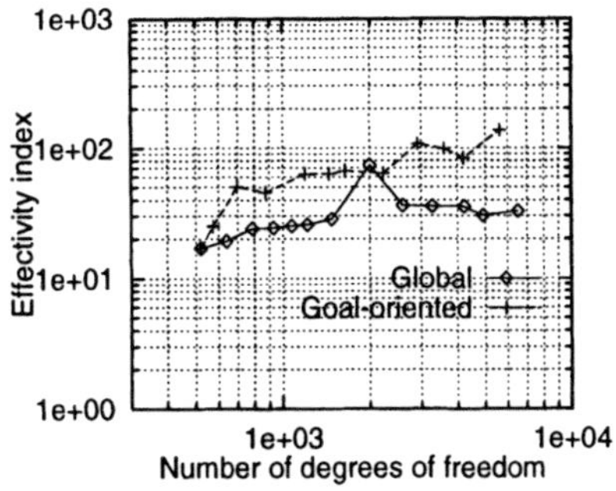

Fig. 19. Effectivity indices λ^L of the error bound η^L in the case of the obstacle problem.

The goal-oriented strategy is again superior to the global adaptation strategy by more than an order of magnitude. Finally, the effectivity indices λ^L for the error bound η^L (i.e. the ratio $\lambda^L = |\eta^L|/|L(e_u, e_p)|$) lie in the range ten to hundred as shown in Fig. 19.

3 Time-Dependent Problems: Stability and Error Control

3.1 Introduction

Developing reliable and practical *a posteriori* error estimators for time dependent problems represent the next challenge in error estimation. In addition to the error due to the discretization in space as in steady-state problems, one also needs to deal with estimating the error accumulation and the error due to the discretization in time. A careful analysis of the equations must then be carried out in order to identify and isolate the different sources of errors.

We also believe that it is necessary to develop specific approaches depending on the problem under investigation and the type of schemes which are chosen to compute the solution. We present in this section of the paper two different methodologies for *a posteriori* error estimation of time-dependent problems: the first deals with error estimation for the heat equation while the second treats the incompressible Navier-Stokes equations. We emphasize here that we focus on constructing *a posteriori* estimators for the error due only to the discretization in space.

3.2 The Heat Equation

Preliminaries. In this section, we consider the heat equation as a model problem. We suppose that one requirement of the analysis is to solve the problem using an explicit scheme due to some constraints in computer resources. This is often the case when one intends to simulate complex or multiphysics time-dependent phenomena.

Let Ω be an open bounded domain in $\mathbb{R}^d$, $d \geq 1$, and $T > 0$ a real number. The heat equation is given by:

$$\begin{aligned} \partial_t u - \kappa \Delta u &= 0, && \text{in } \Omega \times (0,T), \\ u(\boldsymbol{x},t) &= 0, && \forall\, \boldsymbol{x} \in \partial\Omega, \quad \forall\, t \in (0,T), \\ u(\boldsymbol{x},0) &= u^0(\boldsymbol{x}), && \forall\, \boldsymbol{x} \in \Omega. \end{aligned} \tag{57}$$

where $u = u(\boldsymbol{x},t)$ is the temperature field and κ the constant of diffusivity. Let $V = H_0^1(\Omega)$. The strong solution is known to be unique in the space $H^1(0,T;L^2(\Omega)) \cap L^2(0,T;V)$. The corresponding weak formulation of the heat equation consists in finding $u \in L^2(0,T;V)$, such that,

$$\begin{aligned} (\partial_t u, v) &= -\kappa(\nabla u, \nabla v), && \forall\, v \in V, \text{ for almost all } t \in (0,T), \\ (u,v) &= (u^0, v), && \forall\, v \in V, \text{ at } t = 0, \end{aligned} \tag{58}$$

where $(\cdot,\cdot)$ denotes the L^2 inner product.

Discretization schemes. The heat conduction problem is supposed to be solved using the explicit forward Euler scheme. Let the time interval $(0,T)$ be divided into subintervals (t^{n-1}, t^n), $n = 1, \ldots, N_T$, where N_T is the total number of subintervals. The time step is defined as $\Delta t = t^n - t^{n-1}$ and is allowed to vary with time. The discrete solution $u^n \in V$ is advanced in time for $n = 1, 2, \ldots, N_T$, such that, given u^0,

$$(u^n, v) = (u^{n-1}, v) - \Delta t\, \kappa(\nabla u^{n-1}, \nabla v), \qquad \forall\, v \in V. \tag{59}$$

The equation is then discretized in space using the finite element Galerkin method. Let V^h denotes a conforming finite element space of V. A finite element approximation $u_h^n \in V^h$ of u^n, for $n = 1, 2, \ldots, N_T$ is obtained, given u_h^0, by solving at each time step:

$$(u_h^n, v) = (u_h^{n-1}, v) - \Delta t\ \kappa(\nabla u_h^{n-1}, \nabla v), \qquad \forall\, v \in V^h, \tag{60}$$

where u_h^0 is the L^2 projection of u^0 on V^h.

Error estimation. We now investigate the numerical error in u_h^n due to the finite element discretization; in other words, the objective here is to estimate the error $e^n = u^n - u_h^n$, for all $n = 1, 2, \ldots, N_T$. Replacing in (59) u^n by $u_h^n + e^n$ and u^{n-1} by $u_h^{n-1} + e^{n-1}$, the error is shown to be governed by the error equation:

$$(e^n, v) = \mathcal{R}_h(v) + \mathcal{R}_e(v), \qquad \forall\, v \in V, \tag{61}$$

where $\mathcal{R}_h(\cdot)$ and $\mathcal{R}_e(\cdot)$ denote the residual functionals in V:

$$\begin{aligned}\mathcal{R}_h(v) &= -\ (u_h^n - u_h^{n-1}, v) - \Delta t\ \kappa(\nabla u_h^{n-1}, \nabla v),\\ \mathcal{R}_e(v) &= (e^{n-1}, v) - \Delta t\ \kappa(\nabla e^{n-1}, \nabla v).\end{aligned}$$

The residual $\mathcal{R}_h$ is identified with the source of error due to the finite element discretization, while the residual $\mathcal{R}_e$ governs the error accumulation in time. In this work, we investigate the effect of $\mathcal{R}_h$ only, so that we seek the component of the error e^n which satisfies:

$$(e^n, v) = \mathcal{R}_h(v). \tag{62}$$

Note that e^n refers now to one component of the error only. We choose to keep the same notation for the sake of simplicity.

We recall here that the equation is assume to be solved in an explicit manner because of some constraints in computer resources. It therefore makes sense to compute the error using inexpensive error estimators. We also choose here to employ the explicit residual method. As before, let r_K and $j_{\gamma,K}$ denote the interior and boundary residuals for each element K of the partition:

$$r_K = -\ \frac{u_h^n - u_h^{n-1}}{\Delta t} + \kappa \Delta u_h^{n-1}$$

and

$$j_{\gamma,K} = \begin{cases} \kappa(\mathbf{n} \cdot \nabla u_h^{n-1})_K, & \text{if } \gamma \in \partial K \cap \partial \Omega,\\ \dfrac{\kappa}{2}(\mathbf{n} \cdot \nabla u_h^{n-1})_K + \dfrac{\kappa}{2}(\mathbf{n} \cdot \nabla u_h^{n-1})_L, & \text{if } \gamma \in \partial K \backslash \partial \Omega,\end{cases}$$

where the subscript L stands for the neighbor element L sharing the edge γ with K. It follows that the residual $\mathcal{R}_h$ can be split into elementwise contributions as:

$$\mathcal{R}_h(v) = \Delta t \sum_{K \in \mathcal{P}_h} \left(\int_K r_K v \,\mathrm{d}x - \int_{\partial K} j_{\gamma,K} v \,\mathrm{d}s \right). \tag{63}$$

Now, let $\mathcal{I}_h$ denote an arbitrary operator mapping elements of V into elements of V^h. Observing that the residual $\mathcal{R}_h$ vanishes on V^h, i.e. $\mathcal{R}_h(v) = 0, \forall\, v \in V^h$ (orthogonality relation), it follows that $\mathcal{R}_h(v) = \mathcal{R}_h(v - \mathcal{I}_h v)$ so that:

$$\mathcal{R}_h(v) = \Delta t \sum_{K \in \mathcal{P}_h} \left(\int_K r_K (v - \mathcal{I}_h v)\, \mathrm{d}x - \int_{\partial K} j_{\gamma,K} (v - \mathcal{I}_h v)\, \mathrm{d}s \right). \tag{64}$$

Then, using the Cauchy-Schwartz inequality, classical interpolation estimates (see [4]), and the discrete Schwartz inequality, we are able to derive the following inequalities:

$$\begin{aligned} \mathcal{R}_h(v) &\leq \Delta t \sum_{K \in \mathcal{P}_h} \|r_K\|_{0,K} \|v - \mathcal{I}_h v\|_{0,K} + \|j_{\gamma,K}\|_{0,\partial K} \|v - \mathcal{I}_h v\|_{0,\partial K} \\ &\leq C \Delta t \sum_{K \in \mathcal{P}_h} \|r_K\|_{0,K} \|v\|_{0,\widetilde{K}} + h_K^{1/2} \|j_{\gamma,K}\|_{0,\partial K} \|v\|_{1,\widetilde{K}} \\ &\leq C \Delta t \left\{ \sum_{K \in \mathcal{P}_h} \|r_K\|_{0,K}^2 + h_K \|j_{\gamma,K}\|_{0,\partial K}^2 \right\}^{1/2} \times \sqrt{\|v\|_0^2 + \|v\|_1^2} \end{aligned}$$

where C is a positive constant independent of h, and $\widetilde{K}$ a patch of elements surrounding one of the vertices of the element K. Observing from (62) that:

$$\|e\|_0^2 = (e, e) = \mathcal{R}_h(e), \tag{65}$$

and replacing v by e^n in the above inequality, we obtain:

$$\|e^n\|_0^2 \leq C \Delta t \left\{ \sum_{K \in \mathcal{P}_h} \|r_K\|_{0,K}^2 + h_K \|j_{\gamma,K}\|_{0,\partial K}^2 \right\}^{1/2} \times \sqrt{\|e^n\|_0^2 + \|e^n\|_1^2}.$$

At this stage, in order to derive an L^2 estimate of the error, we need to be able to bound the term $\|e^n\|_1$ by $\|e^n\|_0$. Unfortunately, such a result is not necessarily true. However, one remarks that the time discretization scheme is explicit in time and therefore should be conditionally stable. We prove in the next subsection that such a stability condition holds.

Stability condition of the time discretization scheme. We show here that the explicit forward Euler scheme is conditionally stable.

Theorem 3. *Let $\psi \in V$ be a perturbation to a solution of (59). The perturbation ψ decays in time, in other words, the explicit forward Euler scheme is stable, if:*

$$|\psi|_1^2 < \frac{2}{\Delta t\, \kappa} \|\psi\|_0^2. \tag{66}$$

Proof. Let $\delta^n \in V$ be a perturbation to the solution u^n of the time discretized equation, that is, $\delta^n = \hat{u}^n - u^n$, where $\hat{u}^n$ is also a solution of the discrete heat equation. Such a perturbation results for example in a perturbation in the initial condition. Since both $\hat{u}^n$ and u^n satisfy the discrete heat equation, the evolution of the perturbation is then governed by:

$$(\delta^n, v) = (\delta^{n-1}, v) - \Delta t\ \kappa(\nabla\delta^{n-1}, \nabla v), \qquad \forall\, v \in V.$$

Suppose that the perturbation can be written in the form: $\delta^n = A^n\psi$, where $A^n \in \mathbb{R}$ and $\psi \in V$, $\psi \neq 0$. Only A^n may change along the time integration. Then

$$(A^n\psi, v) = (A^{n-1}\psi, v) - \Delta t\ \kappa(\nabla A^{n-1}\psi, \nabla v), \qquad \forall\, v \in V,$$

and replacing v by ψ in the equation above, we get:

$$A^n(\psi,\psi) = A^{n-1}(\psi,\psi) - \Delta t\ \kappa A^{n-1}(\nabla\psi, \nabla\psi),$$

which yields:

$$\frac{A^n}{A^{n-1}} = 1 - \Delta t\ \kappa \frac{(\nabla\psi, \nabla\psi)}{(\psi,\psi)}.$$

The perturbation δ^n is stable, i.e. the scheme is stable, if $|A^n/A^{n-1}| < 1$, or equivalently if

$$\left|1 - \Delta t\ \kappa \frac{(\nabla\psi, \nabla\psi)}{(\psi,\psi)}\right| < 1.$$

which yields the desired stability condition. □

Remark 10. The implicit backward Euler scheme is unconditionally stable since the same analysis yields:

$$\frac{A^n}{A^{n-1}} = \frac{1}{1 + \Delta t\ \kappa\, |\psi|_1^2 / ||\psi||_0^2},$$

and the ratio A^n/A^{n-1} is necessarily less than one. □

Explicit error estimator in the L^2 norm. We now apply the above result to derive an explicit *a posteriori* error estimator for the heat equation. Suppose that the error e^n satisfies the above stability condition in the sense that:

$$|e^n|_1^2 < \frac{2}{\Delta t\ \kappa}||e^n||_0^2. \tag{67}$$

Then, we get

$$\sqrt{||e^n||_0^2 + ||e^n||_1^2} = \sqrt{2||e^n||_0^2 + |e^n|_1^2} \leq \sqrt{2\,||e^n||_0^2 + \left(\frac{2}{\Delta t\ \kappa}\right)||e^n||_0^2}$$
$$\leq \sqrt{2\left(1 + \frac{1}{\Delta t\ \kappa}\right)} \times ||e^n||_0,$$

so that

$$\|e^n\|_0 \leq C\Delta t \sqrt{2\left(1+\frac{1}{\Delta t\ \kappa}\right)} \left\{ \sum_{K\in\mathcal{P}_h} \|r_K\|_{0,K}^2 + h_K \|j_{\gamma,K}\|_{0,\partial K}^2 \right\}^{1/2} . \quad (68)$$

This provides us with an explicit error estimator in the L^2 norm. Denoting the elementwise contributions η_K where:

$$\eta_K = \left\{ \|r_K\|_{0,K}^2 + h_K \|j_{\gamma,K}\|_{0,\partial K}^2 \right\}^{1/2}, \quad (69)$$

the global error estimator η would be then defined as:

$$\eta = \Delta t \sqrt{2\left(1+\frac{1}{\Delta t\ \kappa}\right)} \left\{ \sum_{K\in\mathcal{P}_h} \eta_K^2 \right\}. \quad (70)$$

In the case in which the space V^h consists of continuous piecewise linear basis functions, the term Δu_h^n in the interior residual r_K naturally vanishes, so the predominant term in the indicator above should be given by $j_{\gamma,K}$. We also remark that the elementwise contributions η_K can naturally be used to construct a refinement indicator.

3.3 The Time-Dependent Navier-Stokes Equations

Preliminaries. The following work on *a posteriori* error estimation for the Navier-Stokes equations is partly based on our recent work [52]. The flow of a viscous fluid in an open bounded domain Ω in $\mathbb{R}^d$, $d \geq 2$, with boundary $\partial\Omega$, is modeled by the Navier-Stokes equations,

$$\left.\begin{aligned} \partial_t \boldsymbol{u} + \boldsymbol{u}\cdot\nabla\boldsymbol{u} - \mathrm{Re}^{-1}\Delta\boldsymbol{u} + \nabla p = \boldsymbol{f} \\ \nabla\cdot\boldsymbol{u} = 0 \end{aligned}\right\} \quad \text{in } \Omega\times(0,T) \quad (71)$$

with boundary and initial conditions:

$$\begin{aligned} \boldsymbol{u} &= \boldsymbol{0}, && \text{on } \partial\Omega, \\ \boldsymbol{u} &= \boldsymbol{u}_0, && \text{at } t = 0, \end{aligned} \quad (72)$$

where $\boldsymbol{u} = \boldsymbol{u}(\boldsymbol{x},t)$ and $p = p(\boldsymbol{x},t)$ respectively denote the velocity vector and the pressure at $\boldsymbol{x} \in \Omega$ and $t \in (0,T)$, Re is the Reynolds number, and $\boldsymbol{f} = \boldsymbol{f}(\boldsymbol{x},t)$ is a prescribed body force. The spaces of trial velocities and pressures are given by:

$$\begin{aligned} V &= [H_0^1(\Omega)]^d, \\ Q &= \{q \in L^2(\Omega) : \int_\Omega q\,\mathrm{d}x = 0\}, \end{aligned}$$

and let $\|v\|_0$ and $|v|_1$ denote the usual $[L^2]^d$ norm and $[H_0^1]^d$ norm of functions in V and $\|q\|_0$ the L^2 norm of functions in Q.

A weak formulation of the Navier-Stokes problem consists, for $\boldsymbol{f}$ and $\boldsymbol{u}_0$ given, in finding $\boldsymbol{u} \in L^2(0,T;V)$ and $p \in L^2(0,T;Q)$, such that $\boldsymbol{u} = \boldsymbol{u}_0$ at $t = 0$, and, for almost every time $t \in (0,T)$:

$$\begin{aligned} (\partial_t \boldsymbol{u}, \boldsymbol{v}) + C(\boldsymbol{u}, \boldsymbol{u}, \boldsymbol{v}) + \mathrm{Re}^{-1} A(\boldsymbol{u}, \boldsymbol{v}) + B(\boldsymbol{v}, p) &= F(\boldsymbol{v}), \quad \forall\, \boldsymbol{v} \in V, \\ B(\boldsymbol{u}, q) &= 0, \qquad \forall\, q \in Q, \end{aligned} \tag{73}$$

where the forms A, B, C and F were defined in Section 1. We assume here that there exists a unique solution $(\boldsymbol{u}, p)$ of the Navier-Stokes problem. This has been actually proven for $d = 2$ under mild conditions on $\boldsymbol{f}$ and $\boldsymbol{u}_0$, but has not yet been proved or disproved for $d = 3$ (see Temam [59]).

Discretizations. The above equations are discretized in space using *h-p* finite element spaces $V^h \subset V$ and $Q^h \subset Q$. The finite element solution $(\boldsymbol{u}_h, p_h) \in L^2(0,T;V^h) \times L^2(0,T;Q^h)$ is then governed by the semi-discrete system of equations, for a.e. $t \in (0,T)$:

$$\begin{aligned} (\partial_t \boldsymbol{u}_h, \boldsymbol{v}) + C(\boldsymbol{u}_h, \boldsymbol{u}_h, \boldsymbol{v}) + \mathrm{Re}^{-1} A(\boldsymbol{u}_h, \boldsymbol{v}) + B(\boldsymbol{v}, p_h) &= F(\boldsymbol{v}), \quad \forall\, \boldsymbol{v} \in V^h, \\ B(\boldsymbol{u}_h, q) &= 0, \qquad \forall\, q \in Q^h, \end{aligned}$$

A fully discrete problem of the Navier-Stokes equations is obtained using for instance the Adams-Bashforth Crank-Nicholson time discretization scheme (ABCN). Given some initial conditions $\boldsymbol{u}_h^0$ and $\boldsymbol{u}_h^1$, the discrete solution $(\boldsymbol{u}_h^n, p_h^n) \in V^h \times Q^h$ at $t^n = n\Delta t$, $n = 2, 3, \ldots, N_T$ is advanced in time by solving

$$\begin{aligned} \frac{1}{\Delta t}(\boldsymbol{u}_h^n, \boldsymbol{v}) + \frac{1}{2}\mathrm{Re}^{-1} A(\boldsymbol{u}_h^n, \boldsymbol{v}) + B(\boldsymbol{v}, p_h^n) &= \mathcal{F}_h(\boldsymbol{v}), \quad \forall\, \boldsymbol{v} \in V^h, \\ B(\boldsymbol{u}_h^n, q) &= 0, \qquad \forall\, q \in Q^h, \end{aligned} \tag{74}$$

where $\mathcal{F}_h$ is a functional which depends on the solutions at the previous discrete times, i.e.

$$\begin{aligned} \mathcal{F}_h(\boldsymbol{v}) = \frac{1}{2}F^n(\boldsymbol{v}) + \frac{1}{2}F^{n-1}(\boldsymbol{v}) + \frac{1}{\Delta t}(\boldsymbol{u}_h^{n-1}, \boldsymbol{v}) - \frac{1}{2}\mathrm{Re}^{-1} A(\boldsymbol{u}_h^{n-1}, \boldsymbol{v}) \\ - \frac{3}{2}C(\boldsymbol{u}_h^{n-1}, \boldsymbol{u}_h^{n-1}, \boldsymbol{v}) + \frac{1}{2}C(\boldsymbol{u}_h^{n-2}, \boldsymbol{u}_h^{n-2}, \boldsymbol{v}). \end{aligned} \tag{75}$$

The ABCN scheme is implicit with respect to the linear terms and explicit for the nonlinear convective terms, and is therefore conditionally stable. The resulting system of equations is therefore linear. Note that $\boldsymbol{u}_h^0$ is taken equal to a projection of $\boldsymbol{u}_0$ on the finite element space V^h and that $\boldsymbol{u}_h^1$ is obtained using a first-order scheme.

Numerical Stability. The Navier-Stokes equations (71) or (73) define a dissipative autonomous dynamical system with control parameter Re. Assuming that the flow evolves to an attractor at any Reynolds number (see i.e. [23,24]), the type of attractor is expected to change as Re increased. However, the fully-discrete system of equations (74) defines a new dynamical system with h and Δt as additional control parameters. The fundamental issue which we want to address here is whether the attractors of the Navier-Stokes equations and of the discrete system are the same at a given Re. In other words, can we expect the bifurcations to occur at the same critical Reynolds numbers? In [29,26,53], numerical simulations of the Navier-Stokes equations were performed with the objective of showing the existence of strange attractors and of unveiling a possible *route to chaos*. Our motivation here is rather to show how and why attractors are sensitive to the mesh discretization in order to design a strategy for the control of the error and stability. We emphasize that here the primary goal of the following numerical experiments is not to obtain accurate solutions but rather to analyze their stability in critical situations.

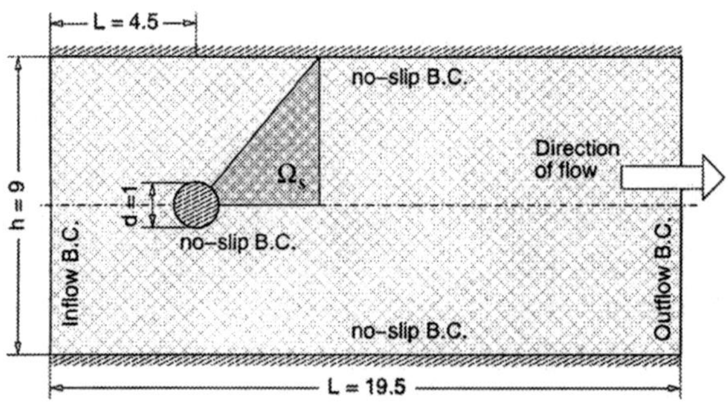

Fig. 20. Channel flow past a cylinder: geometry and boundary conditions.

The numerical study is performed on channel flows past a cylinder since these undergo the first bifurcation from steady-state to periodic vortex shedding flows at low Reynolds numbers. The Reynolds number is defined here as Re $= U_c d/\nu$, where d is the diameter of the cylinder and U_c the inflow velocity at the centerline of the channel. Boundary conditions are prescribed as shown in Fig. 20, and $\boldsymbol{u}_0$ is set to zero everywhere in Ω. Four meshes using the Taylor-Hood element are constructed. In view of approaching the attractors, i.e. the long-time behavior of flows, simulations are run during the time interval $[0, 500]$ and the timestep is chosen sufficiently small ($\Delta t = 0.02$) so that the stability of the ABCN scheme is satisfied and the numerical errors due to the discretization in time are minimized. To identify the type of attractors, we extract from the output data *time series signals* based on the kinetic energy computed in the "triangular" subregion Ω_s as shown in

Fig. 20. The time series signals are then postprocessed into *time delay reconstruction diagrams* to classify the attractors (see Table 4). We observe that the attractor depends on the spatial discretization as expected (at fixed Δt and Re). For instance, for Re = 75, the flow evolves to a steady-state on the coarsest mesh, and to a periodic-state on the other meshes. On the other hand, for Re = 200, the flow is still periodic on the two finer meshes while it has changed to a more complex state on the other two. Finally, the numerical flow on the coarsest mesh does not reach any attractors at Re = 250 and 300 as the solution "blows up" after a finite time.

Table 4. Types of attractor versus the Reynolds number and the number of mesh elements: (·) fixed point, (O) periodic orbit with on fundamental, (X) other, (-) no attractor.

	Reynolds Number							
# Elements	1	50	75	100	150	200	250	300
112	·	·	·	O	X	X	-	-
160	·	·	O	O	X	X	X	X
192	·	·	O	O	O	O	O	X
262	·	·	O	O	O	O	O	X

In the next set of experiments, we study the influence of the choice V^h and Q^h on the flow stability. We take advantage of the *h-p* data structure of our code to independently increase the spectral orders of the velocity and pressure (while preserving the *inf-sup* condition (8)), and run the experiments on the coarsest of the meshes used previously. Let $\varrho = (\varrho_u, \varrho_p)$ denote the pair of polynomial degrees ϱ_u and ϱ_p for the velocity and the pressure respectively (we use here the notation ϱ for the polynomial degree to avoid any confusion with the pressure denoted by p). We then consider $\varrho = (3,1)$ and $\varrho = (3,2)$.

We display in Fig. 21 the time delay reconstruction diagrams determined from the signals based on the kinetic energy computed in Ω_s. At Re = 75, in the case $\varrho = (2,1)$, the solution evolves towards a steady-state. When the dimension of V^h is increased, taking $\varrho = (3,1)$, we observe a periodic state, but with frequency $f = 0.095$. When enriching Q^h using $\varrho = (3,2)$, the new periodic signal has frequency $f = 0.145$, which is a value that compares better with the one obtained on the finest mesh. At Re = 200, the signal looks chaotic for $\varrho = (2,1)$. When we employ the discretization $\varrho = (3,1)$, we obtain a periodic signal with $f = 0.084$. On the other hand, the periodic signal obtained with $\varrho = (3,2)$ has frequency $f = 0.173$. Finally, the solution at Re = 250 blows up at even earlier time when using $\varrho = (3,1)$ instead of $\varrho = (2,1)$. A stable solution is finally obtained when the dimension of Q^h is eventually increased to $\varrho = (3,2)$ and is periodic with $f = 0.171$. We also observe that the amplitude of the signals varies from one discretization to the other.

In conclusion, these experiments reveal that the solutions have better stability properties when the dimension of Q^h is increased. Indeed, the larger $\dim Q^h$, the better the incompressibility constraint is enforced. In other words, the divergence of $\boldsymbol{u}_h$, which is not necessarily zero in classical finite element approximations, can be viewed as a source of unphysical perturbations artificially generated by the spatial discretization. It suggests that this quantity should be carefully controlled in order to control the numerical stability and hence the overall accuracy of the solutions. The control of the error essentially relies on *a posteriori* error estimation and mesh adaptation during the flow evolution. We shall see in the next section that the quantity $\nabla \cdot \boldsymbol{u}_h$ naturally appears in the error estimates.

Error and Residual. The numerical errors at time t^n, due to the spatial discretization, are defined as the pair $(\boldsymbol{e}_u^n, e_p^n)$ in $V \times Q$, $\boldsymbol{e}_u^n = \boldsymbol{u}^n - \boldsymbol{u}_h^n$ and $e_p^n = p^n - p_h^n$, and are governed by, for all $n = 2, 3, \ldots, N_T$,

$$
\begin{aligned}
\frac{1}{\Delta t}(\boldsymbol{e}_u^n, \boldsymbol{v}) + \frac{1}{2}\mathrm{Re}^{-1} A(\boldsymbol{e}_u^n, \boldsymbol{v}) + B(\boldsymbol{v}, e_p^n) &= \mathcal{R}_h^m(\boldsymbol{v}) + \mathcal{R}_e(\boldsymbol{v}), &\quad \forall\, \boldsymbol{v} \in V, \\
B(\boldsymbol{e}_u^n, q) &= \mathcal{R}_h^c(q), &\quad \forall\, q \in Q,
\end{aligned}
\tag{76}
$$

where $\mathcal{R}_h^m$ and $\mathcal{R}_h^c$ are simply the residual functionals with respect to the momentum and continuity equations:

$$
\begin{aligned}
\mathcal{R}_h^m(\boldsymbol{v}) = {} & \frac{1}{2} F^n(\boldsymbol{v}) + \frac{1}{2} F^{n-1}(\boldsymbol{v}) \\
& - \frac{1}{\Delta t}(\boldsymbol{u}_h^n - \boldsymbol{u}_h^{n-1}, \boldsymbol{v}) - \frac{1}{2}\mathrm{Re}^{-1} A(\boldsymbol{u}_h^n + \boldsymbol{u}_h^{n-1}, \boldsymbol{v}) \\
& - B(\boldsymbol{v}, p_h^n) - \frac{3}{2} C(\boldsymbol{u}_h^{n-1}, \boldsymbol{u}_h^{n-1}, \boldsymbol{v}) + \frac{1}{2} C(\boldsymbol{u}_h^{n-2}, \boldsymbol{u}_h^{n-2}, \boldsymbol{v}), \\
\mathcal{R}_h^c(q) = {} & - B(\boldsymbol{u}_h^n, q).
\end{aligned}
$$

The residual $\mathcal{R}_e$ combines all the terms involving $\boldsymbol{e}_u^{n-1}$ and $\boldsymbol{e}_u^{n-2}$, and thus, governs the error accumulation after long periods of time. It is conjectured that the accumulated errors do not reflect the mesh inadequacy since errors created at a given time are later convected away from their source. Consequently, for the purpose of mesh adaptation, we only focus on the errors which are local in time by setting $\mathcal{R}_e = 0$.

The key idea in our approach is to characterize the respective influence of $\mathcal{R}_h^m$ and $\mathcal{R}_h^c$ on the errors $(\boldsymbol{e}_u^n, e_p^n)$. Let $(\boldsymbol{e}_u^m, e_p^m)$ and $(\boldsymbol{e}_u^c, e_p^c)$ denote the errors governed by $\mathcal{R}_h^m$ and $\mathcal{R}_h^c$, respectively. Thanks to the linearity of the ABCN scheme, these pairs of errors can be treated independently, i.e.

$$
\begin{aligned}
\boldsymbol{e}_u &= \boldsymbol{e}_u^m + \boldsymbol{e}_u^c, \\
e_p &= e_p^m + e_p^c.
\end{aligned}
$$

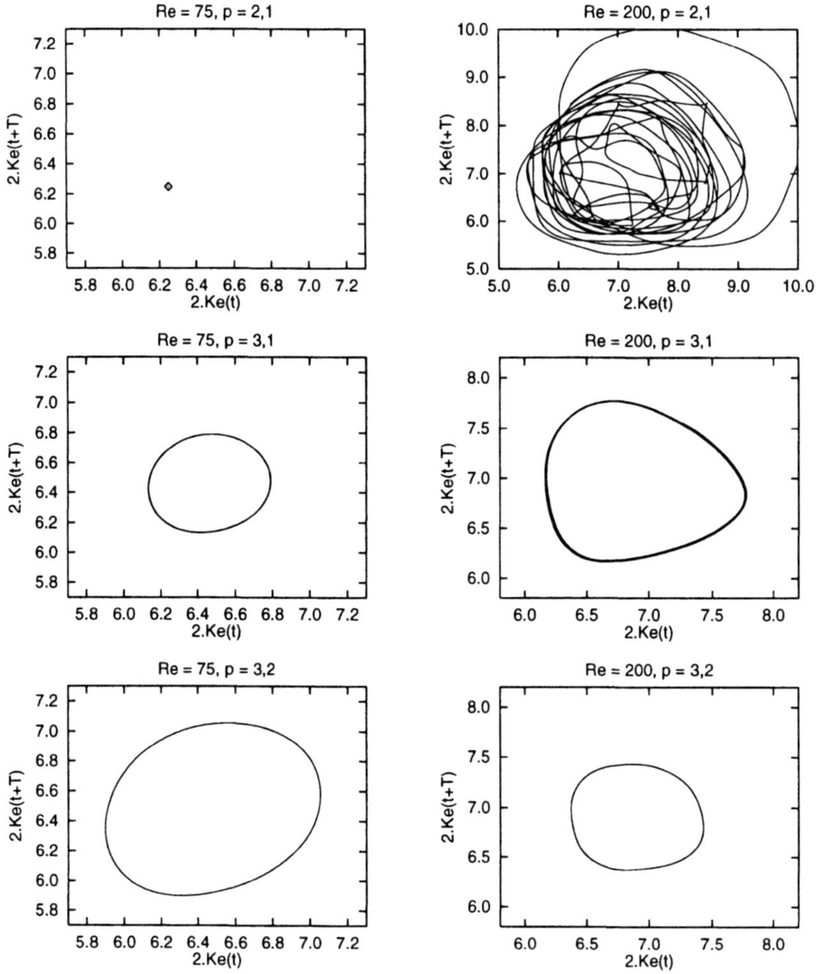

Fig. 21. Time delay reconstruction diagrams.

By the superposition principle, the governing equations for the error component $(\boldsymbol{e}_u^m, e_p^m) \in V \times Q$ are deduced from (76) as:

$$\begin{aligned} \frac{1}{\Delta t}(\boldsymbol{e}_u^m, \boldsymbol{v}) + \frac{1}{2}\mathrm{Re}^{-1}A(\boldsymbol{e}_u^m, \boldsymbol{v}) + B(\boldsymbol{v}, e_p^m) &= \mathcal{R}_h^m(\boldsymbol{v}), \quad \forall\, \boldsymbol{v} \in V, \\ B(\boldsymbol{e}_u^m, q) &= 0, \qquad\quad \forall\, q \in Q. \end{aligned} \tag{77}$$

Defining the norms of the error and residual as:

$$\|v\|_{\Delta t,\mathrm{Re}} = \sqrt{\frac{1}{\Delta t}(v,v) + \frac{1}{2}\mathrm{Re}^{-1}A(v,v)},$$

$$\|\mathcal{R}_h^m\|_* = \sup_{v\in V\setminus\{0\}} \frac{|\mathcal{R}_h^m(v)|}{\|v\|_{\Delta t,\mathrm{Re}}}$$

we can show that

$$\|e_u^m\|_{\Delta t,\mathrm{Re}} \le \|\mathcal{R}_h^m\|_*. \tag{78}$$

Indeed, substituting e_p^m for q in the continuity equation (77), we obtain that $B(e_u^m, e_p^m) = 0$. It follows, after replacing v by e_u^m in the momentum equation, that:

$$\frac{1}{\Delta t}(e_u^m, e_u^m) + \frac{1}{2}\mathrm{Re}^{-1}A(e_u^m, e_u^m) = \mathcal{R}_h^m(e_u^m).$$

Using the definition of the norm and the Cauchy-Schwartz inequality yield:

$$\|e_u^m\|^2_{\Delta t,\mathrm{Re}} \le \|\mathcal{R}_h^m\|_* \|e_u^m\|_{\Delta t,\mathrm{Re}},$$

which is precisely (78).

In a similar manner, the errors $(e_u^c, e_p^c) \in V \times Q$ are governed by:

$$\begin{aligned} \frac{1}{\Delta t}(e_u^c, v) + \frac{1}{2}\mathrm{Re}^{-1}A(e_u^c, v) + B(v, e_p^c) &= 0, \qquad \forall\, v \in V, \\ B(e_u^c, q) &= \mathcal{R}_h^c(q), \ \forall\, q \in Q. \end{aligned} \tag{79}$$

This time however, we cannot directly relate e_u^c to $\mathcal{R}_h^c$. On the other hand, decomposing the vector e_u^c into the divergence-free component e_u^d and its orthogonal vector e_u^p (with respect to the inner product $A(\cdot,\cdot)$), we can show (cf [42,50]) that

$$\beta\, |e_u^p|_1 \le \|\mathcal{R}_h^c\|_* \le |e_u^p|_1 \tag{80}$$

where β is the constant in the *inf-sup* condition for the bilinear form $B(\cdot,\cdot)$ (compare with the discrete condition (8)). We recall that the norm of $\mathcal{R}_h^c$ is defined as:

$$\|\mathcal{R}_h^c\|_* = \sup_{q\in Q\setminus\{0\}} \frac{|\mathcal{R}_h^c(q)|}{\|q\|_0}. \tag{81}$$

Such a result shows that the residual $\mathcal{R}_h^c$ provides a measure equivalent to the norm of a component of the velocity.

Global estimators. We have shown in the previous subsection how to relate the residual to specific norms of the error in the velocity. In other words, we can estimate the norm of the residuals to obtain some error estimators. The norm of the residual $\mathcal{R}_h^c$ can exactly computed as:

$$\|\mathcal{R}_h^c\|_* = \|\nabla \cdot u_h^n\|_0. \tag{82}$$

In other words, we have the error estimator η^c such that

$$\eta^c = \|\nabla \cdot \boldsymbol{u}_h^n\|_0 = \sqrt{\sum_{K \in \mathcal{P}_h} \|\nabla \cdot \boldsymbol{u}_h^n\|_{0,K}^2}. \tag{83}$$

On the other hand, the residual $\mathcal{R}_h^m$ cannot be computed exactly. However it can be estimated using either explicit or implicit residual methods as shown in Section 1. Such an estimate will be denoted η^m.

Adaptation strategy. The residuals represent the source terms for the errors. We then use in our adaptation strategy the information provided by the norms of the residuals in order to control the errors. The global quantities η^m and η^c are decomposed into the elementwise quantities $\eta_{m,K}$ and $\eta_{c,K}$ on each element K. Whenever the relative errors, associated with $|\boldsymbol{e}_u^p|_1$ and $\|\boldsymbol{e}_u^m\|_{\Delta t,\mathrm{Re}}$ and estimated by η^m and η^c, become too large with respect to preset tolerances C_c^{tol} and C_m^{tol}, the elements for which the contributions $\eta_{m,K}$ or $\eta_{c,K}$ are the largest are refined. In this procedure, we always check η^c first, since failing to enforce the divergence-free constraint is supposedly held responsible for stronger instabilities in numerical flows, as seen in the previous section.

Numerical Examples. The adaptation strategy is now applied to the simulation of a flow in an oblique cavity (obtained from the square cavity by rotating the vertical walls by an angle of 20 degrees, while keeping the area of the cavity to one; see Fig. 22). The velocities are zero everywhere on $\partial\Omega$ except on the top part where $\boldsymbol{u} = (1,0)$ and the fluid is initially at rest. Note that the solution may not exist in $(H^1(\Omega))^2$ due to the strong singularity in the upper right corner of the cavity. This test case has actually been studied in [22] for various Reynolds numbers and the authors have identified three possible steady-state solutions in the range $1449.7 \leq \mathrm{Re} \leq 2002.8$. The Reynolds number is set here to $\mathrm{Re} = 2000$ to see whether the adaptive strategy allows the computed flow field to approach the most stable state. We select the timestep $\Delta t = 0.002$ and time range $[0, 40]$. The tolerances are chosen as $C_m^{\mathrm{tol}} = 0.016$ and $C_c^{\mathrm{tol}} = 0.015$. The contour plots of the velocities, pressure and vorticity are shown in Fig. 23 along with the initial and adapted meshes. We also measure the square-root of the kinetic energy K_e and find $K_e = 0.219$, which is close to the value of the most stable of the three possible steady-states (see Fig. 5.2(a) in [22]).

3.4 Estimation of Errors due to the Discretization in Time

So far, we have not considered the error due to the discretization in time. Once the error in space is controlled, one approach would be to advance the solution performing two half-timestep integrations. The error could be

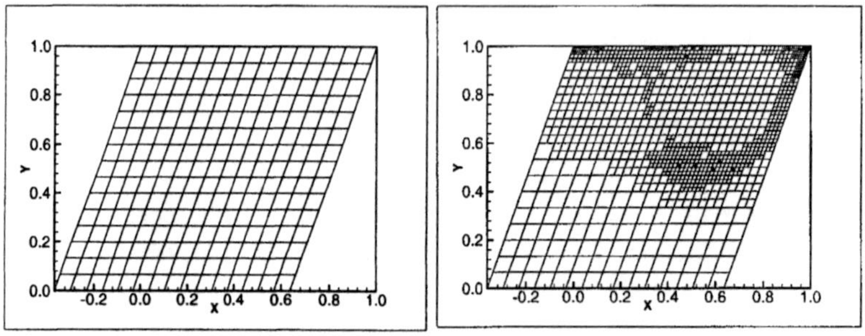

Fig. 22. Initial mesh and adapted mesh.

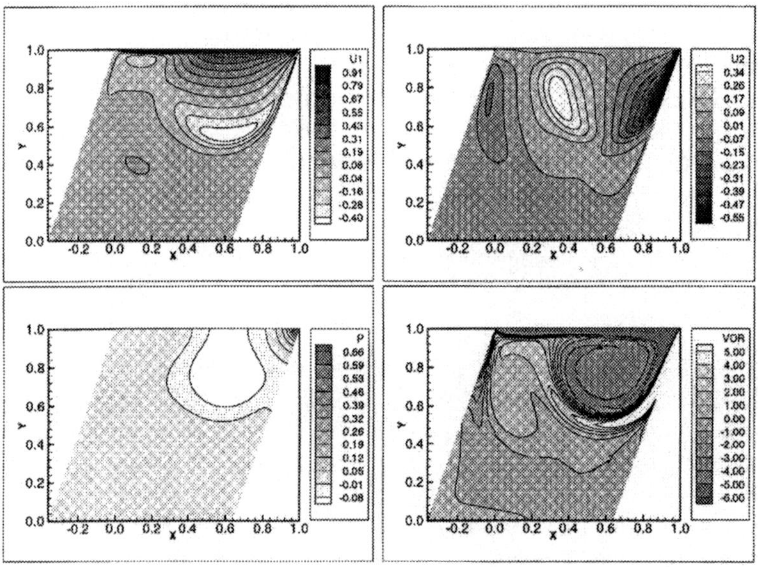

Fig. 23. Shown clockwise: velocities u_1 and u_2, vorticity, pressure.

then estimated by comparing this solution to the solution obtained when performing just one timestep integration. This is an ongoing research effort.

3.5 Estimation of Accumulated Errors

The natural approach to estimate the accumulated errors is to consider the dual problem associated with the time-dependent primal problem. The dual problem is also a time-dependent problem that needs to be solved backward in time from the time t down to $t = 0$. At this stage of our research, we do not believe that such a method is practical when simulating complex phenomena such as, for example, those modeled by the Navier-Stokes equations. The dual problem can also be used to compute error estimates in quantity of interest. In this case, it seems possible to approximate the time-dependent dual problem by a steady-state problem. We refer the reader to the work of Johnson *et al.* [34], Estep *et al.* [25], Becker and Rannacher [17] (and references therein) for complete details.

Acknowledgements The first author would like to thank T.J. Barth and H. Deconinck for their kind invitation to give lectures at NASA Ames and VKI during the NATO RTO/NASA/VKI Lectures Series on error estimation and solution adaptive discretization in computational fluid dynamics. The support of part of this research work through the Office of Naval Research under grant N00014-95-0401 is gratefully acknowledged.

References

1. AINSWORTH, M., KELLY, D. W., SLOAN, I. H., AND WANG, S. L. Post-processing with computable error bounds for the finite element approximation of a nonlinear heat conduction problem. *IMA J. Numer. Anal. 17* (1997), 547–561.
2. AINSWORTH, M., AND ODEN, J. T. A unified approach to a posteriori error estimation using element residual methods. *Numer. Math. 65*, 1 (1993), 23–50.
3. AINSWORTH, M., AND ODEN, J. T. A posteriori error estimation in finite element analysis. *Comp. Meth. in Appl. Mech. and Eng. 142*, 1–2 (1997), 1–88.
4. AINSWORTH, M., AND ODEN, J. T. *A Posteriori Error Estimation in Finite Element Analysis.* John Wiley & Sons, 2000.
5. BABUŠKA, I., AND MILLER, A. The post-processing approach in the finite element method - Part 1: Calculation of displacements, stresses and other higher derivatives of the displacements. *Int. J. Numer. Methods Eng. 20* (1984), 1085–1109.
6. BABUŠKA, I., AND MILLER, A. The post-processing approach in the finite element method - Part 2: The calculation of stress intensity factors. *Int. J. Numer. Methods Eng. 20* (1984), 1111–1129.
7. BABUŠKA, I., AND MILLER, A. The post-processing approach in the finite element method - Part 3: A-posteriori error estimates and adaptive mesh selection. *Int. J. Numer. Methods Eng. 20* (1984), 2311–2324.
8. BABUŠKA, I., AND MILLER, A. A feedback finite element method with *a posteriori* error estimation Part 1. *Comp. Meth. in Appl. Mech. and Eng. 61* (1987), 1–40.
9. BABUŠKA, I., AND RHEINBOLDT, W. A posteriori error estimates for the finite element method. *Int. J. Numer. Methods Eng. 12* (1978), 1597–1615.
10. BABUŠKA, I., AND RHEINBOLDT, W. On the reliability and optimality of the finite element method. *Computers & Structures 10* (1979), 87–94.
11. BABUŠKA, I., AND STROUBOULIS, T. *The Finite Element Method and its reliability.* Oxford University Press, 2001.
12. BABUŠKA, I., STROUBOULIS, T., AND GANGARAJ, S. K. Guaranteed computable bounds for the exact error in the finite element solution Part I: One-dimensional model problem. *Comp. Meth. in Appl. Mech. and Eng. 176* (1999), 51–79.
13. BABUŠKA, I., STROUBOULIS, T., UPADHYAY, C. S., AND GANGARAJ, S. K. A posteriori estimation and adaptive control of the pollution error in the *h*-version of the finite element method. *Int. J. Numer. Methods Eng. 38* (1995), 4207–4235.

14. Babuška, I., Strouboulis, T., Upadhyay, C. S., Gangaraj, S. K., and Copps, K. Validation of a posteriori error estimators by numerical approach. *Int. J. Numer. Methods Eng. 37* (1994), 1073–1123.
15. Bank, R. E., and Weiser, A. Some a posteriori error estimators for elliptic partial differential equations. *Math. Comp. 44*, 170 (1985), 283–301.
16. Becker, R., and Rannacher, R. A feedback approach to error control in finite elements methods: Basic analysis and examples. *East-West J. Numer. Math. 4* (1996), 237–264.
17. Becker, R., and Rannacher, R. An optimal control approach to *a posteriori* error estimation in finite element methods. *Acta Numerica* (2001). To appear.
18. Bernardi, C., and Girault, V. A local regularisation operator for triangular and quadrilateral finite elements. *SIAM J. Numer. Anal. 35*, 5 (1998), 1893–1916.
19. Cao, T., Kelly, D. W., and Sloan, I. H. Local error bounds for post-processed finite element calculations. *Int. J. Numer. Methods Eng. 45* (1999), 1085–1098.
20. Carstensen, C., and Funken, S. A. Fully reliable localized error control in the FEM. *SIAM J. Sci. Comput. 21*, 4 (2000), 1465–1484.
21. Chilton, L., and Suri, M. On the selection of a locking-free *hp* element for elasticity problems. *Int. J. Numer. Methods Eng. 40* (1997), 2045–2062.
22. de Almeida, V. F., and Derby, J. J. Construction of solution curves for large two-dimensional problems of steady-state flows of incompressible fluids. *SIAM J. Sci. Comput. 22*, 1 (1999), 285–311.
23. Doering, C. R., and Gibbon, J. D. *Applied Analysis of the Navier-Stokes Equations.* Cambridge University Press, 1995.
24. Eckmann, J.-P., and Ruelle, D. Ergodic theory of chaos and strange attractors. *Review of Modern Physics 57* (1985), 617–656.
25. Estep, D. J., Larson, M. G., and Williams, R. D. *Estimating the Error of Numerical Solutions of Systems of Reaction-Difffusion Equations.* No. 696 in Memoirs of the American Mathematical Society. AMS, Providence, Rhode Island, 2000.
26. Fortin, A., Fortin, M., and Gervais, J. A numerical simulation of the transition to turbulence in a two-dimensional flow. *J. Comp. Phys. 70* (1987), 295–310.
27. Fragos, V. P., Psychoudaki, S. P., and Malamataris, N. A. Computer-aided analysis of flow past a surface-mounted obstacle. *Int. J. Numer. Methods Fluids 25* (1997), 495–512.
28. Gartland, Jr., E. C. Computable pointwise error bounds and the ritz method in one dimension. *SIAM J. Numer. Anal. 21*, 1 (1984), 84–100.
29. Ghia, K. N., Osswald, G. A., and Ghia, U. Simulation of self-induced unsteady motion in the near wake of a Joukowski airfoil. In *Supercomputers and fluid dynamics: proceedings of the First Nobeyama Workshop, September 3-6, 1985* (1986), K. Kuwahara, R. Mendez, and S. A. Orszag, Eds., no. 24 in Lecture Notes in Engineering, Springer-Verlag, pp. 118–132.
30. Girault, V., and Raviart, P.-A. *Finite Element Methods for Navier-Stokes Equations.* Springer-Verlag, 1986.
31. Gunzburger, M. D. *Finite element methods for viscous incompressible flows: a guide to theory, practice and algorithms.* Academic Press, 1989.

32. Huerta, A., Díez, P., Rodríguez-Ferran, A., and Pijaudier-Cabot, G. Error estimation and adaptive finite element analysis of softening solids. In *Advances in Adaptive Computational Methods in Mechanics* (1998), P. Ladevèze and J. T. Oden, Eds., Elsevier, Amsterdam, pp. 333–347.
33. Jin, H., and Prudhomme, S. A posteriori error estimation of finite element solutions of the Navier-Stokes equations. *Comp. Meth. in Appl. Mech. and Eng. 159* (1998), 19–48.
34. Johnson, R., Rannacher, R., and Boman, M. Numerics and hydrodynamic stability: Towards error control in CFD. *SIAM J. Numer. Anal. 32*, 4 (1995), 1058–1079.
35. Ladevèze, P., and Leguillon, D. Error estimate procedure in the finite element method and applications. *SIAM J. Numer. Anal. 20* (1983), 485–509.
36. Ladevèze, P., and Maunder, E. A. W. A general method for recovering equilibrating element tractions. *Comp. Meth. in Appl. Mech. and Eng. 137* (1996), 111–151.
37. Ladevèze, P., and Pelle, J.-P. *La maîtrise du calcul en mécanique linéaire et non linéaire.* Hermès, 2001.
38. Morin, P., Nochetto, R. H., and Siebert, K. G. Local problems on stars: A posteriori error estimators, convergence, and performance. *Website:* `http://www.math.umd.edu/~pmorin/` (2001).
39. Niu, Q., and Shephard, M. S. Superconvergent extraction techniques for finite element analysis. *Int. J. Numer. Methods Eng. 36* (1993), 811–836.
40. Niu, Q., and Shephard, M. S. Superconvergent boundary stress extraction and some experiments with adaptive pointwise error control. *Int. J. Numer. Methods Eng. 37* (1994), 877–891.
41. Oden, J. T., Demkowicz, L., Westermann, T., and Rachowicz, W. Toward a universal *h-p* adaptive finite element strategy. Part 2. A posteriori error estimates. *Comp. Meth. in Appl. Mech. and Eng. 77* (1989), 79–120.
42. Oden, J. T., and Prudhomme, S. A technique for a posteriori error estimation of h-p approximations of the Stokes equations. In *Advances in Adaptive Computational Methods in Mechanics* (1998), P. Ladevèze and J. T. Oden, Eds., Elsevier, Amsterdam, pp. 43–63.
43. Oden, J. T., and Prudhomme, S. New appoaches to error estimation and adaptivity for the Stokes and Oseen equations. *Int. J. Numer. Methods Fluids 31* (1999), 3–15.
44. Oden, J. T., and Prudhomme, S. Goal-oriented error estimation and adaptivity for the finite element method. *Computers Math. Applic. 41*, 5–6 (2001), 735–756.
45. Paraschivoiu, M., and Patera, A. T. A hierarchical duality approach to bounds for the outputs of partial differential equations. *Comp. Meth. in Appl. Mech. and Eng. 158* (1998), 389–407.
46. Paraschivoiu, M., and Patera, A. T. A posteriori bounds for linear functional outputs of Crouzeix-Raviart finite element discretizations of the incompressible Stokes problem. *Int. J. Numer. Methods Fluids 32*, 7 (2000), 823–849.
47. Paraschivoiu, M., Peraire, J., and Patera, A. T. A posteriori finite element bounds for linear-functional outputs of elliptic partial differential equations. *Comp. Meth. in Appl. Mech. and Eng. 150* (1997), 289–312.
48. Peraire, J., and Patera, A. T. Bounds for linear-functional outputs of coercive partial differential equations: local indicators and adaptive refine-

ment. In *Advances in Adaptive Computational Methods in Mechanics* (1998), P. Ladevèze and J. T. Oden, Eds., Elsevier, Amsterdam, pp. 199–215.
49. PRUDHOMME, S. Adaptive finite element simulation of a supersonic underexpanded jet. Master's thesis, University of Virginia, 1992.
50. PRUDHOMME, S. *Adaptive Control of Error and Stability of h-p Approximations of the Transient Navier-Stokes Equations.* PhD thesis, The University of Texas at Austin, 1999.
51. PRUDHOMME, S., AND ODEN, J. T. On goal-oriented error estimation for elliptic problems: Application to the control of pointwise errors. *Comp. Meth. in Appl. Mech. and Eng. 176* (1999), 313–331.
52. PRUDHOMME, S., AND ODEN, J. T. Numerical stability and error analysis for the incompressible Navier-Stokes equations. *Communications in Numerical Methods in Engineering* (Submitted).
53. PULLIAM, T. H., AND VASTANO, J. A. Transition to chaos in an open unforced 2D flow. *J. Comp. Phys. 105* (1993), 133–149.
54. SCHWAB, C. *p- and hp-Finite Element Methods: Theory and Applications in Solid and Fluid Mechanics.* Oxford University Press, New York, 1998.
55. STENBERG, R., AND SURI, M. Mixed *hp* finite element methods for problems in elasticity and Stokes flow. *Numer. Math. 72* (1996), 367–389.
56. STROUBOULIS, T., BABUŠKA, I., DATTA, D. K., COPPS, K., AND GANGARAJ, S. K. A posteriori estimation and adaptive control of the error in the quantity of interest. Part I: A posteriori estimation of the error in the von Mises stress and the stress intensity factor. *Comp. Meth. in Appl. Mech. and Eng. 181* (2000), 261–294.
57. STROUBOULIS, T., BABUŠKA, I., AND GANGARAJ, S. K. Guaranteed computable bounds for the exact error in the finite element solution – Part II: bounds for the energy norm of the error in two dimensions. *Int. J. Numer. Methods Eng. 47* (2000), 427–475.
58. SZABÓ, B., AND BABUŠKA, I. *Finite Element Analysis.* John Wiley & Sons, 1991.
59. TEMAM, R. *Navier-Stokes Equations: Theory and Numerical Analysis*, 2nd ed. North Holland, 1984.
60. VERFÜRTH, R. *A Review of A Posteriori Error Estimation and Adaptive Mesh-refinement Techniques.* Wiley-Teubner, Stuttgart, 1996.
61. WANG, S., SLOAN, I. H., AND KELLY, D. W. Computable error bounds for pointwise derivatives of a neumann problem. *IMA J. Numer. Anal. 18* (1998), 251–271.
62. YANG, J. D., KELLY, D. W., AND ISLES, J. D. A posteriori pointwise upper bound error estimates in the finite element method. *Int. J. Numer. Methods Eng. 36* (1993), 1279–1298.
63. ZIENKIEWICZ, O. C., AND ZHU, J. Z. A simple error estimator and adaptive procedure for practical engineering analysis. *Int. J. Numer. Methods Eng. 24* (1987), 337–357.
64. ZIENKIEWICZ, O. C., AND ZHU, J. Z. The superconvergent patch recovery and a posteriori error estimates. Part 1: the recovery technique. *Int. J. Numer. Methods Eng. 33* (1992), 1331–1364.
65. ZIENKIEWICZ, O. C., AND ZHU, J. Z. The superconvergent patch recovery and a posteriori error estimates. Part 2: error estimates and adaptivity. *Int. J. Numer. Methods Eng. 33* (1992), 1365–1382.

Adaptive Finite Element Approximation of Hyperbolic Problems

Endre Süli[1] and Paul Houston[2,*]

[1] University of Oxford,
Computing Laboratory,
Wolfson Building, Parks Road,
Oxford OX1 3QD, UK
email: endre@comlab.ox.ac.uk

[2] University of Leicester,
Department of Mathematics and Computer Science,
University Road,
Leicester LE1 7RH, UK
email: Paul.Houston@mcs.le.ac.uk

Abstract. We review some recent developments concerning the *a posteriori* error analysis of $h-$ and $hp-$version finite element approximations to hyperbolic problems. The error bounds stem from an error representation formula which equates the error in an output functional of interest to the inner product of the finite element residual with the solution of a dual (adjoint) problem whose data is the density function of the target functional. *Type I a posteriori* error bounds are derived which, unlike the cruder *Type II* bounds, retain the dual solution in the bound as a local weight-function. The relevance of Type I *a posteriori* bounds is argued by showing that the local size of the error in a hyperbolic problem may be only very weakly correlated to the local size of the residual; consequently, adaptive refinement algorithms based on the size of the local residual alone can be ineffective. The sharpness of Type I *a posteriori* error bounds is demonstrated on both structured and adaptively refined meshes.

Contents

* Paul Houston acknowledges the financial support of the EPSRC (GR/N24230).

1 Introduction

Over the last decade, tremendous progress has been made in the area of *a posteriori* error estimation and adaptive finite element approximation of partial differential equations (PDEs); for a review of some of the main developments in the subject we refer to the recent monographs [4,56,57], and the articles [14,8]. Despite a number of significant advances in the field, much of the research to date has focused on partial differential equations of elliptic type, while for hyperbolic partial differential equations the theory of *a posteriori* error estimation and adaptivity has remained in its infancy.

This state of affairs is perhaps not accidental: the mechanisms of error propagation in finite element approximations of hyperbolic PDEs are more complex and less well understood than in elliptic problems; a further hindrance is that the theory of well-posedness and regularity for hyperbolic PDEs is in a less satisfactory state than the corresponding theoretical framework for elliptic problems due, among other things, to the fact that unlike elliptic differential operators, first-order hyperbolic operators have no inherent smoothing properties in isotropic Sobolev spaces; finally, in engineering applications, finite element methods have been traditionally employed for the numerical approximation of elliptic boundary value problems and their use for the solution of hyperbolic PDEs is more recent and less widespread.

The aim of these notes is to provide a brief introduction to some recent advances in the field of *a posteriori* error estimation and adaptivity for hyperbolic PDEs. The focus of this work and its scope is, necessarily, influenced by our own interests in the subject and we do not attempt to give an exhaustive survey. For further developments in the field we refer to [11,24,29,52,53], and [25–28,30,35,40,42,54] for some more recent contributions. In order to keep the presentation simple, we shall concentrate on two popular finite element methods: the streamline–diffusion finite element method and the discontinuous Galerkin finite element method; nevertheless, the ideas developed here apply more generally, to other partial differential equations and to other finite element discretizations. The *a posteriori* error analyses for the two methods will be pursued within a single, general theoretical framework based on the idea of duality.

The relevance of duality arguments in *a posteriori* error estimation has been highlighted in the review articles [14] and [8] (see, also, [7,24,30,35,42,44–47,49,53,54]); concerning the use of duality arguments in post-processing and design, we refer to [17,18,20,21,48], and references therein. The key ingredient in duality-based error estimation is an auxiliary PDE problem, the dual problem, involving the formal adjoint of the partial differential operator under consideration. The data for the dual problem is the quantity of interest: in engineering applications, this is typically an output functional of the analytical solution such as a mean value, point value, boundary flux, for example, or, in fluid dynamics, it may be the pressure–drop between inflow and outflow, or the drag or lift coefficients of a body immersed into the fluid (see, [7,8,17,18,20,21,27,42,45–48], for example).

The relevance and generality of duality-based error estimation has been powerfully argued in the work of Johnson and his collaborators; see, for example, [14] for an excellent survey. The *a posteriori* error bounds resulting from this analysis involve the finite element residual which is obtained by inserting the computed finite element solution into the partial differential equation; the residual measures the extent to which the finite element approximation to the analytical solution fails to satisfy the underlying PDE. In the framework of [14], the error bounds are arrived at by exploiting Galerkin orthogonality (a fundamental property of all finite element methods expressing the fact that the residual is orthogonal to the finite element space), and strong stability (well-posedness/regularity in isotropic Sobolev norms of positive index) of the dual problem. In view of the remarks at the end of the second paragraph concerning the inherent lack of isotropic smoothing properties in hyperbolic problems, it is this latter step in the process of *a posteriori* error estimation that causes difficulties when dealing with hyperbolic PDEs: the underlying problems will be explained in more detail in the next section.

Stimulated by our findings, in this paper we revisit the problem of *a posteriori* error estimation and adopt a minimalistic variant of duality–based error estimation which avoids explicitly invoking strong stability of the dual problem through retention of the dual solution in the bound; the approach in question is stimulated by the work of Becker and Rannacher [7] on weighted-residual-based *a posteriori* error estimation; see also the recent review article [8] and references therein. As in [14] and [8], the starting point in our error analysis is an *error representation formula* which equates the error in the output functional of interest with the inner product over the computational domain of the finite element residual and the analytical solution to the dual problem. In our approach, the inner product over the computational domain contained in the error representation formula is decomposed as a sum of inner products over elements in the finite element partition, and hence the absolute value of the error in the output functional is bounded by the sum of absolute values of elemental inner products; this process is referred to as *localization*. However, unlike [14] and [8], we shall refrain from applying the Cauchy–Schwarz inequality to these elemental inner products; rather, we shall compute them numerically through replacing the analytical dual solution by a finite element approximation. If necessary, the error committed in the course of this additional approximation can be controlled through an auxiliary *a posteriori* error estimate. It will be shown that the proposed approach leads to sharp *a posteriori* error bounds, in the sense that the *effectivity index* (the ratio of the error bound and the true error) is typically very close to unity.

The paper is structured as follows. We begin, in Section 2, by outlining the fundamental problems associated with residual-based error estimation in the context of hyperbolic problems. We then continue, in Section 3, with a brief excursion into the theory of function spaces to introduce the notational conventions. In Section 4, we discuss some of the key model problems whose numerical approximations we shall consider. In this work, we shall concentrate on two established finite element methods for hyperbolic problems: the streamline–diffusion finite element method (SDFEM) and the discontinuous Galerkin finite element method (DGFEM). Thus, Section 5 is devoted to the formulation of the SDFEM which uses continuous piecewise polynomial functions to approximate the underlying analytical solution. For the sake of simplicity we shall confine ourselves to a scalar linear hyperbolic model problem. We shall be concerned with the derivation of *a posteriori* error bounds for the SDFEM within the framework of h–version finite element methods (i.e., with variable mesh size h, but fixed polynomial degree p, $p \geq 1$); in particular, we shall focus on situations when the quantity of interest is a linear functional of the analytical solution rather than the solution itself. The derivation of the *a posteriori* error bound is based on a duality argument which may involve either the formal adjoint of the partial differential operator or the formal adjoint of a perturbed partial differential operator which stems from the bilinear form of the stabilized finite element method. It will be shown in Section 5 that for stabilized finite element methods, such as the SDFEM, the second approach leads to sharper *a posteriori* error bounds. In Section 6, we shall consider the *hp*–version of the discontinuous Galerkin finite element method, based on the use of discontinuous piecewise polynomial approximations, which, in addition to local variation of the granularity h of the computational mesh, also allows local adjustment of the degree p of the approximating piecewise polynomial. We shall discuss the *a posteriori* error analysis of the *hp*–version of the discontinuous Galerkin finite element method and consider the implementation of our *a posteriori* error bounds into an *hp*–adaptive algorithm. For *hp*–adaptive finite element methods, a key question is to decide when to h–refine and when to p–refine. The decision concerning which of the two refinement strategies to adopt will be based on monitoring the local smoothness of the solutions to the primal and dual problems and will involve an algorithm for the estimation of their local Sobolev regularity indices. In Section 7, we conclude with a brief summary of the main results.

2 So, what is the problem, then?

In order to describe the fundamental problem in the simplest possible terms, let us suppose that $A : D(A) \subset \mathcal{H} \to \mathcal{H}$ is a linear differential operator defined on its domain $D(A)$ contained in a certain Hilbert space $\mathcal{H}$. Given $f \in \mathcal{H}$, suppose that $u \in D(A)$ solves

$$Au = f\,.$$

In this simple discussion we shall assume that any (homogeneous) boundary condition that u has to satisfy is imposed through the definition of $D(A)$.

Now, let us seek a numerical approximation u_h to u by means of a Galerkin method. Thereby, we consider a sequence of finite-dimensional linear subspaces $\{\mathcal{H}^h\}$ of $D(A)$ parametrized by $h \in (0,1]$, and look for $u_h \in \mathcal{H}^h$ such that

$$P_h A u_h = P_h f\,, \tag{1}$$

where P_h is the orthogonal projector in $\mathcal{H}$ onto $\mathcal{H}^h$. On defining the *residual* $R(u_h) = f - Au_h$, (1) can be restated in the following equivalent form:

$$P_h R(u_h) = P_h(f - Au_h) = 0\,;$$

this identity is usually referred to as the *Galerkin orthogonality property*, as it expresses the fact that $R(u_h)$ is orthogonal to $\mathcal{H}^h$ in $\mathcal{H}$.

Suppose, for example, that Ω, the computational domain, is a bounded open set in $\mathbb{R}^n$, $\mathcal{H} = L_2(\Omega)$ is the set of all real-valued square-integrable functions defined on Ω, and let ζ be a given nonnegative weight function in $L_2(\Omega)$. [A summary of some standard notation and a collection of basic results from the theory of function spaces will be provided in the next section.] Then, with $(\cdot,\cdot)$ denoting the inner product in $L_2(\Omega)$, $v \mapsto J(v) = (v,\zeta)$ is a bounded linear functional on $L_2(\Omega)$, and $J(u)$ represents the weighted mean value of u over Ω.

A typical practical question one might encounter is the following: given a certain tolerance $\mathtt{TOL} > 0$, find $u_h \in \mathcal{H}^h$ such that

$$|J(u) - J(u_h)| \le \mathtt{TOL}\,. \tag{2}$$

In the field of *a posteriori* error estimation a problem of this kind is approached by deriving a bound on $|J(u) - J(u_h)|$ in terms of the residual $R(u_h)$ of the form

$$|J(u) - J(u_h)| \le \mathtt{Computable\ expression}(R(u_h))\,, \tag{3}$$

where the right-hand side is a computable expression involving available data and the residual, $R(u_h) = f - Au_h$, which can be calculated *after* u_h has been computed; hence the terminology *a posteriori error bound.* Once such an *a posteriori* error bound is available, the task of satisfying (2) is reduced to finding $\mathcal{H}^h$ and $u_h \in \mathcal{H}^h$ such that

$$\texttt{Computable expression}(R(u_h)) \leq \texttt{TOL},$$

– an objective that is usually achieved through a feed-back process, by adaptive adjustment of $\mathcal{H}^h$. It is clear from this discussion that the fundamental theoretical problem, therefore, is to derive a bound of the kind (3).

A general approach to the derivation of *a posteriori* error bounds is based on the use of a *duality argument* involving the adjoint A^* of the linear operator A, defined on its domain $D(A^*) \subset \mathcal{H} = L_2(\Omega)$. For this purpose, we consider the *dual problem* of finding z in $D(A^*) \subset \mathcal{H}$ such that

$$A^* z = \zeta, \tag{4}$$

whose right-hand side is the given weight-function ζ. Thus the data for the dual problem is the density function (Riesz representer) of the linear functional of interest $J(\cdot)$. Hence,

$$\begin{aligned} J(u) - J(u_h) &= (u - u_h, \zeta) \\ &= (Au - Au_h, z) \\ &= (f - Au_h, z) \\ &\equiv (R(u_h), z)\,. \end{aligned} \tag{5}$$

As $P_h R(u_h) = 0$, it follows that

$$\begin{aligned} J(u) - J(u_h) &= (R(u_h), z) \\ &= (R(u_h), z - z_h) \qquad \forall z_h \in \mathcal{H}^h\,; \end{aligned} \tag{6}$$

this identity is referred to as the *error representation formula.* A possible choice of z_h may be $P_h z$, although there are other viable alternatives. For example, if (1) represents a finite element method, then z_h may be chosen as a finite element interpolant or quasi-interpolant of z. It should be borne in mind, however, that z is the analytical solution to the dual problem, and even though $R(u_h)$ will become 'computable' after u_h has been calculated, $z - z_h$ will not be available. Hence, the right-hand side of the error representation formula (6) is *not* computable.

One approach to obtaining a computable error bound from (6) (see, e.g., [14]) proceeds by attempting to eliminate $z - z_h$. Briefly, the argument is as follows. Suppose that $\mathcal{H}_s$ is a Hilbert subspace of $\mathcal{H}$ with norm $\|\cdot\|_{\mathcal{H}_s}$; here s is a positive index related to smoothness (for example, when $\mathcal{H} = L_2(\Omega)$, $\mathcal{H}_s$ might be the Sobolev space $H^s(\Omega) \subset L_2(\Omega)$ of index $s > 0$); assume that $z \in \mathcal{H}_s$ and that we have found a constant $C_{\mathrm{stab}} = C_{\mathrm{stab}}(\zeta)$ such that

$$\|z\|_{\mathcal{H}_s} \leq C_{\mathrm{stab}}\,.$$

Suppose also that there exists a positive constant C_{int} and $z_h \in \mathcal{H}^h$ such that

$$\|h^{-s}(z - z_h)\|_{\mathcal{H}} \leq C_{\text{int}} \|z\|_{\mathcal{H}_s} . \tag{7}$$

Then, on applying the Cauchy-Schwarz inequality in $\mathcal{H}$, we get

$$\begin{aligned} |J(u) - J(u_h)| &\leq \|h^s R(u_h)\|_{\mathcal{H}} \, \|h^{-s}(z - z_h)\|_{\mathcal{H}} \\ &\leq C_{\text{int}} \, \|h^s R(u_h)\|_{\mathcal{H}} \, \|z\|_{\mathcal{H}_s} \\ &\leq C_{\text{int}} \, C_{\text{stab}} \, \|h^s R(u_h)\|_{\mathcal{H}} . \end{aligned}$$

The constants C_{int} and C_{stab} are referred to as the *interpolation constant* and *strong stability constant*, respectively. Typical values of C_{int} are available from approximation theory, while C_{stab} has to be determined either analytically or computationally by studying the well-posedness of the dual problem (4) in the norm $\|\cdot\|_{\mathcal{H}_s}$. Thus we have established the following *a posteriori* error bound:

$$\boxed{|J(u) - J(u_h)| \leq C_{\text{int}} C_{\text{stab}} \|h^s R(u_h)\|_{\mathcal{H}}} \qquad s \geq 0 . \tag{8}$$

An error bound of this kind which does not directly involve the dual solution z will be referred to here as a Type II *a posteriori* error bound, cf. [35].

Let us pause for a moment at this point to reflect on some of the steps in our argument and examine the validity of the hypotheses we have made. Our first assumption concerned the existence of a Hilbert subspace $\mathcal{H}_s$ contained in $\mathcal{H}$ such that the dual solution z is contained in $\mathcal{H}_s$; in particular, when $\mathcal{H} = L_2(\Omega)$, it would be convenient to choose $\mathcal{H}_s$ as a Sobolev space $H^s(\Omega)$ of positive index $s > 0$ for it is in these spaces that approximation properties of the kind (7) are readily available. We shall consider two examples.

Example A. Suppose that $\Omega = (0,1)^2$, the unit square in the $x = (x_1, x_2)$ plane, and consider the following Dirichlet boundary value problem for Poisson's equation

$$\Delta u = 1 \quad \text{in } \Omega , \qquad u = 0 \quad \text{on } \partial\Omega .$$

Here the dual problem is

$$\Delta z = \zeta \quad \text{in } \Omega , \qquad z = 0 \quad \text{on } \partial\Omega .$$

As $\zeta \in L_2(\Omega)$, it follows from regularity theory for elliptic partial differential equations in polygonal domains that $z \in H^s(\Omega) \cap H^1_0(\Omega)$, $1 \leq s < 3$; thus we may select $\mathcal{H}_s = H^s(\Omega) \cap H^1_0(\Omega)$, $1 \leq s < 3$.

Example B. Suppose, again, that $\Omega = (0,1)^2$, the unit square in the $x = (x_1, x_2)$ plane, and consider the following boundary value problem for a scalar first-order hyperbolic equation for a function u of the variable $(x_1, x_2) \in \Omega$:

$$u_{x_1} = 1 \quad \text{in } \Omega, \qquad u = 0 \quad \text{on } \Gamma_- ,$$

where $\Gamma_- = \{(0, x_2) : 0 \le x_2 \le 1\}$. This time, the dual problem is

$$-z_{x_1} = \zeta \quad \text{in } \Omega\,, \qquad z = 0 \quad \text{on } \Gamma_+ = \{(1, x_2) : 0 \le x_2 \le 1\}.$$

Unlike Example A, here z need not belong to $H^s(\Omega)$ for any $s > 0$. This is easily seen by choosing ζ as a function of x_2 only; say, $\zeta(x_1, x_2) = \varphi(x_2)$ where φ belongs to $L_2(0,1)$ but does not belong to $H^s(0,1)$ for any $s > 0$ (e.g., $\varphi(x_2) = 1/(|x_2 - (1/2)|^{1/2} \log|x_2 - (1/2)|)$ belongs to $L_2(0,1)$ but not to $L_p(0,1)$ for $p > 2$; by the Sobolev embedding theorem, φ cannot belong to any $H^s(0,1)$, $s > 0$, for otherwise it would be an element of $L_p(0,1)$ for some $p > 2$, which is not the case). While this is a slightly extreme example, it nevertheless highlights the stark difference between the regularity properties of solutions to elliptic and hyperbolic boundary value problems.

We see that while in Example A our hypothesis about the existence of a Hilbert subspace $\mathcal{H}_s$, $s > 0$, of $\mathcal{H} = L_2(\Omega)$ containing the dual solution z can be satisfied, for Example B this hypothesis has failed. Although our examples are trivial, the observed problem is generic, and is due to the fact that unlike elliptic partial differential operators, first-order hyperbolic operators have no inherent smoothing properties in isotropic Sobolev spaces.

In fact, our examples allow us to gain further insight into the difficulties. Let us select a point $\hat{x}$ in $\Omega = (0,1)^2$ and suppose that the aim of the computation is now to ensure that

$$|u(\hat{x}) - u_h(\hat{x})| \le \mathtt{TOL}\,,$$

where u is the solution to the boundary value problem from either Example A or B.

Example A′. For the elliptic boundary value problem from Example A, the appropriate dual problem associated with $J(u) = u(\hat{x})$ is

$$\Delta z = \delta(x - \hat{x}) \qquad \text{for } x \in \Omega\,, \qquad z = 0 \quad \text{on } \partial\Omega\,;$$

i.e., z is the classical Green's function. It is well known that z is a smooth function in $\Omega \setminus \{\hat{x}\}$ and it exhibits a logarithmic singularity near $\hat{x}$. A simple calculation reveals that $z \in H^s(\Omega)$ for $s \in [0,1)$.

Example B′. For the hyperbolic boundary value problem from Example B, the appropriate dual problem associated with $J(u) = u(\hat{x})$ is now

$$-z_{x_1} = \delta(x - \hat{x}) \qquad \text{for } x \in \Omega\,, \qquad z = 0 \quad \text{on } \Gamma_+.$$

Hence,

$$z(x_1, x_2) = (1 - H(x_1 - \hat{x}_1))\,\delta(x_2 - \hat{x}_2)\,.$$

Clearly, z is not a locally integrable function but a measure supported on the line-segment $x_2 = \hat{x}_2$, $0 \le x_1 \le \hat{x}_1$; in particular, z does not belong

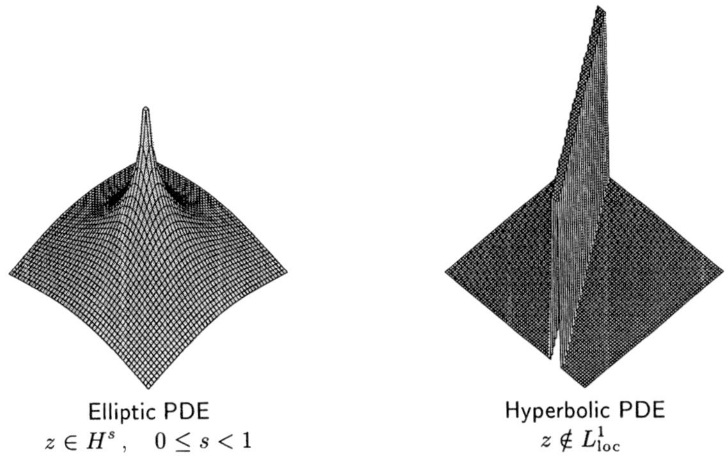

Fig. 1. Green's function for an elliptic (left) and a hyperbolic problem (right)

to any Sobolev space on Ω of index $s \geq 0$. Incidentally, had we chosen to consider the more general scalar linear first-order hyperbolic differential equation $a_1 u_{x_1} + a_2 u_{x_2} = 1$, where $a_1 > 0$ and $a_2 \geq 0$ are smooth functions of $x = (x_1, x_2)$, a similar conclusion would have resulted except that z would have been concentrated on a piece of the characteristic curve that passes through the point $(\hat{x}_1, \hat{x}_2)$. Thus, while the Green's function for an elliptic boundary value problem on Ω has singularity at a single point within Ω and decays away from that point, for a scalar linear hyperbolic boundary value problem the Green's function (or, *Green's measure*, to be more precise) has nonlocal singularity concentrated along a characteristic and does not, in general, decay along this characteristic, cf. Fig. 1, for example, for a sketch of z.

Remark 1. Strictly speaking, Examples A′ and B′ do not fit into the general theoretical framework we had presented given that $\zeta = \delta(\cdot - \hat{x})$ is not an element of $\mathcal{H} = L_2(\Omega)$, so (4) is meaningless for this choice of ζ. However, the effects which we observed in Examples A′ and B′ can be emulated within the theory presented above by approximating, for $\varepsilon > 0$ small, $J(u) = u(\hat{x})$ by $J_\varepsilon(u) = (u, \zeta_\varepsilon(\cdot - \hat{x}))$ and $\delta(\cdot - \hat{x})$ with $\zeta_\varepsilon(\cdot - \hat{x})$, where $\zeta_\varepsilon(x) = \varepsilon^{-2}\zeta(x/\varepsilon)$ and ζ is a nonnegative function whose support is contained in the unit circle centered at the origin and such that the integral of ζ over its support is equal to 1.

If now, with the observations from Examples A′ and B′ in mind, we return to (5) we conclude, at least formally[1], that for an elliptic operator the size

[1] The argument can be made rigorous by local mollification of $\delta(\cdot - \hat{x})$ as indicated in the previous Remark.

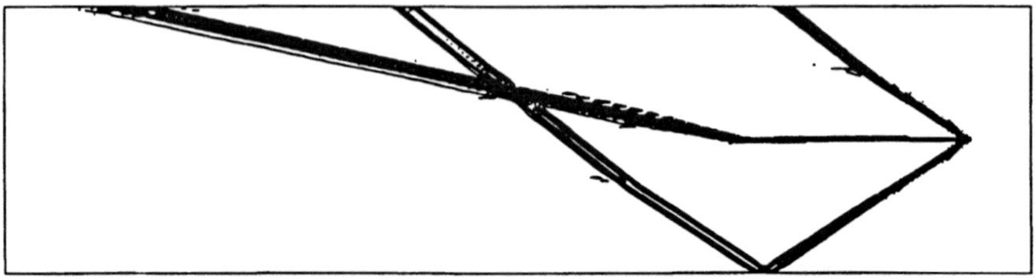

Fig. 2. Green's function for the linearized adjoint compressible Euler equations of gas dynamics

of the error $(u - u_h)$ at $\hat{x}$ will be strongly dependent on the size of $R(u_h)$ in the neighborhood of the point $\hat{x}$ and, given that z decays away from $\hat{x}$, values of $R(u_h)$ at more distant points will have less influence on $(u-u_h)(\hat{x})$. On the other hand, for a hyperbolic operator, the singularity of z is nonlocal and z does not, in general, decay along the characteristic curve on which the singularity is concentrated, so the size of $(u - u_h)$ at $\hat{x}$ can be strongly influenced by values of $R(u_h)$ at points that are quite distant from $\hat{x}$. This is further demonstrated in Fig. 2 where we show the Green's function for the linearized adjoint compressible Euler equations of gas dynamics in the case of a shock reflecting off a solid wall, cf. Kröner [39], p. 401.

In order to give an experimental illustration of the issue raised in the last paragraph, we note that, letting $e = u - u_h$ denote the *global error*, the residual and the error are related via

$$Ae = R(u_h)\,.$$

Now suppose that A is a scalar first-order linear hyperbolic operator. Following [29], we decompose the error $e_\kappa = e|_\kappa$ on an open subset κ of the computational domain Ω as

$$e_\kappa = e_\kappa^{\text{trans}} + e_\kappa^{\text{cell}}\,.$$

We shall refer to e_κ^{cell} as the *cell error* and e_κ^{trans} will be called the *transmitted error*; they are defined as the respective solutions to the following problems:

$$\begin{aligned} Ae_\kappa^{\text{cell}} &= R(u_h) \quad \text{on } \kappa; \qquad & e_\kappa^{\text{cell}} &= 0 \quad \text{on the inflow boundary of } \kappa\,,\\ Ae_\kappa^{\text{trans}} &= 0 \quad \text{on } \kappa; \qquad & e_\kappa^{\text{trans}} &= e_\kappa \quad \text{on the inflow boundary of } \kappa\,. \end{aligned}$$

The cell error e_κ^{cell} is the part of the error $e_\kappa = e|_\kappa$ that is governed by the residual $R(u_h)|_\kappa$ and is not influenced by pollution effects from outside κ. On the other hand, the transmitted error e_κ^{trans} is independent of the residual on κ and is driven completely by the error that has been created outside κ and transmitted into κ through the inflow boundary of κ.

Table 1. Correlation coefficients under mesh refinement

Mesh	Elements	$\rho(e, e^{\mathrm{trans}})$	$\rho(e, e^{\mathrm{cell}})$	$\rho(e, R(u_h))$
17×17	512	0.99	0.31	0.34
33×33	2048	1.00	0.42	0.43
65×65	8192	1.00	0.44	0.44
129×129	32768	1.00	0.45	0.45
257×257	131072	1.00	0.44	0.44

Let us consider the model scalar hyperbolic problem $a_1 u_{x_1} + a_2 u_{x_2} = 0$, with $\mathbf{a} = (a_1, a_2) = (1/10 + \sin(\pi x_2), 2)$, on the unit square $\Omega = (0,1)^2$ in the (x_1, x_2)–plane, with inflow boundary condition

$$u(x_1, 0) = \begin{cases} \mathrm{e}^{-300(0.4-x_1)^4}, & \text{for } 0.0 \le x_1 \le 0.4, \\ 1, & \text{for } 0.4 \le x_1 \le 0.6, \\ 0, & \text{for } 0.6 < x_1 \le 1.0, \end{cases}$$

with the compatible boundary condition $u(0, x_2) = \mathrm{e}^{-300(0.4)^4}$ $(= \mathrm{e}^{-7.68})$ along $x_1 = 0$, $0 \le x_2 \le 1$. This hyperbolic problem has been solved numerically using the streamline-diffusion finite element method on a sequence of uniform meshes consisting of triangular elements κ. The streamline-diffusion parameter was selected according to the standard formula: $\delta|_\kappa = 0.5h/|\bar{\mathbf{a}}_\kappa|$, where $\bar{\mathbf{a}}_\kappa$ denotes the average of $\mathbf{a}$ on element κ and $|\bar{\mathbf{a}}_\kappa|$ is the Euclidean norm of $\bar{\mathbf{a}}_\kappa$. Since the exact solution u can be calculated using the method of characteristics, we may compute $\|e_\kappa\|_{L_2(\kappa)}$, $\|e_\kappa^{\mathrm{cell}}\|_{L_2(\kappa)}$ and $\|e_\kappa^{\mathrm{trans}}\|_{L_2(\kappa)}$ on each triangle κ in the mesh, thus obtaining three sets of data. Our aim is now to compute the coefficients of correlation[2] $\rho(e, e^{trans})$, $\rho(e, e^{cell})$, $\rho(e, R(u_h))$ between the data set $\{\|e_\kappa\|_{L_2(\kappa)}\}_\kappa$ and each of the three data sets $\{\|e_\kappa^{\mathrm{trans}}\|_{L_2(\kappa)}\}_\kappa$, $\{\|e_\kappa^{\mathrm{cell}}\|_{L_2(\kappa)}\}_\kappa$ and $\{\|R(u_h)|_\kappa\|_{L_2(\kappa)}\}_\kappa$, respectively.

In Table 1 we display the correlation coefficients $\rho(e, e^{trans})$, $\rho(e, e^{cell})$ and $\rho(e, R(u_h))$ for these data sets, on a sequence of uniform structured triangular meshes: in each case the mesh is constructed from a uniform $N \times N$ mesh by connecting the bottom–left corner of each mesh square with its top–right corner. We observe that there is a very strong correlation between the error

---

2

$$\rho(X, Y) = \frac{\mathrm{cov}(X, Y)}{\sqrt{\mathrm{cov}(X, X)}\ \sqrt{\mathrm{cov}(Y, Y)}},$$

where the covariance of the data sets X and Y is defined by

$$\mathrm{cov}(X, Y) = \mathrm{E}([X - \mathrm{E}(X)]\,[Y - \mathrm{E}(Y)]);$$

here $\mathrm{E}(X)$ signifies the arithmetic mean of the data set X (i.e., the expectation of X, assuming that X is uniformly distributed).

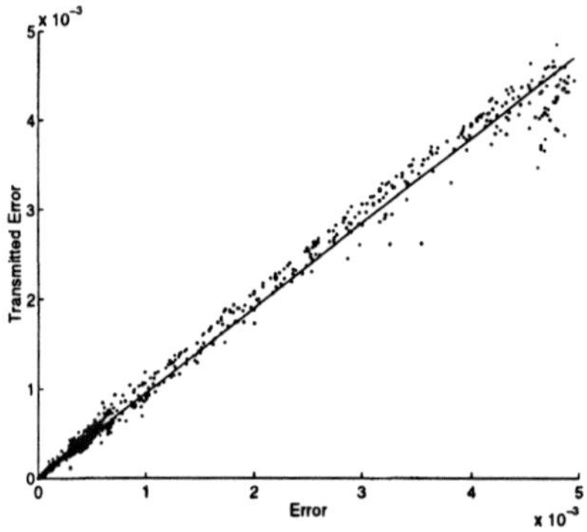

Transmitted Error vs. Error

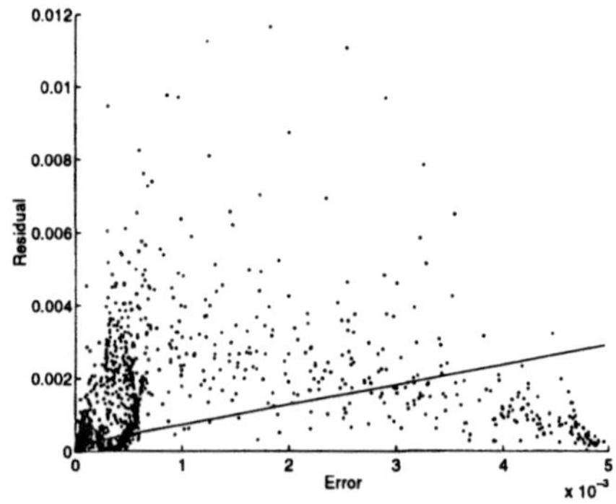

Residual vs. Error

Fig. 3. Scatter plots for the data-set-pairs $(\|e_\kappa\|_{L_2(\kappa)}, \|e_\kappa^{\mathrm{trans}}\|_{L_2(\kappa)})_\kappa$ (left) and $(\|e_\kappa\|_{L_2(\kappa)}, \|R(u_h)|_\kappa\|_{L_2(\kappa)})_\kappa$ (right). Each dot in the left scatter plot corresponds to a triangle κ in the mesh. In the plot on the left, the first co-ordinate of a point is $\|e_\kappa\|_{L_2(\kappa)}$ and the second co-ordinate is $\|e_\kappa^{\mathrm{trans}}\|_{L_2(\kappa)}$; similarly, in the right-hand figure, the first co-ordinate of a point in the scatter plot is $\|e_\kappa\|_{L_2(\kappa)}$ and the second co-ordinate is $\|R(u_h)|_\kappa\|_{L_2(\kappa)}$

and the transmitted error, while there is only very weak correlation between the error and the cell error, as well as between the error and the residual.

This is further confirmed in the scatter diagrams presented in Fig. 3; here we plot each of $\|e_\kappa^{\mathrm{trans}}\|_{L_2(\kappa)}$ and $\|R(u_h)|_\kappa\|_{L_2(\kappa)}$ against the error $\|e_\kappa\|_{L_2(\kappa)}$ as κ ranges through the triangulation of the computational domain, together with the linear regression line, on the third mesh with 8192 elements. In particular, we see from the right-hand figure that there is a cluster of points lying on the horizontal axis; i.e., as the error increases, the residual remains small in magnitude. Clearly, this has serious implications in the context of mesh design; an adaptive algorithm based on refining elements κ where the *local* residual is large may fail to refine all the elements where $\|e_\kappa\|_{L_2(\kappa)}$ is large, hence leading to poor resolution of localized structures in the solution.

One can draw a number conclusions from this discussion. First, the dual solution measures the sensitivity of the error in the quantity of interest to local perturbations in the residual. Second, the smoothness and decay properties of the dual solution in a hyperbolic problem may be completely different than in an elliptic problem; indeed, the lack of decay of the dual solution to a hyperbolic problem induces nonlocal dependence between the error and the residual; hence the local size of the residual may give little or no information about the local size of the error. From the point of view of sharpness of the *a posteriori* error bound, it therefore appears beneficial to retain the dual solution in the bound as a local weight function. In Sections 5 and 6, we shall revisit the error representation formula (6) that gave rise to the Type II *a posteriori* error bound (8), and attempt to derive a new bound which still includes z (or, rather, a numerical approximation to z); such bounds will be referred to henceforth as Type I *a posteriori* error bounds, cf. [35]. Before doing so, however, we shall make a brief diversion in order to introduce some basic results from function space theory and formulate the mathematical models of fluid dynamics whose discretizations we shall consider.

3 Elements of function space theory

The aim of this section is to provide a brief overview of some elementary results from the theory of function spaces and to introduce the notation which will be used throughout. For proofs and further technical details on classical function spaces the reader is referred to the monograph of Adams [1], for example.

Spaces of continuous functions. Let $\mathbb{N}$ denote the set of all nonnegative integers. An n–tuple $\alpha = (\alpha_1, \ldots, \alpha_n)$ in $\mathbb{N}^n$ will be referred to as a *multi-index*; the nonnegative integer $|\alpha| = |\alpha_1| + \ldots + |\alpha_n|$ is called the length of the multi-index α. We define $\partial^\alpha = \partial_1^{\alpha_1} \ldots \partial_n^{\alpha_n}$, where $\partial_j = \partial/\partial x_j$ for $j = 1, \ldots, n$.

Suppose that Ω is an open set in $\mathbb{R}^n$. For $k \in \mathbb{N}$, we denote by $C^k(\Omega)$ the set of all continuous real-valued functions u defined on Ω such that $\partial^\alpha u$ is continuous on Ω for every multi-index α, $|\alpha| \leq k$. When $k = 0$, we shall write $C(\Omega)$ in place of $C^0(\Omega)$. For $k = \infty$, $C^\infty(\Omega)$ will denote the intersection $\bigcap_{k \geq 0} C^k(\Omega)$.

We shall also require spaces of functions defined over the closure $\bar{\Omega}$ of an open set $\Omega \subset \mathbb{R}^n$. For $k \in \mathbb{N}$, $C^k(\bar{\Omega})$ will signify the set of all $u \in C^k(\Omega)$ such that $\partial^\alpha u$ can be continuously extended from Ω onto $\bar{\Omega}$ for every multi-index α, $|\alpha| \leq k$. Further, we define $C^\infty(\bar{\Omega})$ as the intersection $\bigcap_{k \geq 0} C^k(\bar{\Omega})$. The notation $C^0(\bar{\Omega})$ is abbreviated to $C(\bar{\Omega})$.

Assuming that Ω is a *bounded* open set in $\mathbb{R}^n$ and $k \in \mathbb{N}$, the linear space $C^k(\bar{\Omega})$ is a Banach space equipped with the norm

$$\|u\|_{C^k(\bar{\Omega})} = \max_{|\alpha| \leq k} \sup_{x \in \Omega} |\partial^\alpha u(x)| .$$

For $k \in \mathbb{N}$ we denote by $C^{k,1}(\bar{\Omega})$ the set of all $u \in C^k(\bar{\Omega})$ such that the quantity

$$|u|_{C^{k,1}(\bar{\Omega})} = \max_{|\alpha| = k} \sup_{x \neq y,\ x,y \in \Omega} \frac{|\partial^\alpha u(x) - \partial^\alpha u(y)|}{|x - y|}$$

is finite. $C^{k,1}(\bar{\Omega})$ is a Banach space with the norm

$$\|u\|_{C^{k,1}(\bar{\Omega})} = \|u\|_{C^k(\bar{\Omega})} + |u|_{C^{k,1}(\bar{\Omega})} .$$

Clearly, $C^{k+1}(\bar{\Omega}) \subset C^{k,1}(\bar{\Omega})$. When u belongs to $C^{0,1}(\bar{\Omega})$, it is said to be *Lipschitz continuous* on $\bar{\Omega}$.

The *support*, $\operatorname{supp} u$, of a continuous function u defined on an open set Ω is the closure in Ω of the set $\{x \in \Omega \ : \ u(x) \neq 0\}$; in other words, $\operatorname{supp} u$ is the smallest closed subset of Ω such that $u = 0$ on $\Omega \setminus \operatorname{supp} u$. For $k = 0, 1, \ldots, \infty$, $C_0^k(\Omega)$ denotes the set of all $u \in C^k(\Omega)$ whose support is a bounded (and, by definition, closed) subset of Ω.

Spaces of integrable functions. For $p \geq 1$ and an open set $\Omega \subset \mathbb{R}^n$, $L_p(\Omega)$ will denote the set of all real-valued Lebesgue measurable functions u defined on Ω such that $|u|^p$ is integrable on Ω with respect to the Lebesgue measure $\mathrm{d}x = \mathrm{d}x_1 \ldots \mathrm{d}x_n$; it is implicitly assumed that any two functions which are equal almost everywhere (i.e., equal, except maybe on a set of zero Lebesgue measure) are identified. $L_p(\Omega)$ is a Banach space equipped with the norm

$$\|u\|_{L_p(\Omega)} = \left(\int_\Omega |u(x)|^p \, \mathrm{d}x \right)^{1/p} .$$

When $p = 2$, $L_2(\Omega)$ is a Hilbert space with the inner product

$$(u, v) = \int_\Omega u(x)\, v(x) \, \mathrm{d}x \, .$$

$L_\infty(\Omega)$ denotes the set of all real-valued Lebesgue measurable functions u defined on Ω such that $|u|$ has finite essential supremum; the essential supremum of $|u|$ is defined as the infimum of the set of all positive real numbers M such that $|u| \leq M$ almost everywhere on Ω. Again, any two functions that are equal almost everywhere on Ω are identified. $L_\infty(\Omega)$ is a Banach space with norm

$$\|u\|_{L_\infty(\Omega)} = \text{ess.sup}_{x \in \Omega} |u(x)| \, .$$

Hölder's Inequality. Let $u \in L_p(\Omega)$ and $v \in L_q(\Omega)$, where $1/p + 1/q = 1$, $1 \leq p, q \leq \infty$. Then $uv \in L_1(\Omega)$ and

$$\left| \int_\Omega u(x)\, v(x) \, \mathrm{d}x \right| \leq \|u\|_{L_p(\Omega)} \|v\|_{L_q(\Omega)} \, .$$

In the special case when $p = q = 2$, this inequality is referred to as the *Cauchy-Schwarz Inequality*.

Sobolev spaces. Given that Ω is an open set in $\mathbb{R}^n$, k a non-negative integer and $1 \leq p \leq \infty$, we define the *Sobolev space*

$$W_p^k(\Omega) = \{u \in L_p(\Omega) \, : \, \partial^\alpha u \in L_p(\Omega) \, , \quad |\alpha| \leq k\} \, ,$$

and equip it with the *Sobolev norm* defined by

$$\|u\|_{W_p^k(\Omega)} = \begin{cases} \left(\sum_{|\alpha| \leq k} \|\partial^\alpha u\|_{L_p(\Omega)}^p \right)^{1/p} , & \text{if } 1 \leq p < \infty \, , \\ \|u\|_{W_\infty^k(\Omega)} = \max_{|\alpha| \leq k} \|\partial^\alpha u\|_{L_\infty(\Omega)} \, , & \text{if } p = \infty \, . \end{cases}$$

The associated *Sobolev seminorm* is defined by

$$|u|_{W_p^k(\Omega)} = \begin{cases} \left(\sum_{|\alpha|=k} \|\partial^\alpha u\|^p_{L_p(\Omega)}\right)^{1/p}, & \text{if } 1 \le p < \infty\,, \\ \max_{|\alpha|=k} \|\partial^\alpha u\|_{L_\infty(\Omega)}\,, & \text{if } p = \infty\,. \end{cases}$$

In these definitions the derivatives are to be understood in the sense of distributions. The Sobolev space $W_p^k(\Omega)$ is a Banach space with the norm $\|\cdot\|_{W_p^k(\Omega)}$, $1 \le p \le \infty$, $k \ge 0$. Specifically, for $p = 2$, the normed linear space $W_2^k(\Omega)$ is a Hilbert space with the inner product

$$(u, v)_{W_2^k(\Omega)} = \sum_{|\alpha|\le k} (\partial^\alpha u, \partial^\alpha v)\,,$$

where $(\cdot,\cdot)$ denotes the inner product in $L_2(\Omega)$.

Finer smoothness properties of integrable functions can be detected by considering fractional-order Sobolev spaces. Given that s is a positive real number, $s \notin \mathbb{N}$, let us write $s = m + \sigma$, where $0 < \sigma < 1$ and $m = [s]$ is the integer part of s. The fractional-order Sobolev space $W_p^s(\Omega)$, $1 \le p < \infty$, is the set of all $u \in W_p^m(\Omega)$ such that

$$|u|_{W_p^s(\Omega)} = \left\{ \sum_{|\alpha|=m} \int_\Omega \int_\Omega \frac{|D^\alpha u(x) - D^\alpha u(y)|^p}{|x-y|^{n+\sigma p}}\, \mathrm{d}x\, \mathrm{d}y \right\}^{1/p} < \infty\,,$$

with the usual modification when $p = \infty$. The fractional-order Sobolev norm of index s is defined by

$$\|u\|_{W_p^s(\Omega)} = \begin{cases} \left\{\|u\|^p_{W_p^m(\Omega)} + |u|^p_{W_p^s(\Omega)}\right\}^{1/p}, & \text{if } 1 \le p < \infty\,, \\ \|u\|_{W_\infty^m(\Omega)} + |u|_{W_\infty^s(\Omega)}\,, & \text{if } p = \infty\,. \end{cases}$$

The fractional-order Sobolev space $W_p^s(\Omega)$ is a Banach space with this norm.

The function spaces defined above will be helpful for characterizing the smoothness properties of a solution to a partial differential equation. In physical applications, however, a PDE is rarely considered in isolation, but is coupled with side–conditions which typically take the form of boundary conditions. The PDE in tandem with boundary conditions is referred to as a *boundary value problem.* In order to guarantee that the boundary condition is mathematically meaningful, it is necessary to ensure that the boundary is sufficiently smooth (in the sense of the following definition, for example).

Definition 1. Suppose that Ω is an open set in $\mathbb{R}^n$. The boundary $\partial\Omega$ of Ω is said to be Lipschitz continuous if, for every $x \in \partial\Omega$, there is an open set

$\mathcal{O} \subset \mathbb{R}^n$ with $x \in \mathcal{O}$ and a local orthogonal coordinate system with coordinate $\zeta = (\zeta_1, \ldots, \zeta_n) \equiv (\zeta', \zeta_n)$ and $a \in \mathbb{R}^n$, such that

$$\mathcal{O} = \{\zeta \, : \, -a_j < \zeta_j < a_j, \quad 1 \le j \le n\},$$

and there is a Lipschitz continuous function φ defined on

$$\mathcal{O}' = \{\zeta' \in \mathbb{R}^{n-1} \, : \, -a_j < \zeta_j < a_j, \quad 1 \le j \le n-1\},$$

with

$$|\varphi(\zeta')| \le a_n/2, \qquad \zeta' \in \mathcal{O}',$$

$$\Omega \cap \mathcal{O} = \{\zeta : \zeta_n < \varphi(\zeta'), \; \zeta' \in \mathcal{O}'\} \text{ and } \partial\Omega \cap \mathcal{O} = \{\zeta : \zeta_n = \varphi(\zeta'), \; \zeta' \in \mathcal{O}'\}.$$

A bounded open set with a Lipschitz continuous boundary is called a *Lipschitz domain.* [Analogously, if φ above belongs to $C^\infty(\mathcal{O}')$ for each $\mathcal{O}'$ then we say that Ω has a C^∞ boundary. A bounded open set with C^∞ boundary is called a C^∞ *domain.*]

An important property of a Lipschitz domain Ω is that the unit outward normal vector to $\partial\Omega$ is defined almost everywhere with respect to the $(n-1)$–dimensional measure on $\partial\Omega$; thus, for example, one may meaningfully talk about the normal flux of a physical quantity across $\partial\Omega$. A simple example of a Lipschitz domain is a bounded open polyhedron in $\mathbb{R}^n$, $n \ge 2$.

Having characterized the smoothness of the boundary, next we need to make sure that the boundary data are sufficiently smooth so as to be meaningful. In order to do so, we give a brief description of Sobolev spaces on the boundary $\partial\Omega$ of a Lipschitz domain Ω. Let us first recall from Definition 1 that for every x on $\partial\Omega$ there exists a Lipschitz continuous function $\varphi : \mathcal{O}' \subset \mathbb{R}^{n-1} \to \mathbb{R}$ such that, using the notation from Definition 1,

$$\partial\Omega \cap \mathcal{O} = \{\zeta = (\zeta', \varphi(\zeta')) \, : \, \zeta' \in \mathcal{O}'\},$$

so that, locally, $\partial\Omega$ is an $(n-1)$–dimensional hypersurface in $\mathbb{R}^n$. We define the mapping ϕ by

$$\phi(\zeta') = (\zeta', \varphi(\zeta')).$$

Then ϕ^{-1} exists and it is Lipschitz continuous on $\phi(\mathcal{O}')$, which leads us to the following definition.

Definition 2. Let Ω be a Lipschitz domain in $\mathbb{R}^n$. For $0 \le s \le 1$ and $1 \le p < \infty$ we denote by $W_p^s(\partial\Omega)$ the set of all $u \in L_p(\partial\Omega)$ such that the composition $u \circ \phi$ belongs to $W_p^s(\mathcal{O}' \cap \phi^{-1}(\partial\Omega \cap \mathcal{O}))$ for all possible $\mathcal{O}'$ and φ satisfying the conditions of Definition 1, where $\phi(\zeta') = (\zeta', \varphi(\zeta'))$ for $\zeta' \in \mathcal{O}'$.

Next, we shall equip $W_p^s(\partial\Omega)$ with a norm; for this purpose we consider any *atlas* $(\mathcal{O}_j, \varphi_j)_{j=1}^J$ for $\partial\Omega$ such that $\mathcal{O}_j$ and φ_j, $j = 1, \ldots, J$, satisfy the conditions of Definition 1. We define $\|\cdot\|_{W_p^s(\partial\Omega)}$ by

$$\|u\|_{W_p^s(\partial\Omega)} = \left(\sum_{j=1}^J \|u \circ \phi_j\|^p_{W_p^s(\mathcal{O}_j' \cap \phi_j^{-1}(\partial\Omega \cap \mathcal{O}_j))} \right)^{1/p},$$

where $\phi_j(\zeta') = (\zeta', \varphi_j(\zeta'))$ for $\zeta' \in \mathcal{O}'_j$, $j = 1, \ldots, J$. The defining expression for $\|\cdot\|_{W^s_p(\partial\Omega)}$ is not particularly convenient from the practical point of view as it explicitly involves the atlas $(\mathcal{O}_j, \varphi_j)_{j=1}^J$. Fortunately, for $0 < s < 1$ the norm $\|\cdot\|_{W^s_p(\partial\Omega)}$ defined above is equivalent to the following norm

$$\|u\|_{W^s_p(\partial\Omega)} = \left(\int_{\partial\Omega} |u|^p \,\mathrm{d}\sigma + \int_{\partial\Omega}\int_{\partial\Omega} \frac{|u(x) - u(y)|^p}{|x - y|^{n-1+sp}} \,\mathrm{d}\sigma(x)\,\mathrm{d}\sigma(y)\right)^{1/p},$$

where $\mathrm{d}\sigma$ denotes the $(n-1)$–dimensional surface measure on $\partial\Omega$. If Ω has a C^∞ boundary then the restriction $s \leq 1$ is redundant; in fact, if $\Gamma_0 \subset \partial\Omega$ and Γ_0 is part of an $(n-1)$–dimensional C^∞ hypersurface in $\mathbb{R}^n$, then $\|u\|_{W^s_p(\Gamma_0)}$ can be defined, analogously as above, for all $s \geq 0$ and $1 \leq p < \infty$.

Proposition 1. *Suppose that Ω is a Lipschitz domain contained in $\mathbb{R}^n$ and let $1 \leq p < \infty$. Then $C^\infty(\bar{\Omega})$ is dense in $W^s_p(\Omega)$ for $s \geq 0$; in other words, the closure of $C^\infty(\bar{\Omega})$ in the norm of $W^s_p(\Omega)$ is equal to $W^s_p(\Omega)$.*

We note that while $C^\infty(\bar{\Omega})$ is dense in $W^s_p(\Omega)$ for $s \geq 0$ and $1 \leq p < \infty$, $C^\infty_0(\Omega)$ is not dense in $W^s_p(\Omega)$ for $s > 1/p$; i.e., for $s > 1/p$, the closure of $C^\infty_0(\Omega)$ in the norm of $W^s_p(\Omega)$ is a proper closed linear subspace of $W^s_p(\Omega)$.

Now we are ready to discuss the notion of trace of a function on the boundary $\partial\Omega$ of a Lipschitz domain $\Omega \subset \mathbb{R}^n$. If ψ belongs to $C^\infty(\bar{\Omega})$ then we put

$$\gamma_0(\psi) = \psi|_{\partial\Omega} \,. \tag{9}$$

The *trace* of a function u in $W^s_p(\Omega)$ with $s > 1/p$, $1 < p < \infty$, is then defined by extending the operator γ_0 from the dense subspace $C^\infty(\bar{\Omega})$ to the whole of $W^s_p(\Omega)$.

Proposition 2. *Suppose that Ω is a Lipschitz domain in $\mathbb{R}^n$, and let $1 < p < \infty$. Assuming that $1/p < s \leq 1$, the mapping γ_0 defined on $C^\infty(\bar{\Omega})$ by (9) has a unique continuous extension to a linear operator, called the trace operator, still denoted γ_0, from $W^s_p(\Omega)$ onto $W^{s-(1/p)}_p(\partial\Omega)$. If Γ_0 is an open subset of $\partial\Omega$ and Γ_0 is part of an $(n-1)$–dimensional C^∞ hypersurface in $\mathbb{R}^n$, then $\gamma_0 : W^s_p(\Omega) \to W^{s-(1/p)}_p(\Gamma_0)$ for $1 < p < \infty$ and all $s > 1/p$ (the restriction $s \leq 1$ being redundant).*

When $p = 2$ we shall write H^s in place of W^s_2 to signify the fact that we are dealing with a Hilbert space. We denote by $H^s_0(\Omega)$ the closure of $C^\infty_0(\Omega)$ in the norm of $H^s(\Omega)$; when Ω is a Lipschitz domain and $1/2 < s < 3/2$, this space coincides with the set of all those functions in $H^s(\Omega)$ whose trace on $\partial\Omega$ is equal to zero.

4 Mathematical models of compressible fluid dynamics

The purpose of this section is to give a brief overview of mathematical models of compressible fluid dynamics and to list some simple model hyperbolic problems derived from these equations whose numerical approximations will be considered here. The computational tools developed in this work, however, are by no means limited to problems in compressible fluid dynamics, and the partial differential equations listed in this section should only be thought of as typical continuum models to which these computational tools apply.

4.1 Compressible Navier–Stokes equations

The motion of a compressible fluid occupying an open set $\Omega \subset \mathbb{R}^n$, $1 \leq n \leq 3$, is governed by the *compressible Navier–Stokes equations.* The equations express key physical properties, such as the conservation of mass, momentum and energy in the flow, and involve the density $\rho : \Omega \to \mathbb{R}$, the velocity field $\mathbf{u} : \Omega \to \mathbb{R}^n$, and the total energy per unit mass $e : \Omega \to \mathbb{R}$.

In Eulerian co-ordinates, the compressible Navier–Stokes equations have the following form:

Conservation of mass

$$\frac{\partial \rho}{\partial t} + \sum_{j=1}^{n} \frac{\partial}{\partial x_j}(\rho u_j) = 0\,;$$

Conservation of momentum

$$\frac{\partial}{\partial t}(\rho u_i) + \sum_{j=1}^{n} \frac{\partial}{\partial x_j}(\rho u_i u_j + p\delta_{ij}) = \sum_{j=1}^{n} \frac{\partial}{\partial x_j}\tau_{ij} + \hat{S}_i\,, \qquad \text{for } i = 1, \ldots, n\,;$$

Conservation of energy

$$\frac{\partial}{\partial t}(\rho e) + \sum_{j=1}^{n} \frac{\partial}{\partial x_j}((\rho e + p)u_j) = \sum_{j=1}^{n} \frac{\partial}{\partial x_j}\left(k\frac{\partial T}{\partial x_j} + \sum_{\ell=1}^{n} \tau_{jl} u_l\right).$$

In these equations, $k > 0$ denotes the thermal diffusivity, τ is the stress tensor, p is the pressure, $\hat{\mathbf{S}} \in [L_2(\Omega)]^n$ is a vector of given sources, δ_{ij} is the Kronecker delta, and $T > 0$ is the temperature.

4.2 Compressible Euler equations

If viscous effects are negligible, the compressible Navier–Stokes equations can be simplified. For example, in the case of an inviscid compressible ideal gas, they reduce to:

$$\frac{\partial \mathbf{U}}{\partial t} + \sum_{j=1}^{n} \frac{\partial}{\partial x_j}\mathbf{F}_j(\mathbf{U}) = \mathbf{S}\,, \tag{10}$$

where

$$\mathbf{U} = [\rho, \rho u_1, \ldots, \rho u_n, \rho e]^{\mathrm{T}}$$

is the vector of *conserved variables*,

$$\mathbf{F}_j = [\rho u_j, \rho u_1 u_j + \delta_{1j} p, \ldots, \rho u_n u_j + \delta_{nj} p, (\rho e + p) u_j]^{\mathrm{T}}\,, \qquad j = 1, \ldots, n\,,$$

are the fluxes, and

$$\mathbf{S} = \left[0, \hat{S}_1, \ldots, \hat{S}_n, 0\right]^{\mathrm{T}} \in [L_2(\Omega)]^{n+2}$$

is a source vector. The range of the conserved variables is a subset $\Sigma \subset \mathbb{R}^{n+2}$ called the *state space*. For an ideal gas, density and pressure are related through the *equation of state*

$$p = (\kappa - 1)\rho \left(e - \frac{1}{2}|\mathbf{u}|^2\right)\,,$$

where $\mathbf{u} = (u_1, \ldots, u_n)^{\mathrm{T}}$ and κ is the ratio of specific heats; for dry air, $\kappa = 1.405$. The nonlinear partial differential equations (10) are called the *compressible Euler equations.*

If $\mathbf{U}$ is sufficiently smooth, say $\mathbf{U} \in [C^1(\Omega)]^{n+2}$, then, by applying the chain rule, (10) can be restated in the equivalent quasilinear (nonconservative) form:

$$\frac{\partial \mathbf{U}}{\partial t} + \sum_{j=1}^{n} A_j(\mathbf{U}) \frac{\partial \mathbf{U}}{\partial x_j} = \mathbf{S}\,,$$

where $A_j(\mathbf{U}) = \nabla_{\mathbf{U}} \mathbf{F}_j(\mathbf{U}) \in \mathbb{R}^{(n+2)\times(n+2)}$ denotes the Jacobi matrix associated with the flux vector $\mathbf{F}_j$, $j = 1, \ldots, n$. The system of compressible Euler equations is *hyperbolic*, i.e., the matrix

$$\sum_{j=1}^{n} \mu_j A_j(\mathbf{U}) \in \mathbb{R}^{(n+2)\times(n+2)}$$

possesses $n + 2$ real eigenvalues λ_i, $i = 1, \ldots, n + 2$, and a complete set of linearly independent eigenvectors for all values of $\mathbf{U} \in \Sigma$ and for all $\boldsymbol{\mu} = (\mu_1, \ldots, \mu_n)^{\mathrm{T}} \in \mathbb{R}^n$.

4.3 Symmetrization of the compressible Euler equations

Symmetric hyperbolic systems frequently arise in mathematical models of continuum mechanics; later on in these notes we shall consider their finite element approximation in more detail. Here, as a motivating example, we show how the compressible Euler equations can be restated as a symmetric system of partial differential equations using entropy variables. Our exposition of the symmetrization process follows [52] and is based on the work of

Hughes, Franca and Mallet [36] and Mock [43]; see also the paper of Barth [6] where symmetric forms of conservation laws have been extensively used in numerical computations. For the sake of simplicity and brevity, we shall confine our presentation to the case of two space dimensions ($n = 2$).

The starting point of the symmetrization process is to hypothesize the existence of a scalar *entropy function*

$$\eta \,:\, \mathbf{U} \in \Sigma \mapsto \eta(\mathbf{U}) \in \mathbb{R}$$

and the existence of the *entropy fluxes*

$$q_i \,:\, \mathbf{U} \in \Sigma \to q_i(\mathbf{U}) \in \mathbb{R}, \qquad i = 1, 2\,,$$

which are compatible with η in the sense that

$$\nabla_{\mathbf{U}}\eta(\mathbf{U}) \cdot A_i(\mathbf{U}) = \nabla_U q_i(\mathbf{U})\,, \qquad i = 1, 2\,.$$

The entropy η and the entropy fluxes q_i, $i = 1, 2$, are assumed to satisfy the entropy inequality

$$\frac{\partial}{\partial t}\eta(\mathbf{U}) + \sum_{j=1}^{2} \frac{\partial}{\partial x_j} q_j(\mathbf{U}) \leq 0$$

in the sense of distributions. For the Euler equations, the thermodynamical entropy density

$$\eta(\mathbf{U}) = -\rho s$$

with entropy

$$s = \log \frac{p}{\rho^{\kappa}}$$

is a candidate for an entropy function; $\mathbf{U} \mapsto \eta(\mathbf{U})$ is then a convex function of

$$\mathbf{U} = [\rho, \rho u_1, \rho u_2, \rho e]^{\mathrm{T}}\,.$$

The compressible Euler equations (10) can be transformed into a symmetric system of nonlinear partial differential equations using the invertible change of variables

$$\mathbf{U} \mapsto \mathbf{V}(\mathbf{U}) = \nabla_{\mathbf{U}}\eta(\mathbf{U})\,.$$

Indeed, on writing

$$\frac{\partial}{\partial t}\mathbf{U}(\mathbf{V}) + \sum_{j=1}^{2} \frac{\partial}{\partial x_j} \mathbf{F}_j(\mathbf{U}(\mathbf{V})) = \mathbf{S}$$

and applying the chain rule, we deduce that

$$(\nabla_{\mathbf{V}}\mathbf{U})\frac{\partial \mathbf{V}}{\partial t} + \sum_{j=1}^{2} (\nabla_U \mathbf{F}_j(\mathbf{U}(\mathbf{V})))\,(\nabla_{\mathbf{V}}\mathbf{U})\frac{\partial \mathbf{V}}{\partial x_j} = \mathbf{S}\,.$$

On defining $A^0(\mathbf{V}) = \nabla_{\mathbf{V}}\mathbf{U}$, the last equation can be rewritten in the following, more compact, form:

$$A^0(\mathbf{V})\frac{\partial \mathbf{V}}{\partial t} + \sum_{j=1}^{2} A_j(\mathbf{U}(\mathbf{V}))A^0(\mathbf{V})\frac{\partial \mathbf{V}}{\partial x_j} = \mathbf{S}\,.$$

After some calculations we find that the mapping $\mathbf{U} \mapsto \mathbf{V}(\mathbf{U}) = \nabla_{\mathbf{U}}\eta(\mathbf{U})$ is given by

$$\mathbf{V}(\mathbf{U}) = \frac{\kappa-1}{p}\begin{bmatrix} \frac{p}{\kappa-1}(\kappa+1-s) - \rho e \\ \rho u_1 \\ \rho u_2 \\ -\rho \end{bmatrix} \equiv \begin{bmatrix} V_1 \\ V_2 \\ V_3 \\ V_4 \end{bmatrix},$$

while the inverse $\mathbf{V} \mapsto \mathbf{U}(\mathbf{V})$ of this mapping can be expressed as

$$\mathbf{U}(\mathbf{V}) = \frac{p}{\kappa-1}\begin{bmatrix} -V_4 \\ V_2 \\ V_3 \\ 1 - \frac{1}{2}\frac{V_2^2+V_3^2}{V_4} \end{bmatrix}.$$

Consequently,

$$A^0 = \frac{1}{\kappa-1}\begin{bmatrix} \rho & \rho u_1 & \rho u_2 & \frac{p}{\kappa-1} + \frac{1}{2}\rho|\mathbf{u}|^2 \\ & p+\rho u_1^2 & \rho u_1 u_2 & \frac{1}{2}\rho u_1|\mathbf{u}|^2 + \frac{\kappa u_1 p}{\kappa-1} \\ & & p+\rho u_2^2 & \frac{1}{2}\rho u_2|\mathbf{u}|^2 + \frac{\kappa u_2 p}{\kappa-1} \\ \text{symm} & & & \frac{1}{4}\rho|\mathbf{u}|^4 - \frac{\kappa|\mathbf{u}|^2 p}{\kappa-1} + \frac{\kappa p^2}{\rho(\kappa-1)^2} \end{bmatrix}.$$

On the other hand,

$$A_1(\mathbf{U}) = \begin{bmatrix} 0 & 1 & 0 & 0 \\ \frac{\kappa-3}{2}u_1^2 + \frac{\kappa-1}{2}u_2^2 & (3-\kappa)u_1 & (1-\kappa)u_2 & \kappa-1 \\ -u_1u_2 & u_2 & u_1 & 0 \\ (\kappa-1)u_1|\mathbf{u}|^2 - \kappa u_1 e & \kappa e - \frac{\kappa-1}{2}(u_2^2+3u_1^2) & (1-\kappa)u_1u_2 & \kappa u_1 \end{bmatrix},$$

$$A_2(\mathbf{U}) = \begin{bmatrix} 0 & 0 & 1 & 0 \\ -u_1u_2 & u_2 & u_1 & 0 \\ \frac{\kappa-3}{2}u_2^2 + \frac{\kappa-1}{2}u_1^2 & (1-\kappa)u_1 & (3-\kappa)u_2 & \kappa-1 \\ (\kappa-1)u_2|\mathbf{u}|^2 - \kappa u_2 e & (1-\kappa)u_1u_2 & \kappa e - \frac{\kappa-1}{2}(u_1^2+3u_2^2) & \kappa u_2 \end{bmatrix}.$$

Thus, we have transformed the compressible Euler equations (10) into the form

$$\sum_{j=0}^{2} B_j(\mathbf{V})\frac{\partial \mathbf{V}}{\partial x_j} = \mathbf{S}\,, \tag{11}$$

with $x_0 = t$, where $B_0(\mathbf{V}) = A^0(\mathbf{V})$ and $B_j(\mathbf{V}) = A_j(\mathbf{U}(\mathbf{V}))A^0(\mathbf{V})$, $j = 1, 2$, are symmetric 4×4 matrices[3].

[3] The symmetry of the matrices B_1 and B_2 may be verified by evaluating the matrix products $A_j(\mathbf{U}(\mathbf{V}))A^0(\mathbf{V})$, $j = 1, 2$.

4.4 Symmetric linear hyperbolic systems

Suppose that the solution to the nonlinear system (11) can be decomposed as

$$\mathbf{V} = \hat{\mathbf{V}} + \mathbf{W},$$

where $\hat{\mathbf{V}}$ is a (4–component vector) mean state and $\mathbf{W}$ is a small nonconstant perturbation to $\hat{\mathbf{V}}$. In many physical situations the mean state is known and one is only interested in modeling the evolution of the small perturbations $\mathbf{W}$. After some simple algebraic manipulations (see [52], p. 627), we deduce that

$$\sum_{j=0}^{2} B_j(\hat{\mathbf{V}}) \frac{\partial \mathbf{W}}{\partial x_j} = \mathbf{S},$$

with $x_0 = t$, which is a symmetric linear hyperbolic system; equivalently, the last set of equations can be rewritten in the following *conservative* (or *conservation*) *form*:

$$\sum_{j=0}^{2} \frac{\partial}{\partial x_j}(B_j(\hat{\mathbf{V}})\mathbf{W}) + C\mathbf{W} = \mathbf{S}, \tag{12}$$

where

$$C = -\sum_{j=0}^{2} \frac{\partial}{\partial x_j} B_j(\hat{\mathbf{V}}).$$

In particular, when the mean state $\hat{\mathbf{V}}$ is a constant vector then C is the 4×4 zero-matrix.

We explained how the fundamental equations of compressible fluid dynamics lead to symmetric (linear or nonlinear) hyperbolic systems. However, symmetric linear hyperbolic systems also arise in many other physical applications such as acoustics, electromagnetism and linear elasticity. This motivates us to adopt, as linear model problem, the following symmetric hyperbolic system where the principal part of the operator is in conservation form:

$$\sum_{j=1}^{n} \frac{\partial}{\partial x_j}(\mathbf{B}_j \mathbf{u}) + \mathbf{C}\mathbf{u} = \mathbf{f} \qquad \text{on } \Omega;$$

here Ω is a bounded open polyhedral domain in $\mathbb{R}^n$. This system of partial differential equation will be considered in tandem with appropriate boundary conditions which will be stated below after we have made appropriate assumptions on the smoothness of the data.

Let Γ denote the union of open faces of the polyhedral domain Ω and suppose that $\mathbf{B} = (\mathbf{B}_1, \ldots, \mathbf{B}_n)$ is an n–component matrix function defined on $\bar{\Omega}$ with $\mathbf{B}_j \in [W^1_\infty(\Omega)]^{m\times m}_{\mathrm{symm}}$, $j = 1,\ldots,n$. Let $\boldsymbol{\mu} = (\mu_1,\ldots,\mu_n)$ denote the unit outward normal vector to Γ, and consider the symmetric matrix

$$\mathbf{B}(\boldsymbol{\mu}) \equiv \mathbf{B}\cdot\boldsymbol{\mu} = \mu_1\mathbf{B}_1 + \ldots + \mu_n\mathbf{B}_n\,.$$

Suppose that $\mathbf{B}$ is diagonalized *via* $\mathbf{B}(\boldsymbol{\mu}) = X(\boldsymbol{\mu})^{-1}\Lambda(\boldsymbol{\mu})X(\boldsymbol{\mu})$, where $\Lambda(\boldsymbol{\mu})$ is a diagonal matrix, with the (real) eigenvalues of $\mathbf{B}(\boldsymbol{\mu})$ appearing along its diagonal. We shall suppose that Γ is nowhere characteristic in the sense that none of the diagonal entries of $\Lambda(\boldsymbol{\mu})$ is zero for any choice of the unit outward normal vector $\boldsymbol{\mu}$ on Γ. We decompose $\Lambda(\boldsymbol{\mu})$ as

$$\Lambda(\boldsymbol{\mu}) = \Lambda_-(\boldsymbol{\mu}) + \Lambda_+(\boldsymbol{\mu})\,,$$

where $\Lambda_-(\boldsymbol{\mu})$ is diagonal and negative semidefinite, and $\Lambda_+(\boldsymbol{\mu})$ is diagonal and positive semidefinite. Thus we define

$$\mathbf{B}_-(\boldsymbol{\mu}) = X(\boldsymbol{\mu})^{-1}\Lambda_-(\boldsymbol{\mu})X(\boldsymbol{\mu}) \quad\text{and}\quad \mathbf{B}_+(\boldsymbol{\mu}) = X(\boldsymbol{\mu})^{-1}\Lambda_+(\boldsymbol{\mu})X(\boldsymbol{\mu})\,.$$

Clearly, for each unit outward normal vector $\boldsymbol{\mu}$ on Γ,

$$\mathbf{B}(\boldsymbol{\mu}) = \mathbf{B}_-(\boldsymbol{\mu}) + \mathbf{B}_+(\boldsymbol{\mu})\,.$$

Given $\mathbf{C} \in [L_\infty(\Omega)]^{m\times m}$, $\mathbf{f} \in [L_2(\Omega)]^m$ and $\mathbf{g} \in [L_2(\Gamma)]^m$, we consider the following hyperbolic boundary value problem: find $\mathbf{u} \in H(\mathcal{L},\Omega)$ such that

$$\mathcal{L}\mathbf{u} \equiv \sum_{j=1}^{n} \frac{\partial}{\partial x_j}(\mathbf{B}_j\mathbf{u}) + \mathbf{C}\mathbf{u} = \mathbf{f} \quad \text{in } \Omega\,, \quad \mathbf{B}_-(\boldsymbol{\mu})\mathbf{u} = \mathbf{B}_-(\boldsymbol{\mu})\mathbf{g} \quad \text{on } \Gamma\,, \tag{13}$$

where $H(\mathcal{L},\Omega) = \{\mathbf{v} \in [L_2(\Omega)]^m : \mathcal{L}\mathbf{v} \in [L_2(\Omega)]^m\}$ denotes the *graph space* of the partial differential operator $\mathcal{L}$ in $[L_2(\Omega)]^m$. The analysis of the well-posedness of this linear system is technical; for details, the interested reader is referred to the papers [16,41,50]. When $m = 1$, the multidimensional symmetric linear hyperbolic system (13) collapses to a scalar linear first-order hyperbolic PDE, an equation that is of great physical importance in its own right.

4.5 Scalar linear hyperbolic equations

As a scalar hyperbolic model problem, we consider the hyperbolic boundary-value problem

$$\mathcal{L}u \equiv \operatorname{div}(\mathbf{b}u) + cu = f \quad \text{in } \Omega\,, \qquad u = g \quad \text{on } \Gamma_-\,, \tag{14}$$

where Ω is a bounded open polyhedron in $\mathbb{R}^n$, with *inflow boundary*

$$\Gamma_- = \{x \in \Gamma\, :\, \mathbf{b}(x)\cdot\boldsymbol{\mu}(x) < 0\}\,;$$

here Γ denotes the union of open faces of Ω, $\boldsymbol{\mu}(x)$ denotes the unit outward normal vector at $x \in \Gamma$. The complement of Γ_- with respect to Γ will be denoted Γ_+ and will be referred to as the *outflow boundary*. We shall suppose that $\mathbf{b} = (b_1, \ldots, b_n) \in [W^1_\infty(\Omega)]^n$, $c \in L_\infty(\Omega)$, $f \in L_2(\Omega)$, and $g \in L_2(\Gamma_-)$.

For the sake of simplicity we adopt the following additional hypothesis.

Hypothesis 1 *Suppose that $\Omega = (0,1)^n$ and the components $b_1, \ldots, b_n$ of the vector field* $\mathbf{b}$ *are strictly positive functions on $\bar{\Omega}$.*

This assumption ensures that Γ is non-characteristic for the differential operator $v \mapsto \mathcal{L}v$, and leads to the following well-posedness result whose proof, in the case of $g = 0$, may be found in [53]; in the case of $g \in L_2(\Gamma_-)$ nonzero, it is a special case of a general result of Rauch [50] (see also, [5] and [13]).

Proposition 3. *Under Hypothesis 1 and given that $f \in L_2(\Omega)$, $g \in L_2(\Gamma_-)$, and $c \in L_\infty(\Omega)$, problem (14) has a unique solution u in the graph-space $H(\mathcal{L}, \Omega) = \{v \in L_2(\Omega) \, : \, \mathcal{L}v \in L_2(\Omega)\}$.*

We note that an analogous result holds for the scalar multi-dimensional linear hyperbolic equation written in *nonconservative form*:

$$\mathbf{b} \cdot \nabla u + c\, u = f \quad \text{in } \Omega\,, \qquad u = g \quad \text{on } \Gamma_-\,,$$

where $\mathbf{b}$, c, f, g and Ω are as in Proposition 3. Scalar hyperbolic PDEs in nonconservative form arise in the mathematical modeling of transport phenomena.

5 Stabilized FE methods for transport problems

In this section we return to the difficulties raised at the beginning of the paper concerning the *a posteriori* error analysis of hyperbolic problems. As we have seen there, the dual solution carries important information about the sensitivity of the error to local perturbations of the residual. It is therefore essential that the local variation of the dual solution is reflected by the error bound. Guided by this observation, here we shall develop Type I *a posteriori* error bounds for a class of stabilized finite element approximations of first-order scalar linear hyperbolic PDEs in nonconservative form. The discussion in this section is a brief survey of our paper with Rolf Rannacher [30].

Let us consider the transport problem:

$$\mathcal{L}u \equiv \mathbf{b} \cdot \nabla u + cu = f\,, \quad x \in \Omega\,, \qquad u = g\,, \quad x \in \Gamma_-\,, \tag{15}$$

where $\Omega = (0,1)^n$. With a slight abuse of notation, here we have used the symbol $\mathcal{L}$ to denote the scalar hyperbolic differential operator $\mathbf{b} \cdot \nabla + c$ in nonconservative form, even though in the previous section $\mathcal{L}$ signified the related partial differential operator in conservation form. Since $\mathbf{b} \in [W^1_\infty(\Omega)]^n$, we can safely move back and forth between the two forms, only having to alter the coefficient c if necessary; therefore, we have chosen not to distinguish between the two forms in terms of $\mathcal{L}$. Of course, if $\nabla \cdot \mathbf{b} = 0$ then the conservative- and nonconservative-form differential operators are identical.

As in the previous section, we assume that the entries $b_1, \ldots, b_n$ of the n-component vector function $\mathbf{b} \in [W^1_\infty(\Omega)]^n$ are positive on $\bar{\Omega} = [0,1]^n$. This hypothesis ensures that Γ, the union of open faces of Ω, is noncharacteristic for the operator $\mathcal{L}$ at each point x of Γ. Also, we shall suppose that $c \in L_\infty(\Omega)$, $f \in L_2(\Omega)$, and $g \in L_2(\Gamma_-)$. In addition, it will be assumed that there exists $\gamma > 0$ such that

$$c_0^2(x) \equiv c(x) - \frac{1}{2}\nabla \cdot \mathbf{b}(x) \geq \gamma^2 \quad \text{for all } x \in \bar{\Omega}\,. \tag{16}$$

In order to set up the variational formulation of the boundary value problem (15), we associate with $\mathcal{L}$ the graph space

$$H(\mathcal{L},\Omega) = \{v \in L_2(\Omega) \, : \, \mathcal{L}v \in L_2(\Omega)\}\,.$$

We consider the bilinear form $B(\cdot,\cdot) \, : \, H(\mathcal{L},\Omega) \times H(\mathcal{L},\Omega) \to \mathbb{R}$ defined by

$$B(w,v) = (\mathcal{L}w, v) - ((\mathbf{b}\cdot\boldsymbol{\mu})w, v)_{\Gamma_-}$$

and introduce the linear functional $\ell \, : \, H(\mathcal{L},\Omega) \to \mathbb{R}$ by

$$\ell(v) = (f,v) - ((\mathbf{b}\cdot\boldsymbol{\mu})g, v)_{\Gamma_-}\,.$$

In these definitions, $(\cdot,\cdot)$ denotes the $L_2(\Omega)$ inner product and $(\cdot,\cdot)_{\Gamma_-}$ is the $L_2(\Gamma_-)$ inner product with respect to the surface measure $d\sigma$ (with analogous definition of $(\cdot,\cdot)_{\Gamma_+}$). In terms of this notation, the boundary value problem (15) can be restated in the following variational form: find $u \in H(\mathcal{L},\Omega)$ such that

$$B(u,v) = \ell(v) \qquad \forall v \in H(\mathcal{L},\Omega)\,. \tag{17}$$

Suppose that $\mathcal{T}_h$ is a finite element partition of the computational domain Ω into open element domains κ. It will be assumed that the family $\mathcal{T}_h$ is shape regular (see, [10], for example); namely, there exists a positive constant α such that $h_\kappa/\rho_\kappa \leq \alpha$ for all $\kappa \in \bigcup_h \mathcal{T}_h$, where $h_\kappa = \mathrm{diam}(\bar{\kappa})$ and ρ_κ is the diameter of the largest ball contained in $\bar{\kappa}$. Let us denote by $h(x)$ the mesh function defined on the union of the elements κ in $\mathcal{T}_h$ whose value on element κ is identically equal to h_κ. We then consider on $\mathcal{T}_h$ the finite element space $S_h \subset H^1(\Omega) \subset H(\mathcal{L},\Omega)$ consisting of continuous piecewise polynomial functions of maximum degree p, $p \geq 1$. It will be assumed that S_h possesses the following approximation property:

(H) Given that $v \in H^{s+1}(\Omega)$ and $v|_{\Gamma_-} \in H^{s+1}(\Gamma_-)$ for some s, $0 \leq s \leq p$, there exists $P_h v$ in S_h and a positive constant C_{int}, independent of v and the mesh function h, such that

$$\|v - P_h v\|_{L_2(\kappa)} + h_\kappa |v - P_h v|_{H^1(\kappa)} \leq C_{\mathrm{int}} h_\kappa^{s+1} |v|_{H^{s+1}(\hat{\kappa})} \quad \forall \kappa \in \mathcal{T}_h\,,$$
$$\|v - P_h v\|_{L_2(\partial\kappa\cap\Gamma_-)} \leq C_{\mathrm{int}} h_\kappa^{s+1} |v|_{H^{s+1}(\partial\hat{\kappa}\cap\Gamma_-)} \; \forall \kappa \in \mathcal{T}_h \text{ s.t. } \partial\kappa \cap \Gamma_- \neq \emptyset\,.$$

In this hypothesis $\hat{\kappa}$ denotes the union of all such elements (including κ itself) whose closure has nonempty intersection with the closure of κ. Hypothesis (H) may be satisfied by choosing $P_h v$ as a finite element quasi-interpolation operator based on local averaging that involves the neighbors of κ. A further possibility is to define $P_h v \in S_h$ at the degrees of freedom interior to $\Omega \cup \Gamma_+$ as indicated in the previous sentence, while on Γ_- one can define $P_h v|_{\Gamma_-}$ as the orthogonal projection of $v|_{\Gamma_-}$ with respect to the inner product

$$\langle w, v\rangle_- = ((\mathbf{b}\cdot\boldsymbol{\mu})w, v)_{\Gamma_-}\,;$$

the inner product $\langle w, v\rangle_+$ is defined analogously.

We note that, according to Proposition 2, the regularity hypothesis $v \in H^{s+1}(\Omega)$ implies $v \in H^{s+1/2}(\Gamma_-)$ only; therefore the regularity of v on Γ_- assumed in (H) marginally exceeds the regularity on Γ_- that is implied by Proposition 2.

We are now ready to consider stabilized finite element approximations of our model problem. Let δ be a positive function contained in $L_\infty(\Omega)$; δ will be referred to as the ***stabilization parameter***. The stabilized finite element approximation of (15) is then defined as follows: find $u_h \in S_h$ such that

$$B_\delta(u_h, v_h) = l_\delta(v_h) \quad \forall v_h \in S_h\,, \tag{18}$$

where the bilinear functional $B_\delta : H(\mathcal{L},\Omega) \times H(\mathcal{L},\Omega) \to \mathbb{R}$ and the linear functional $\ell_\delta : H(\mathcal{L},\Omega) \to \mathbb{R}$ are defined by

$$B_\delta(w,v) = (\mathcal{L}w, v + \delta\hat{\mathcal{L}}v) - \langle w, v\rangle_-\,, \qquad l_\delta(v) = (f, v + \delta\hat{\mathcal{L}}v) - \langle g, v\rangle_-\,,$$

respectively, with $\mathcal{L}w = \mathbf{b}\cdot\nabla w + cw$ and $\hat{\mathcal{L}}w = \mathbf{b}\cdot\nabla w + \hat{c}w$. Depending on the choice of $\hat{c}$ we obtain different stabilization techniques; two typical choices are listed below:

$$\hat{c} = \begin{cases} 0 & \text{Streamline-Diffusion FEM (SDFEM)}\,, \\ c & \text{Least-Squares FEM (LSFEM)}\,. \end{cases}$$

The standard choice of the stabilization parameter, based on *a priori* error analysis, is $\delta = C_\delta h$, where C_δ is a positive constant which should be selected by the user. Condition (16) implies that $B_\delta(v,v) > 0$ for all $v \in S_h \setminus \{0\}$. As (18) is a linear problem over a finite-dimensional space S_h, the existence of a unique solution u_h to (18) follows from the positivity of $B_\delta(v,v)$, $v \in S_h \setminus \{0\}$.

5.1 *A posteriori* error estimation for linear functionals

In many problems of physical interest the quantity of interest is an output or target functional of the solution to the partial differential equation rather than the solution itself. Relevant physical examples have already been discussed in the Introduction. For simplicity, here we restrict ourselves to linear target functionals; the extension to general nonlinear output functionals will be discussed in Section 6.4.

Let us suppose, for example, that we wish to control the discretization error in some linear functional $J(\cdot)$ defined on $H(\mathcal{L},\Omega)+S_h$. A typical situation might be that a certain tolerance $\mathtt{TOL}>0$ is given and the aim of the computation is to find a subdivision $\mathcal{T}_h$ of the computational domain Ω and u_h in the finite element space S_h associated with $\mathcal{T}_h$ such that

$$|J(u)-J(u_h)|<\mathtt{TOL}.$$

Problems of this kind are referred to as *measurement problems.* In order to solve a measurement problem, we resort to *a posteriori* error analysis to derive a 'computable' bound on $|J(u)-J(u_h)|$ and then perform adaptive mesh refinement until the *a posteriori* error bound drops below the specified tolerance.

Here we shall discuss the derivation of an *a posteriori* error bound on $|J(u)-J(u_h)|$ in the case when a stabilized finite element method is used for the computation of u_h. The derivation of the *a posteriori* error bound will be based on a duality argument. Two approaches will be considered: the first approach involves the formal adjoint of the partial differential operator, while the second approach exploits the adjoint of the stabilized primal operator. Thus the two alternative dual problems are as follows:

$$\text{Find } z\in H(\mathcal{L},\Omega) \text{ such that} \quad B(w,z)=J(w) \quad \forall w\in H(\mathcal{L},\Omega)\,; \tag{19}$$

$$\text{Find } z_\delta\in H(\mathcal{L},\Omega) \text{ such that} \quad B_\delta(w,z_\delta)=J(w) \quad \forall w\in H(\mathcal{L},\Omega)\,. \tag{20}$$

These two dual problems will lead to two different *a posteriori* error bounds, and one of our aims will be to assess which of the two provides a sharper estimate of the error. The starting point for the error analysis is the following result which states the error representation formulas that arise from the two dual problems.

Theorem 1. *The dual problems (19), (20) give rise to the following error representation formulas, respectively:*

$$J(u)-J(u_h)=-\langle r_h^-,z-z_h\rangle_-+(r_h,z-z_h)-(r_h,\delta\hat{\mathcal{L}}z_h)\,, \tag{21}$$

$$J(u)-J(u_h)=-\langle r_h^-,z_\delta-z_{\delta,h}\rangle_-+(r_h,z_\delta-z_{\delta,h})+(r_h,\delta\hat{\mathcal{L}}(z_\delta-z_{\delta,h}))\,, \tag{22}$$

where $r_h=f-\mathcal{L}u_h$ *is the* internal residual, $r_h^-=g-u_h$ *is the boundary residual,* $z_h=P_hz$ *and* $z_{\delta,h}=P_hz_\delta$.

Proof. In order to prove (22), we begin by noting the following *Galerkin orthogonality property*:

$$B_\delta(u-u_h,v_h)=(r_h,v_h+\delta\hat{\mathcal{L}}v_h)-\langle r_h^-,v_h\rangle_-=0 \quad \forall v_h\in S_h\,.$$

Then, by virtue of the linearity of $J(\cdot)$ and the definition of the dual problem (20), we have that

$$J(u) - J(u_h) = J(u - u_h) = B_\delta(u - u_h, z_\delta) = B_\delta(u - u_h, z_\delta - z_{\delta,h}),$$

where $z_{\delta,h} = P_h z_\delta$. Hence,

$$J(u) - J(u_h) = \ell_\delta(z_\delta - z_{\delta,h}) - B_\delta(u_h, z_\delta - z_{\delta,h}).$$

We then deduce (22) by recalling the definitions of the linear functional $\ell_\delta(\cdot)$ and the bilinear form $B_\delta(\cdot,\cdot)$.

The proof of (21), using the standard dual problem (19), is completely analogous, except that Galerkin orthogonality does not enter into the argument. □

It follows from Theorem 1 that

$$J(u) - J(u_h) = \sum_{\kappa \in \mathcal{T}_h} \eta_\kappa,$$

where, letting $z_h = P_h z$ and $z_{h,\delta} = P_h z_\delta$, we have the following definitions of the *local refinement indicator* η_κ corresponding to the dual problems (19) and (20), respectively:

$$\eta_\kappa = \begin{cases} -((\mathbf{b}\cdot\boldsymbol{\mu})\, r_h^-, z - z_h)_{\partial\kappa\cap\Gamma_-} + (r_h, z - z_h)_\kappa - (r_h, \delta\hat{\mathcal{L}} z_h)_\kappa, \\ -((\mathbf{b}\cdot\boldsymbol{\mu})\, r_h^-, z_\delta - z_{\delta,h})_{\partial\kappa\cap\Gamma_-} + (r_h, z_\delta - z_{\delta,h})_\kappa + (r_h, \delta\hat{\mathcal{L}}(z_\delta - z_{\delta,h}))_\kappa. \end{cases}$$

We adopt the following notational convention: $\eta_\kappa^{(1)}$ will denote the first line in the definition of η_κ corresponding to the standard dual problem, while $\eta_\kappa^{(2)}$ will signify the second line in the definition of η_κ associated with the stabilization-dependent dual problem.

We see from this discussion that a necessary and sufficient condition for ensuring that

$$|J(u) - J(u_h)| < \mathtt{TOL} \tag{23}$$

is that

$$|\mathcal{E}_\Omega(u_h)| < \mathtt{TOL}, \qquad \text{where } \mathcal{E}_\Omega(u_h) \equiv \textstyle\sum_{\kappa\in\mathcal{T}_h} \eta_\kappa. \tag{24}$$

If (24) holds then $J(u_h)$ is accepted as an accurate representation of $J(u)$; otherwise, u_h is discarded and a new (hopefully more accurate) approximation is computed on a new partition which is obtained from $\mathcal{T}_h$ by refining certain elements $\kappa \in \mathcal{T}_h$. In order to ensure that refinement is performed only where it is necessary, a local decision has to be made on each element κ in $\mathcal{T}_h$ as to whether h_κ is acceptable in relation to TOL. A convenient approach to obtaining a refinement criterion that relates the local element size h_κ to the global tolerance TOL is to *localize* $\mathcal{E}_\Omega(u_h)$ by noting that the following is a sufficient condition for (24) to hold (and thereby also for (23)):

$$\mathcal{E}_{|\Omega|}(u_h) < \mathtt{TOL}, \qquad \text{where } \mathcal{E}_{|\Omega|}(u_h) \equiv \textstyle\sum_{\kappa\in\mathcal{T}_h} |\eta_\kappa|. \tag{25}$$

In the following section we discuss the implementation of this bound into an adaptive mesh refinement algorithm. However, before proceeding, a word of caution is in order: the seemingly harmless transition from (24) to (25) can be detrimental in the sense that the asymptotic rate of convergence of the resulting *a posteriori* error bound (25) may be suboptimal, leading to uneconomical mesh design. Indeed, the presence of such adverse effects in the *a posteriori* error bound is closely related to *global* superconvergence phenomena in the three terms constituting (21) and (22), respectively, which may be destroyed in the course of localization. We end this section by showing that the persistence of superconvergence under localization in the three terms of the error representation formula is closely related to the choice of the dual problem. To this end, let us first introduce some notation. We write I_1, II_1 and III_1 to denote the three terms (in modulus) on the right-hand side of (21), respectively; i.e.

$$\mathrm{I}_1 = |\langle r_h^-, z - z_h\rangle_-|, \quad \mathrm{II}_1 = |(r_h, z - z_h)|, \quad \mathrm{III}_1 = |(r_h, \delta\hat{\mathcal{L}} z_h)|. \qquad (26)$$

Similarly, we write

$$\mathrm{I}_{1,\delta} = |\langle r_h^-, z_\delta - z_{\delta,h}\rangle_-|, \qquad \mathrm{II}_{1,\delta} = |(r_h, z_\delta - z_{\delta,h})|,$$
$$\mathrm{III}_{1,\delta} = |(r_h, \delta\hat{\mathcal{L}}(z_\delta - z_{\delta,h}))| \qquad (27)$$

to denote the moduli of the three terms on the right-hand side of (22), respectively.

Next we recall the following result which is a minor variation on the *a priori* error bound derived in [38].

Lemma 1. *Assume that (H) holds and let u and u_h denote the solutions of (15) and (18), respectively. Suppose that $u \in H^{s+1}(\Omega)$ for some s, $0 \le s \le p$, and there exist positive constants c_δ and C_δ such that $c_\delta h \le \delta \le C_\delta h$ on Ω; then*

$$|||u - u_h||| \le C h^{s+1/2} |u|_{H^{s+1}(\Omega)},$$

where

$$|||v|||^2 = \|\sqrt{\delta}\,\mathcal{L}v\|^2_{L_2(\Omega)} + \|v\|^2_{L_2(\Omega)} + \|v\|^2_{L_2(\Gamma)}$$

and C is a constant independent of the mesh function h.

Thereby, from Hypothesis (H) and Lemma 1, we deduce the following *a priori* bound for the error in the computed linear functional $J(\cdot)$.

Theorem 2. *Assume that (H) holds and let u and u_h denote the solutions of (15) and (18), respectively. Given that $u \in H^{s+1}(\Omega)$ for some s, $0 \le s \le$*

p, the dual solution z satisfying (19) is sufficiently regular, and there exist positive constants c_δ and C_δ such that $c_\delta h \leq \delta \leq C_\delta h$ on Ω, we have that

$$\begin{aligned} \mathrm{I}_1 &\leq C h^{2s+3/2} \, |u|_{H^{s+1}(\Omega)} \, |z|_{H^{s+1}(\Gamma_-)} \,, \\ \mathrm{II}_1 &\leq C h^{2s+1} \, |u|_{H^{s+1}(\Omega)} \, |z|_{H^{s+1}(\Omega)} \,, \\ \mathrm{III}_1 &\leq C h^{2s+1} \, |u|_{H^{s+1}(\Omega)} \, |z|_{H^{s+1}(\Omega)} + C h^2 \, |u|_{H^{s+1}(\Omega)} \, |z|_{H^1(\Omega)} \,, \end{aligned}$$

where C is a constant independent of the mesh function h. Analogously, assuming that the solution z_δ of the stabilization–dependent dual problem (20) is sufficiently regular, we deduce that

$$\begin{aligned} \mathrm{I}_{1,\delta} &\leq C h^{2s+3/2} \, |u|_{H^{s+1}(\Omega)} \, |z_\delta|_{H^{s+1}(\Gamma_-)} \,, \\ \mathrm{II}_{1,\delta} &\leq C h^{2s+1} \, |u|_{H^{s+1}(\Omega)} \, |z_\delta|_{H^{s+1}(\Omega)} \,, \\ \mathrm{III}_{1,\delta} &\leq C h^{2s+1} \, |u|_{H^{s+1}(\Omega)} \, |z_\delta|_{H^{s+1}(\Omega)} \,. \end{aligned}$$

Thereby, Theorem 2 indicates that under suitable smoothness for both primal and dual solutions u and z, respectively, the error in the computed target functional $J(\cdot)$ is of size $\mathcal{O}(h^2)$, when the standard dual problem is employed, *irrespective* of the degree p of the approximating polynomial used in the finite element space S_h. In contrast, employing the error representation formula based on the stabilization–dependent dual problem, cf. (22), we find that

$$|J(u) - J(u_h)| \leq C h^{2s+1} \,,$$

for $0 \leq s \leq p$, provided that $z_\delta \in H^{s+1}(\Omega)$ with $\|z_\delta\|_{H^{s+1}(\Omega)} \leq \mathit{Const.}$, independent of h.

Finally, we note that if we assume that $u \in H^{s+1}(\Gamma_-)$ for some s, $0 \leq s \leq p$, then the *a priori* error bound for the boundary terms I_1 and $I_{1,\delta}$ stated in Theorem 2 may be improved provided that the boundary residual is redefined as $r_h^- = g - P_h g$. In this case, we deduce that

$$I_1 \leq C h^{2s+2} \, |u|_{H^{s+1}(\Gamma_-)} \, |z|_{H^{s+1}(\Gamma_-)} \,,$$

provided that $z \in H^{s+1}(\Gamma_-)$, and

$$I_{1,\delta} \leq C h^{2s+2} \, |u|_{H^{s+1}(\Gamma_-)} \, |z_\delta|_{H^{s+1}(\Gamma_-)} \,,$$

provided that $z_\delta \in H^{s+1}(\Gamma_-)$.

5.2 Mesh refinement criteria and mesh modification strategies

Given a user defined tolerance `TOL`, in this section we provide an outline of mesh refinement criteria and mesh modification strategies that may be employed to ensure the reliable and efficient control of the error in the computed functional $J(\cdot)$.

Local tolerance criterion. A possible mesh refinement criterion might consist of checking whether on each element κ in the partition $\mathcal{T}_h$ the following inequality holds:

$$|\eta_\kappa| \leq \frac{\mathtt{TOL}}{N}\,, \tag{28}$$

where N is the number of elements in $\mathcal{T}_h$. If the inequality (28) is violated on an element $\kappa \in \mathcal{T}_h$, then κ is refined; otherwise κ is accepted as being of adequate size. It is also possible to incorporate derefinement into the algorithm by selecting θ, $0 < \theta \ll 1$, and marking elements κ with

$$|\eta_\kappa| \leq \theta \frac{\mathtt{TOL}}{N}$$

for derefinement.

Fixed fraction criterion. This involves choosing two numbers $\varphi_{\mathbf{ref}}$ and $\varphi_{\mathbf{deref}}$ in the interval $(0, 100)$ with $\varphi_{\mathbf{deref}} + \varphi_{\mathbf{ref}} < 100$, ordering the *local refinement indicators* $|\eta_\kappa|$, $\kappa \in \mathcal{T}_h$, according to their size, and then refining those elements κ which correspond to $\varphi_{\mathbf{ref}}\%$ of the largest entries in the ordered sequence (the top 20%, say), and derefining those elements κ which correspond to the $\varphi_{\mathbf{deref}}\%$ of the smallest entries in this ordered sequence (the bottom 10%, say); further variations on this strategy, with dynamically varying $\varphi_{\mathbf{ref}}$ and $\varphi_{\mathbf{deref}}$, are also possible.

Optimized mesh criterion. Yet a further technique, called the *optimized mesh strategy* [49] (see, also, [19] and [22]) aims to design a subdivision $\mathcal{T}_h$ for the primal problem so that

$$\mathcal{E}_{|\Omega|}(u_h) \leq \mathtt{TOL}$$

subject to the constraint that the number N of elements in the subdivision $\mathcal{T}_h$ is minimized. Suppose, for example, that we have chosen to use the stabilization-parameter-dependent dual problem (20); then,

$$\eta_\kappa = -((\mathbf{b}\cdot\boldsymbol{\mu})\, r_h^-, z_\delta - z_{\delta,h})_{\partial\kappa\cap\Gamma_-} + (r_h, z_\delta - z_{\delta,h})_\kappa + (r_h, \delta\hat{\mathcal{L}}(z_\delta - z_{\delta,h}))_\kappa\,.$$

Let us suppose that the finite element space S_h consists of continuous piecewise polynomials of degree $p = 1$. We then define

$$\begin{aligned} A(x) &= |r_h(z_\delta - z_{\delta,h}) + \delta r_h \hat{\mathcal{L}}(z_\delta - z_{\delta,h})|/h_\Omega^3(x)\,,\\ B(x) &= |(\mathbf{b}\cdot\boldsymbol{\mu})(g - u_h)(z_\delta - z_{\delta,h})|/h_{\Gamma_-}^4(x)\,, \end{aligned}$$

where h_Ω and h_{Γ_-} are the mesh functions on $\Omega \cup \Gamma_+$ and Γ_-, respectively. After an elementary calculation based on the use of Lagrange multipliers, we arrive at the following optimal mesh functions $h_\Omega^{\mathrm{opt}}(x)$ and $h_{\Gamma_-}^{\mathrm{opt}}(x)$:

$$h_\Omega^{\mathrm{opt}} = \left(\frac{2}{3\lambda A}\right)^{1/5}, \qquad h_{\Gamma_-}^{\mathrm{opt}} = \left(\frac{1}{4\lambda B}\right)^{1/5},$$

where λ is the positive root of

$$\left(\frac{2}{3\lambda}\right)^{3/5} \int_\Omega A^{2/5}\,\mathrm{d}x + \left(\frac{1}{4\lambda}\right)^{4/5} \int_{\Gamma_-} B^{1/5}\,\mathrm{d}\sigma = \mathtt{TOL}\,.$$

For $\mathtt{TOL} \ll 1$ we expect $\lambda \gg 1$, so that $(1/\lambda)^{4/5} \ll (1/\lambda)^{3/5}$. Thus, for simplicity, we may neglect the boundary integral term in the last equality. Thereby, we can explicitly determine λ in terms of $\mathtt{TOL}$ and the integral of $A^{2/5}$; substituting the resulting expression into the formulae for h_Ω^{opt}, we obtain

$$h_\Omega^{\mathrm{opt}}(x) \approx \left(\frac{\mathtt{TOL}}{W}\right)^{1/3} \frac{1}{A^{1/5}(x)}, \qquad \text{where} \qquad W = \int_\Omega A^{2/5}(x)\,\mathrm{d}x\,,$$

with a similar expression for $h_{\Gamma_-}^{\mathrm{opt}}$.

Any of these criteria can be coupled with a suitable mesh modification algorithm. For example, in two space dimensions a popular mesh modification strategy is a red-green refinement. Here, the user must first specify a coarse *background mesh* upon which any future refinement will be based. Red refinement corresponds to dividing a certain triangle into four similar triangles by connecting the midpoints of the three sides. Since red refinement is performed only locally (rather than in each element in the triangulation) hanging nodes are created in the mesh; green refinement is then used to remove any hanging nodes in the mesh created in the course of red refinement by connecting a hanging node on an edge to the opposite vertex of the triangle. Green refinement is temporary and is only applied to elements which contain one hanging node; on elements with two or more hanging nodes red refinement is performed. Within this mesh modification algorithm elements may also be removed from the mesh through derefinement provided they do not lie in the original background mesh. It is perhaps worth noting here that the removal of hanging nodes through green refinement is necessary only if S_h is contained in $C(\bar{\Omega})$, as is the case for SDFEM considered here. In the discontinuous Galerkin finite element method to be discussed in the next section it is not assumed that S_h is contained in $C(\bar{\Omega})$, so the existence of hanging nodes in the mesh is perfectly acceptable there.

An important issue concerning practical implementation of the *stopping criterion* (25) and the *refinement criterion*, cf. (28) for example, into the adaptive algorithm is that the calculation of η_κ presupposes knowledge of z or z_δ, depending on which error representation formula is selected. These functions, however, are the analytical solutions to the dual problems (19) and (20), respectively, and therefore they are not available explicitly but have to be computed numerically. The numerical approximation of the dual solution requires special care and will be discussed in more detail in the next section. We shall comment on the numerical approximation of the solution z_δ of (20) only, as problem (19) is a special case of (20) upon formally setting $\delta = 0$ in (20), so identical comments apply to z.

5.3 Numerical approximation of z_δ

Let us recall that the function z_δ is defined as the solution to the following dual problem: find $z_\delta \in H(\mathcal{L}, \Omega)$ such that

$$B_\delta(w, z_\delta) = J(w) \quad \forall w \in H(\mathcal{L}, \Omega). \tag{29}$$

In order to be more specific, let us consider two relevant examples.

- In the case when the measurement problem concerns a local weighted mean value of the analytical solution u, we have

$$J(u) \equiv M_\zeta(u) = \int_\Omega u(x)\zeta(x)\, \mathrm{d}x,$$

where $\zeta \in L_2(\Omega)$ is a given nonnegative weight function defined on Ω.
- In the case when the measurement problem concerns the normal advective flux of u across the outflow boundary Γ_+, we have

$$J(u) \equiv N_\psi(u) = \int_{\Gamma_+} (\mathbf{b} \cdot \boldsymbol{\mu})\, u\psi\, \mathrm{d}\sigma,$$

where $\psi \in L_2(\Gamma_+)$ is a nonnegative weight function defined on Γ_+.

In the case of the first of these examples, (29) can be rewritten in the following strong form:

$$\begin{aligned} \mathcal{L}^*(z_\delta + \delta\hat{\mathcal{L}}z_\delta) &= \zeta, && x \in \Omega, \\ \delta\hat{\mathcal{L}}z_\delta &= 0, && x \in \Gamma_-, \\ z_\delta + \delta\hat{\mathcal{L}}z_\delta &= 0, && x \in \Gamma_+, \end{aligned} \tag{30}$$

whereas in the case of the second example the strong form of (29) is

$$\begin{aligned} \mathcal{L}^*(z_\delta + \delta\hat{\mathcal{L}}z_\delta) &= 0, && x \in \Omega, \\ \delta\hat{\mathcal{L}}z_\delta &= 0, && x \in \Gamma_-, \\ z_\delta + \delta\hat{\mathcal{L}}z_\delta &= \psi, && x \in \Gamma_+. \end{aligned} \tag{31}$$

The differential operator $\mathcal{L}_\delta$ defined by

$$\mathcal{L}_\delta v \equiv \mathcal{L}^*(v + \delta\hat{\mathcal{L}}v) = -\nabla \cdot (a_\delta \nabla v) + \mathbf{b}_\delta \cdot \nabla v + c_\delta v,$$

where

$$a_\delta = \delta \mathbf{b}\mathbf{b}^{\mathrm{T}}, \qquad \mathbf{b}_\delta = (\delta(c - \hat{c}) - 1)\mathbf{b}, \qquad c_\delta = \mathcal{L}^*(1 + \delta\hat{c}),$$

is a second-order partial differential operator with nonnegative characteristic form, given that the matrix a_δ is only negative semidefinite but not positive definite.

Problem (30) or (31), as the case may be, can be solved numerically using the streamline–diffusion finite element method, say, by considering a shape regular subdivision $\mathcal{T}_{h_d}$ of Ω with mesh function h_δ, where, for $\kappa \in \mathcal{T}_{h_d}$, $h_\delta|_\kappa = \mathrm{diam}(\kappa)$ (see, [34]). We note that, typically, $\mathcal{T}_h$ and $\mathcal{T}_{h_d}$ will differ; after all, the primal and dual problems are of completely different nature so there is no reason (other than that of minimizing the overall cost of the computation) to use the same sequence of computational meshes for the primal and dual problems. Further, we define S_{h_d} to be a finite element space consisting of piecewise polynomials of fixed degree p_d, $p_d \geq 1$ on $\mathcal{T}_{h_d}$. Again, in general, $p_d \neq p$; more importantly, however, if for any reason $\mathcal{T}_{h_d}$ has been chosen to be the same as $\mathcal{T}_h$, then it is necessary to select $p_d > p \geq 1$, for otherwise our numerical approximation to η_κ will be equal to zero and the computed error bound will then become meaningless.

A further question is: *What constitutes an appropriate computational mesh $\mathcal{T}_{h_d}$ for the dual problem?* A suitable mesh for the numerical solution of the dual problem may be designed adaptively through the use of an auxiliary *a posteriori* error bound to ensure that the additional error incurred through the replacement of z (respectively, z_δ) by its finite element approximation $\hat{z}_{h_d}$ (respectively, $\hat{z}_{\delta,h_d}$) does not exceed a small fraction $\varepsilon\mathtt{TOL}$ (with $0 < \varepsilon \ll 1$) of the global tolerance TOL. We shall say more about this approach in the next section when we discuss the discontinuous Galerkin finite element method. For the sake of simplicity of exposition, here we shall adopt a pragmatic and completely empirical approach: the sequence of meshes $\{\mathcal{T}_{h_d}\}$ for the finite element approximation of the dual problem will be generated from a triangular background mesh of the computational domain Ω (which is geometrically different from the background mesh that gives rise to the family $\{\mathcal{T}_h\}$); the background mesh for the dual problem will contain about 50% more elements than the background mesh for the primal problem so as to ensure that the error committed in the numerical solution of the dual problem has little or no effect on the local refinement indicators η_κ, $\kappa \in \mathcal{T}_h$. Furthermore, for simplicity, we set $p_d = p = 1$.

5.4 Numerical experiments

In this section we present a series of numerical experiments to both compare the asymptotic rate of convergence of the *a posteriori* error bound (25) using both the standard and stabilization–dependent dual problems (19) and (20), respectively, as well as to demonstrate the performance of an adaptive algorithm employing the local refinement indicator $|\eta_\kappa^{(2)}|$.

Example 1. The purpose of this example is to highlight the difference between using the standard dual problem (19) and the stabilization–dependent dual problem (20). Let us take $\Omega = (0,1)^2$, $\mathbf{b} = (1+x, 1+y)$, $c = 0$ and

Table 2. Example 1: Convergence of $\|u-u_h\|_{L_2(\Omega)}$ and $|N_\psi(u)-N_\psi(u_h)|$ with $\delta = h/4$, and the rate of convergence k

Mesh	$\|u-u_h\|_{L_2(\Omega)}$	k	$\|N_\psi(u)-N_\psi(u_h)\|$	k
17×17	1.723×10^{-3}	-	2.631×10^{-4}	-
33×33	3.541×10^{-4}	2.28	3.581×10^{-5}	2.88
65×65	8.101×10^{-5}	2.13	4.672×10^{-6}	2.94
129×129	1.967×10^{-5}	2.04	5.965×10^{-7}	2.97
257×257	4.871×10^{-6}	2.01	7.534×10^{-8}	2.99

Table 3. Example 1: Convergence of the terms in the error representation formula (21) with $\delta = h/4$, and the rate of convergence k

Mesh	I_1	k	II_1	k	III_1	k
17×17	4.396×10^{-6}	-	5.667×10^{-6}	-	2.644×10^{-4}	-
33×33	2.980×10^{-7}	3.88	3.569×10^{-7}	3.99	3.587×10^{-5}	2.88
65×65	1.951×10^{-8}	3.93	2.238×10^{-8}	4.00	4.675×10^{-6}	2.94
129×129	1.249×10^{-9}	3.97	1.401×10^{-9}	4.00	5.967×10^{-7}	2.97
257×257	7.903×10^{-11}	3.98	8.764×10^{-11}	4.00	7.535×10^{-8}	2.99

$f=0$ with boundary condition

$$u(x,y)=\begin{cases}1-y^6\,, & \text{for } x=0,\ 0\le y\le 1\,,\\ \mathrm{e}^{-10x^4}\,, & \text{for } 0\le x\le 1,\ y=0\,.\end{cases}$$

We set $C_\delta = 1/4$ in the definition of the stabilization parameter δ and let

$$\psi=\begin{cases}1-\sin^2(\pi(1-y)/2)\cos(\pi y/2)\,, & \text{for } x=1,\ 0\le y\le 1\,,\\ 1-(1-x)^3-(1-x)^4/2\,, & \text{for } 0\le x\le 1,\ y=1\,.\end{cases}$$

The goal of the exercise is to compute the weighted outflow normal flux, $J(u)=N_\psi(u)=((\mathbf{b}\cdot\boldsymbol{\mu})u,\psi)_{\Gamma_+}$, of the analytical solution u. For purposes of comparison, the analytical solution u and the dual solutions z and z_δ have been computed to high accuracy using the method of characteristics; in particular, the exact value of the weighted outward normal flux was found to be $N_\psi(u) = 2.641445145716141$.

In Table 2 we have displayed the orders of convergence of the error in the $L_2(\Omega)$ norm as well as in the functional $N_\psi(\cdot)$ as h tends to zero on a sequence of uniform triangular meshes obtained from uniform square meshes by cutting each mesh square into two triangles, and S_h consisting of continuous piecewise polynomials of degree 1 ($p=1$). We observe that $|N_\psi(u)-N_\psi(u_h)|$ converges

Table 4. Example 1: Convergence of the terms in the error representation formula (22) with $\delta = h/4$, and the rate of convergence k

Mesh	$\mathrm{I}_{1,\delta}$	k	$\mathrm{II}_{1,\delta}$	k	$\mathrm{III}_{1,\delta}$	k
17×17	4.396×10^{-6}	-	1.400×10^{-5}	-	2.447×10^{-4}	-
33×33	2.980×10^{-7}	3.88	2.130×10^{-6}	2.72	3.339×10^{-5}	2.87
65×65	1.951×10^{-8}	3.93	2.748×10^{-7}	2.95	4.378×10^{-6}	2.93
129×129	1.249×10^{-9}	3.97	3.414×10^{-8}	3.01	5.611×10^{-7}	2.96
257×257	7.903×10^{-11}	3.98	4.225×10^{-9}	3.01	7.104×10^{-8}	2.98

like $\mathcal{O}(h^3)$ with $\mathcal{O}(h)$ stabilization, while the $L_2(\Omega)$ norm of $u - u_h$ is $\mathcal{O}(h^2)$. In Tables 3 and 4 we show the convergence of each of the terms in the error representation formulas (21) and (22), respectively. Here, we recall the notation

$$\mathrm{I}_1 = |\langle r_h^-, z - z_h \rangle_-|, \quad \mathrm{II}_1 = |(r_h, z - z_h)|, \quad \mathrm{III}_1 = |(r_h, \delta \hat{\mathcal{L}} z_h)|,$$

and

$$\mathrm{I}_{1,\delta} = |\langle r_h^-, z_\delta - z_{\delta,h} \rangle_-|, \qquad \mathrm{II}_{1,\delta} = |(r_h, z_\delta - z_{\delta,h})|,$$
$$\mathrm{III}_{1,\delta} = |(r_h, \delta \hat{\mathcal{L}}(z_\delta - z_{\delta,h}))|$$

introduced in Section 5.1, cf. (26) and (27), respectively.

Table 3 shows that the second term in the error representation formula (21), i.e. term II_1, is superconvergent; here $\mathrm{II}_1 = \mathcal{O}(h^4)$ as h tends to zero. Term III_1 which arises as the result of the stabilization employed exhibits (superconvergent) $\mathcal{O}(h^3)$ convergence and entirely dominates the error in the weighted outward normal flux. On the other hand, Table 4 indicates that while the first term, term $\mathrm{I}_{1,\delta}$, in the error representation formula (22) associated with the stabilization-dependent dual problem (20) is fourth-order convergent as h tends to zero, the second term, term $\mathrm{II}_{1,\delta}$, is now only $\mathcal{O}(h^3)$ as the mesh is refined, as is term $\mathrm{III}_{1,\delta}$. Therefore, by using the stabilization-dependent dual problem (20) the terms involving integration over the entire computational domain (namely, terms $\mathrm{II}_{1,\delta}$ and $\mathrm{III}_{1,\delta}$) have now become 'balanced'. This balancing of the terms in the error representation formula (22) will be essential for ensuring that any estimate obtained by further bounding $\mathrm{I}_{1,\delta}$, $\mathrm{II}_{1,\delta}$ and $\mathrm{III}_{1,\delta}$ from above upon localization exhibits the same asymptotic convergence rate as $|N_\psi(u) - N_\psi(u_h)|$.

To demonstrate that this is so when the stabilization-dependent dual problem (20) is used, but not so when the standard dual problem (19) is employed, we show in Table 5 the effectivity indices:

$$\theta = \mathcal{E}_{|\Omega|}(u_h)/|N_\psi(u) - N_\psi(u_h)| \quad \text{and} \quad \theta_\delta = \mathcal{E}_{|\Omega|,\delta}(u_h)/|N_\psi(u) - N_\psi(u_h)|,$$

Table 5. Example 1: Comparison of the *a posteriori* error bounds $\mathcal{E}_{|\Omega|}(u_h)$ and $\mathcal{E}_{|\Omega|,\delta}(u_h)$ arising from the dual problems (19) and (20), respectively, and the corresponding convergence rates k and effectivity indices θ and θ_δ

Mesh	$\mathcal{E}_{\|\Omega\|}(u_h)$	k	θ	$\mathcal{E}_{\|\Omega\|,\delta}(u_h)$	k	θ_δ
17 × 17	1.921×10^{-3}	-	7.30	2.642×10^{-4}	-	1.00
33 × 33	4.936×10^{-4}	1.96	13.8	3.639×10^{-5}	2.86	1.02
65 × 65	1.246×10^{-4}	1.99	26.7	4.912×10^{-6}	2.89	1.05
129 × 129	3.128×10^{-5}	1.99	52.4	6.420×10^{-7}	2.94	1.08
257 × 257	7.833×10^{-6}	2.00	104.0	8.624×10^{-8}	2.90	1.14

where $\mathcal{E}_{|\Omega|}(u_h)$ (respectively, $\mathcal{E}_{|\Omega|,\delta}(u_h)$) is defined through the use $\eta_\kappa = \eta_\kappa^{(1)}$ (respectively, $\eta_\kappa = \eta_\kappa^{(2)}$). Here, very high accuracy approximations to the dual solutions z and z_δ, computed by means of the method of characteristics, have been used in the calculation of $\eta_\kappa^{(1)}$ and $\eta_\kappa^{(2)}$, respectively, to ensure that the effects observed are not due to additional errors which arise from the numerical solution of the dual problem. We observe that $\mathcal{E}_{|\Omega|}(u_h)$ converges at the suboptimal rate of $\mathcal{O}(h^2)$ as h tends to zero while $\mathcal{E}_{|\Omega|,\delta}(u_h)$ exhibits the same $\mathcal{O}(h^3)$ rate as the true error $|N_\psi(u) - N_\psi(u_h)|$. The deterioration of the asymptotic convergence rate of $\mathcal{E}_{|\Omega|}(u_h)$ results in large effectivity indices which grow under mesh refinement, while θ_δ remains in the range $1 - 1.2$.

The final two tables concerning this example demonstrate that localization adversely affects the *a posteriori* error bound which stems from using the standard dual problem, while the bound that arises from the error representation formula based on the stabilization-dependent dual problem is insensitive to localization. We define

$$\mathrm{I}_2 = \sum_{\kappa \in \mathcal{T}_h} |\langle r_h^-, z - z_h \rangle_{\partial\kappa \cap \Gamma_-}|,$$

$$\mathrm{II}_2 = \sum_{\kappa \in \mathcal{T}_h} |(r_h, z - z_h)_\kappa|, \qquad \mathrm{III}_2 = \sum_{\kappa \in \mathcal{T}_h} |(r_h, \delta \hat{\mathcal{L}} z_h)_\kappa|,$$

and, analogously, we introduce

$$\mathrm{I}_{2,\delta} = \sum_{\kappa \in \mathcal{T}_h} |\langle r_h^-, z_\delta - z_{\delta,h} \rangle_{\partial\kappa \cap \Gamma_-}|,$$

$$\mathrm{II}_{2,\delta} = \sum_{\kappa \in \mathcal{T}_h} |(r_h, z_\delta - z_{\delta,h})_\kappa|, \qquad \mathrm{III}_{2,\delta} = \sum_{\kappa \in \mathcal{T}_h} |(r_h, \delta \hat{\mathcal{L}}(z_\delta - z_{\delta,h}))_\kappa|.$$

Table 6 shows each of the terms I_2, II_2 and III_2 in the error bound $\mathcal{E}_{|\Omega|}(u_h)$ together with their respective rates of convergence. We see that term I_2 arising from the localization of term I_1 still converges like $\mathcal{O}(h^4)$ as h tends to

Table 6. Example 1: Convergence of the terms I_2, II_2 and III_2 in $\mathcal{E}_{|\Omega|}(u_h)$

Mesh	I_2	k	II_2	k	III_2	k
17×17	7.088×10^{-6}	-	5.985×10^{-5}	-	1.954×10^{-3}	-
33×33	4.513×10^{-7}	3.97	7.596×10^{-6}	2.98	4.986×10^{-4}	1.97
65×65	2.923×10^{-8}	3.95	9.502×10^{-7}	3.00	1.253×10^{-4}	1.99
129×129	1.872×10^{-9}	3.96	1.188×10^{-7}	3.00	3.137×10^{-5}	2.00
257×257	1.186×10^{-10}	3.98	1.484×10^{-8}	3.00	7.844×10^{-6}	2.00

Table 7. Example 1: Convergence of the terms $\mathrm{I}_{2,\delta}$, $\mathrm{II}_{2,\delta}$, $\mathrm{III}_{2,\delta}$ in $\mathcal{E}_{|\Omega|,\delta}(u_h)$

Mesh	$\mathrm{I}_{2,\delta}$	k	$\mathrm{II}_{2,\delta}$	k	$\mathrm{III}_{2,\delta}$	k
17×17	7.088×10^{-6}	-	1.497×10^{-4}	-	2.447×10^{-4}	-
33×33	4.513×10^{-7}	3.97	2.297×10^{-5}	2.70	3.339×10^{-5}	2.87
65×65	2.923×10^{-8}	3.95	3.250×10^{-6}	2.82	4.398×10^{-6}	2.92
129×129	1.872×10^{-9}	3.96	4.357×10^{-7}	2.90	5.680×10^{-7}	2.95
257×257	1.186×10^{-10}	3.98	5.657×10^{-8}	2.95	7.609×10^{-8}	2.90

zero, and term II_2 which stems from the localization of term II_1 is now third-order convergent; however, unlike III_1, term III_2 is now only second-order convergent. Table 7 displays the analogous terms $\mathrm{I}_{2,\delta}$, $\mathrm{II}_{2,\delta}$ and $\mathrm{III}_{2,\delta}$; clearly, these exhibit exactly the same convergence rates as $\mathrm{I}_{1,\delta}$, $\mathrm{II}_{1,\delta}$ and $\mathrm{III}_{1,\delta}$, respectively. This insensitivity of the convergence rate of the *a posteriori* error bound based on the dual problem (29) to localization is of great importance from the point of view of adaptive algorithms, given that any standard mesh refinement criterion will require a local quantity, such as $|\eta_\kappa|$, to decide which elements need to be refined/derefined.

Example 2. Let us again consider the transport equation

$$\mathbf{b} \cdot \nabla u + cu = f$$

in $\Omega = (0,1)^2$, but now with $\mathbf{b} = (10y^2 - 12x + 1, 1 + y)$, $c = 0$ and $f = 0$. Here the characteristics enter Ω through the bottom of the square and through the two vertical sides, and exit through the top edge. We impose the following

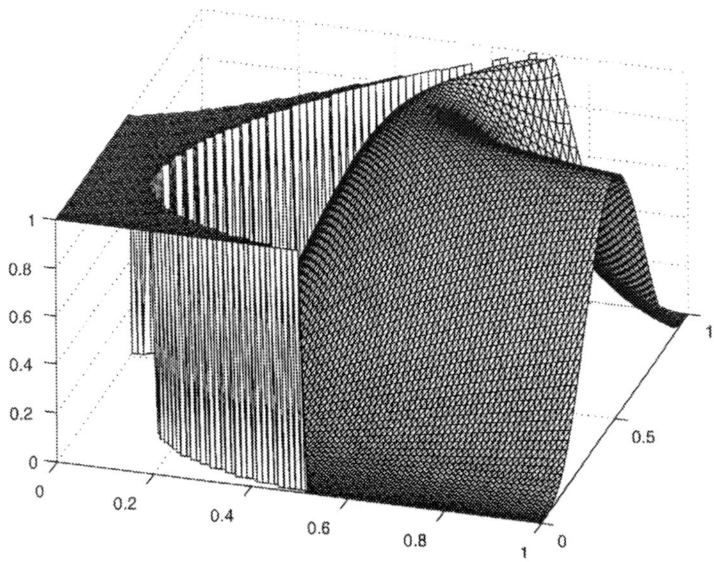

Fig. 4. Example 2: The analytical solution to the primal problem

inflow boundary conditions:

$$u(x,y) = \begin{cases} 0\,, & \text{for } x=0,\ 0.5<y\le 1\,,\\ 1\,, & \text{for } x=0,\ 0<y\le 0.5\,,\\ 1\,, & \text{for } 0\le x\le 0.5,\ y=0\,,\\ 0\,, & \text{for } 0.5<x\le 1,\ y=0\,,\\ \sin^3(\pi y)\,, & \text{for } x=1,\ 0\le y\le 1\,. \end{cases}$$

The analytical solution is depicted in Fig. 4. Let us suppose that the aim of the computation is to calculate the weighted normal flux $N_\psi(u)$ of the analytical solution u through the outflow edge of the square, where the weight-function ψ is defined by

$$\psi(x) = \sin(\pi x/2)\,, \qquad \text{for } 0\le x\le 1,\, y=1\,.$$

For purposes of comparison, using the method of characteristics one may compute a highly accurate approximation to u and thereby deduce that $N_\psi(u) = 0.198826929742166869$. In this example we shall only use the *a posteriori* error bound arising from the stabilization-dependent dual problem (20).

Numerical experiments analogous to those in Example 1 indicate that on both uniform triangular and unstructured triangular meshes the convergence rate of the error measured in the $L_2(\Omega)$ norm is $k \approx 0.4$ due to the presence of the two discontinuities in the solution u. In contrast, the error in the weighted outward normal flux $N_\psi(\cdot)$ is $\mathcal{O}(h^3)$ as h tends to zero, cf. Fig. 5. The high

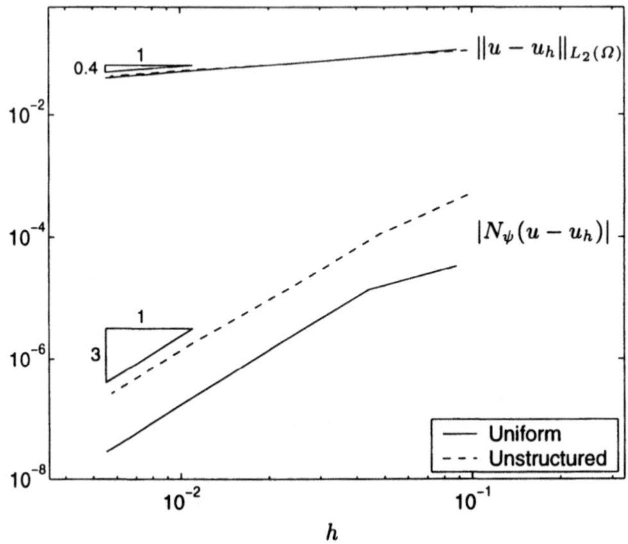

Fig. 5. Example 2: Convergence of $\|u - u_h\|_{L_2(\Omega)}$ and $|N_\psi(u - u_h)|$ with uniform mesh refinement

rate of convergence of $N_\psi(u_h)$ to $N_\psi(u)$ in the presence of a discontinuous primal solution u is attributed to the fact that while the residual terms r_h and r_h^- are large in the neighborhood of the discontinuities in u (cf. Fig. 6(a)), the local weighting terms in $\eta_\kappa^{(2)}$ involving $z_\delta - z_{\delta,h}$ are inactive in these regions of the computational domain, since the dual solution is smooth with 'small' support concentrated near the boundary $y = 1$, cf. Fig. 6(b).

The mesh design in our adaptive algorithm is based on the *optimal mesh criterion* with $\texttt{TOL} = 5.0 \times 10^{-5}$, and $\delta = h/4$. The background meshes for the primal and dual problems as well as the adaptively refined meshes resulting from these are depicted in Fig. 7. Here, Fig. 7(c) shows that most of the nodes in the adaptively refined mesh for the primal problem are concentrated near the outflow boundary where the quantity of interest, $N_\psi(u)$, is defined. It is interesting to note the lack of mesh refinement in the discontinuities as they enter the computational domain from $y = 0$ and $x = 0$. Any refinement indicator based on the residual *alone* will induce heavy (and clearly unnecessary) refinement along the discontinuities in the primal solution, cf. Fig. 8, for example. The adaptively refined mesh for the stabilization–dependent dual problem is, again, finer in the upper half of the computational domain, with most of the elements concentrated near the corners $(x, y) = (0, 1)$ and $(x, y) = (1, 1)$ due to the presence of boundary layers in the dual solution z_δ along Γ_- of thickness $|\mathbf{b} \cdot \boldsymbol{\mu}|\delta/|\mathbf{b}|$.

Finally, in Fig. 9 we show the performance of our adaptive algorithm with $\texttt{TOL} = 5.0 \times 10^{-6}$. The initial meshes are as shown in Fig. 7. It is important to note that even though the stabilization–dependent dual problem has been

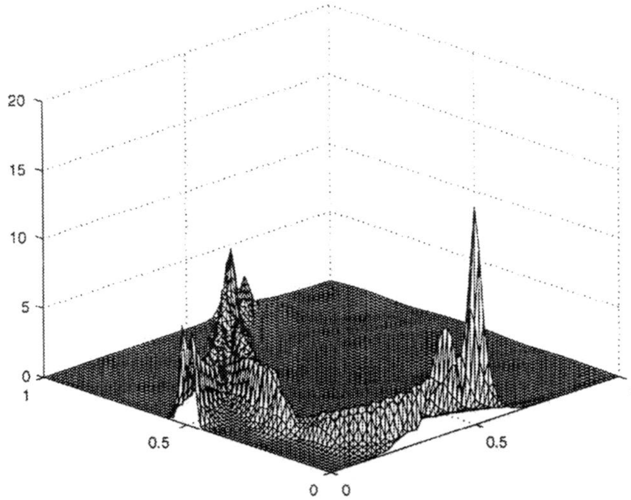

(a)

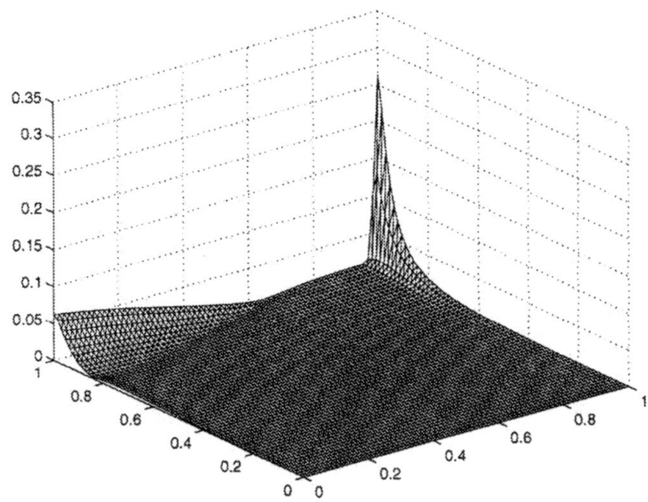

(b)

Fig. 6. Example 2: (a) Scaled residual term $\|r_h\|_{L_2(\kappa)}/h_\kappa$; (b) Scaled weighting term $\|z_\delta - z_{\delta,h}\|_{L_2(\kappa)}/h_\kappa^2$; from [30]

solved numerically, our *a posteriori* error bound $\hat{\mathcal{E}}_{|\Omega|,\delta}$, based on the use of the numerically computed dual solution $\hat{z}_{\delta,h_d}$ and $P_h(\hat{z}_{\delta,h_d})$ in place of z_δ and $z_{\delta,h} = P_h z_\delta$, respectively, remains an upper bound on the true error in our approximation of $N_\psi(u)$. In Fig. 9, $N_\psi(e(\hat{\mathcal{E}}_{|\Omega|,\delta}))$ denotes the true error in the outward normal flux on the sequence of adaptively refined meshes which have been generated using the optimized mesh criterion. Here, Fig. 9 also shows

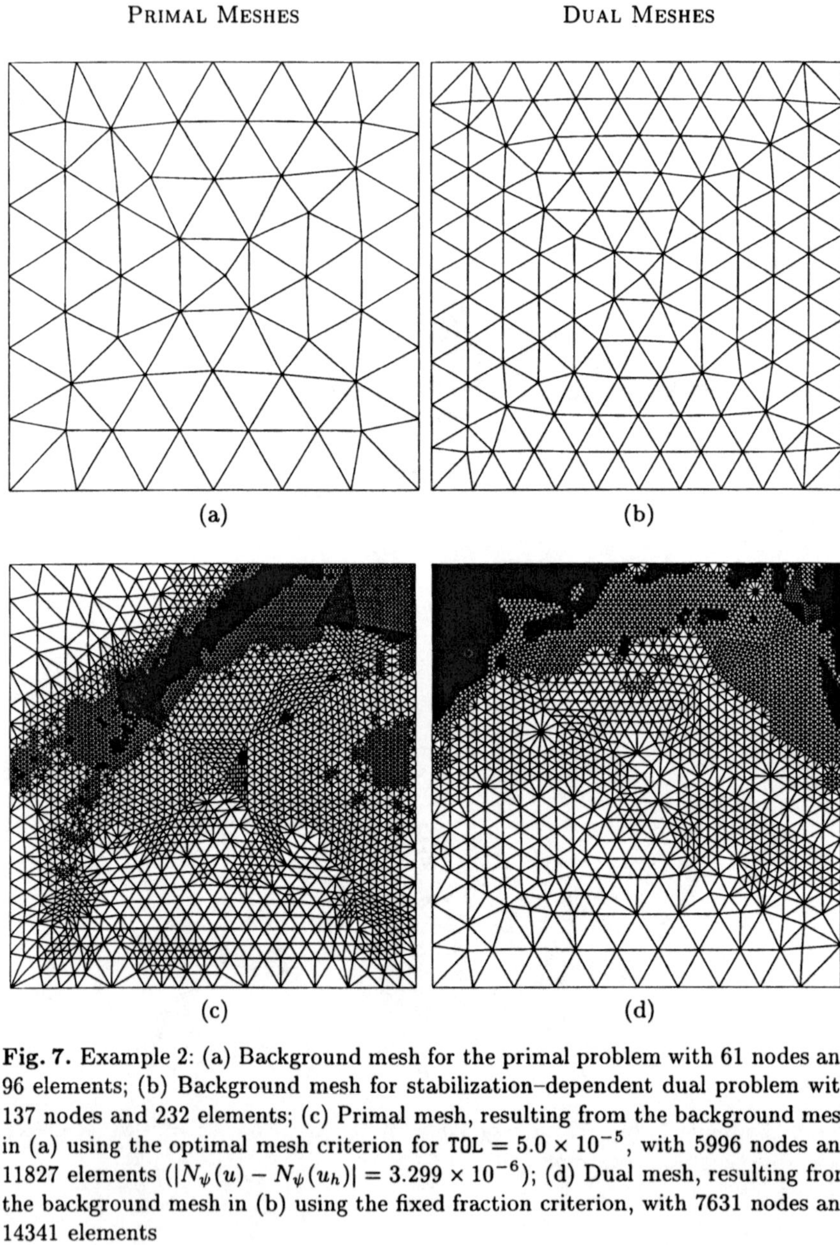

Fig. 7. Example 2: (a) Background mesh for the primal problem with 61 nodes and 96 elements; (b) Background mesh for stabilization–dependent dual problem with 137 nodes and 232 elements; (c) Primal mesh, resulting from the background mesh in (a) using the optimal mesh criterion for $\mathtt{TOL} = 5.0 \times 10^{-5}$, with 5996 nodes and 11827 elements ($|N_\psi(u) - N_\psi(u_h)| = 3.299 \times 10^{-6}$); (d) Dual mesh, resulting from the background mesh in (b) using the fixed fraction criterion, with 7631 nodes and 14341 elements

the true error in the outward normal flux on a sequence of adaptively refined meshes which have been constructed on a sequence of meshes using the energy refinement indicator $\|r_h\|_{L_2(\kappa)}$ on each element κ in $\mathcal{T}_h$ in conjunction with a fixed fraction strategy, cf. Fig. 8. Clearly, the latter approach is inferior:

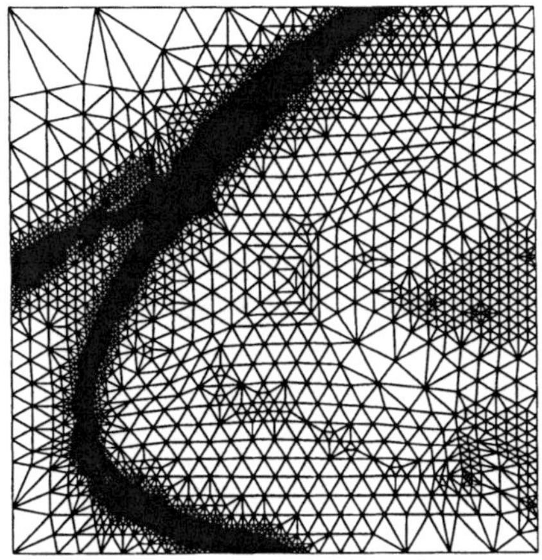

Fig. 8. Example 2: Mesh constructed using energy refinement indicator with 9147 nodes and 18113 elements ($|N_\psi(u-u_h)| = 4.778 \times 10^{-5}$)

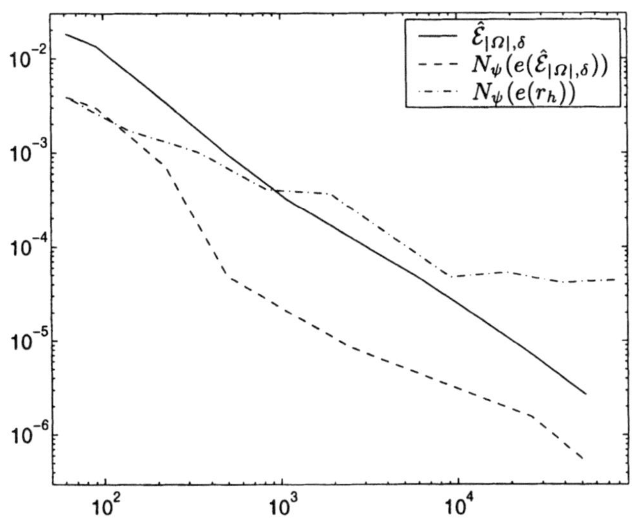

Fig. 9. Example 2: Performance of the adaptive algorithm for TOL $= 5.0 \times 10^{-6}$ and $\delta = h/4$. $N_\psi(e(\hat{\mathcal{E}}_{|\Omega|,\delta}))$ denotes the error in the functional on a sequence of adaptively refined meshes using the stabilization-dependent dual problem; $\hat{\mathcal{E}}_{|\Omega|,\delta}$ is the corresponding Type I *a posteriori* error bound; $N_\psi(e(r_h))$ error denotes the error in the functional when the adaptive meshes are generated by using $\|r_h\|_{L_2(\kappa)}$ only as refinement indicator

stagnation of the error in the functional is observed in the course of the adaptive mesh refinement. The inferiority of the energy refinement indicator

$\|r_h\|_{L_2(\kappa)}$, $\kappa \in \mathcal{T}_h$, is due to the fact that it does not take into account the behavior of the solution to the dual problem.

In conclusion then, the use of a stabilization–dependent dual problem in the *a posteriori* error analysis of a stabilized Galerkin finite element method, such as the streamline–diffusion method, leads to sharper *a posteriori* error bounds than the standard dual problem. This is achieved through 'balancing' the magnitudes of the various terms in the error representation formula, which then leads to economical meshes when implemented into an adaptive algorithm. We have also seen that a Type I *a posteriori* error bound leads to more economical computational meshes than empirical residual-based refinement indicators (such as the energy refinement indicator $\|r_h\|_{L_2(\kappa)}$ that arises from a Type II error bound).

6 *hp*–DGFEM for systems of hyperbolic PDEs

Adaptive finite element methods that exploit both local polynomial degree variation (p-refinement) and local mesh subdivision (h-refinement) offer much greater flexibility and improved efficiency than mesh refinement algorithms which only incorporate h–refinement or p–refinement in isolation.

Since the early analytical paper of Gui and Babuška [23], the benefits of hp–version finite element methods have been clearly established for elliptic boundary value problems (see, for example, the monograph of Schwab [51]), particularly in the field of linear elasticity. The application of hp-version finite element methods to hyperbolic problems is less standard, although their potential in compressible gas dynamics has been demonstrated recently by Flaherty and collaborators; see [2], for example. The argument in favor of using an hp–version finite element method for the numerical solution of a hyperbolic equation rests on the observation that while solutions to these equations may exhibit local singularities and discontinuities, in large parts of the computational domain the solution is typically a real analytic function. Such large variations in the smoothness of the solution can be captured in a particularly simple and flexible manner by using a finite element method based on discontinuous piecewise polynomials, such as the discontinuous Galerkin finite element method (DGFEM). The method admits local variation of the degree of the approximating polynomial and meshes with an arbitrary number of interelement hanging nodes. In its lowest order form, with piecewise polynomials of degree 0, the discontinuous Galerkin finite element method becomes a finite volume method; therefore the hp–version of DGFEM can be thought of as a high-order extension of classical finite volume schemes. For a comprehensive review of discontinuous Galerkin methods, including a detailed historical survey, we refer to [12] and the references cited therein.

The aim of this section is to discuss the *a posteriori* error analysis of the hp–version of the discontinuous Galerkin finite element method; we refer to [9] for earlier work in this area. As in the previous section, we shall be concerned with the derivation of computable error bounds for target functionals of the solution; in Sections 6.1–6.3 we consider first–order symmetric linear hyperbolic systems; Section 6.4 is devoted to the extension of these ideas to nonlinear systems of hyperbolic conservation laws. The work presented here is based on the series of papers [33,35,54,55].

6.1 The model problem and its *hp*–DGFEM approximation

Let Ω be a bounded open polyhedral domain in $\mathbb{R}^n$, $n \geq 2$, and let Γ denote the union of open faces of Ω. We consider the boundary value problem (13); using the notation introduced in Section 4.4, we recall the following symmetric hyperbolic system: find $\mathbf{u} \in H(\mathcal{L}, \Omega)$ such that

$$\mathcal{L}\mathbf{u} \equiv \sum_{j=1}^{n} \frac{\partial}{\partial x_j}(\mathbf{B}_j \mathbf{u}) + \mathbf{C}\mathbf{u} = \mathbf{f} \quad \text{in } \Omega\,, \quad \mathbf{B}_-(\boldsymbol{\mu})\mathbf{u} = \mathbf{B}_-(\boldsymbol{\mu})\mathbf{g} \quad \text{on } \Gamma\,, \tag{32}$$

where $H(\mathcal{L}, \Omega) = \{\mathbf{v} \in [L_2(\Omega)]^m : \mathcal{L}\mathbf{v} \in [L_2(\Omega)]^m\}$ denotes the graph space of the partial differential operator $\mathcal{L}$ in $[L_2(\Omega)]^m$. Here, $\mathbf{B}_-(\boldsymbol{\mu})$ is the negative part of the matrix

$$\mathbf{B}(\boldsymbol{\mu}) \equiv \mathbf{B} \cdot \boldsymbol{\mu} = \mu_1 \mathbf{B}_1 + \ldots + \mu_n \mathbf{B}_n\,,$$

where $\boldsymbol{\mu}$ denotes the unit outward normal vector to the boundary Γ.

We begin by formulating the *hp*–version of the discontinuous Galerkin finite element method (*hp*–DGFEM, for short) for the numerical solution of (32). Let us suppose that $\mathcal{T}_h$ is a regular or 1-irregular subdivision of the computational domain Ω into disjoint open element domains κ such that $\bar{\Omega} = \cup_{\kappa \in \mathcal{T}_h} \bar{\kappa}$. By *1-irregular* we mean that an $(n-1)$–dimensional face of each element κ in $\mathcal{T}_h$ is allowed to contain at most one hanging (irregular) node — typically the barycenter of the face. Further, we shall suppose that the family of subdivisions $\mathcal{T}_h$ is shape–regular and that each $\kappa \in \mathcal{T}_h$ is a smooth bijective image of a fixed master element $\hat{\kappa}$, that is, $\kappa = F_\kappa(\hat{\kappa})$ for all $\kappa \in \mathcal{T}_h$, where $\hat{\kappa}$ is either the open unit simplex or the open unit hypercube in $\mathbb{R}^n$.

On the reference element $\hat{\kappa}$, with $\hat{x} = (\hat{x}_1, \ldots, \hat{x}_n) \in \hat{\kappa}$ and a multi-index $\alpha \in \mathbb{N}^n$, we define spaces of polynomials of degree $p \geq 1$ as follows:

$$\mathcal{Q}_p = \text{span}\,\{\hat{x}^\alpha : 0 \leq \alpha_i \leq p,\ \ 1 \leq i \leq n\}\,, \quad \mathcal{P}_p = \text{span}\,\{\hat{x}^\alpha : 0 \leq |\alpha| \leq p\}\,.$$

With each element κ in $\mathcal{T}_h$ we associate an integer $p_\kappa \geq 1$; collecting the p_κ and F_κ in the vectors $\mathbf{p} = \{p_\kappa : \kappa \in \mathcal{T}_h\}$ and $\mathbf{F} = \{F_\kappa : \kappa \in \mathcal{T}_h\}$, respectively, we introduce the finite element space

$$\begin{aligned} S^{\mathbf{p}}(\Omega, \mathcal{T}_h, \mathbf{F}) = \{\mathbf{v} \in [L_2(\Omega)]^m \ :\ & \mathbf{v}|_\kappa \circ F_\kappa \in [\mathcal{Q}_{p_\kappa}]^m \text{ if } F_\kappa^{-1}(\kappa) \text{ is} \\ & \text{the open unit hypercube and } \ \mathbf{v}|_\kappa \circ F_\kappa \in [\mathcal{P}_{p_\kappa}]^m \\ & \text{if } F_\kappa^{-1}(\kappa) \text{ is the open unit simplex; } \kappa \in \mathcal{T}_h\}\,. \end{aligned} \tag{33}$$

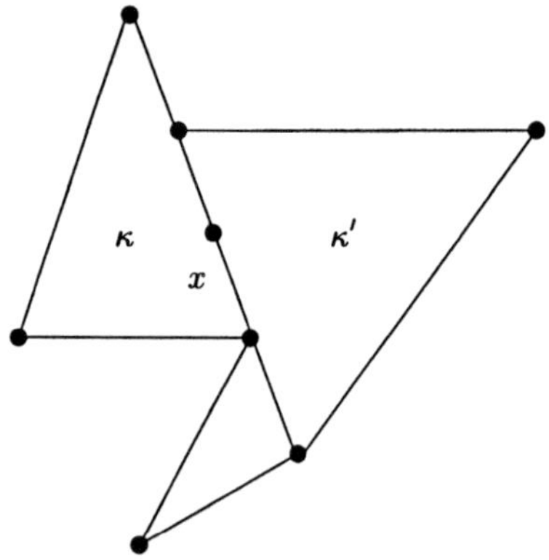

Fig. 10. A point x such that $x \in \partial\kappa$ and $x \in \partial\kappa'$

Assuming that $\mathcal{T}_h$ is a subdivision of Ω, we consider the broken Sobolev space $H^{\mathbf{s}}(\Omega, \mathcal{T}_h)$ of composite index $\mathbf{s}$ with nonnegative components s_κ, $\kappa \in \mathcal{T}_h$, defined by

$$H^{\mathbf{s}}(\Omega, \mathcal{T}_h) = \{\mathbf{v} \in [L_2(\Omega)]^m : \mathbf{v}|_\kappa \in [H^{s_\kappa}(\kappa)]^m \quad \forall \kappa \in \mathcal{T}_h\}.$$

If $s_\kappa = s \geq 0$ for all $\kappa \in \mathcal{T}_h$, we shall simply write $H^s(\Omega, \mathcal{T}_h)$.

For each element κ in $\mathcal{T}_h$, we denote by $\partial\kappa$ the union of $(n-1)$–dimensional open faces of κ. Given that $x \in \partial\kappa$, we label by $\boldsymbol{\mu}_\kappa(x)$ the unit outward normal vector to $\partial\kappa$ at x. With these conventions, we define $\mathbf{B}(\boldsymbol{\mu}_\kappa)$, $\mathbf{B}_-(\boldsymbol{\mu}_\kappa)$ and $\mathbf{B}_+(\boldsymbol{\mu}_\kappa)$ analogously to $\mathbf{B}(\boldsymbol{\mu})$, $\mathbf{B}_-(\boldsymbol{\mu})$ and $\mathbf{B}_+(\boldsymbol{\mu})$ in Section 4.4, respectively.

For each $\kappa \in \mathcal{T}_h$ and any $\mathbf{v} \in [H^1(\kappa)]^m$ we denote by $\mathbf{v}_\kappa^+$ the interior trace of $\mathbf{v}$ on $\partial\kappa$ (i.e., the trace of $\mathbf{v}$ on $\partial\kappa$ taken from within κ). Let us consider an element κ such that the set $\partial\kappa\backslash\Gamma$ is nonempty; then for each $x \in \partial\kappa\backslash\Gamma$ (with the exception of a set of $(n-1)$–dimensional measure zero) there exists a unique element κ', depending on the choice of x, such that $x \in \partial\kappa'$, cf. Fig. 10. Suppose that $\mathbf{v} \in H^1(\Omega, \mathcal{T}_h)$. If $\partial\kappa\backslash\Gamma$ is nonempty for some $\kappa \in \mathcal{T}_h$, then we define the outer trace $\mathbf{v}_\kappa^-$ of $\mathbf{v}$ on $\partial\kappa\backslash\Gamma$ relative to κ as the inner trace $\mathbf{v}_{\kappa'}^+$ relative to those elements κ' for which $\partial\kappa'$ has intersection with $\partial\kappa\backslash\Gamma$ of positive $(n-1)$–dimensional measure. Below, it will always be clear from the context which element κ in the subdivision $\mathcal{T}_h$ the quantities $\boldsymbol{\mu}_\kappa$, $\mathbf{v}_\kappa^+$ and $\mathbf{v}_\kappa^-$ correspond to; therefore, for the sake of notational simplicity, we shall suppress the letter κ in the subscript and write, respectively, $\boldsymbol{\mu}$, $\mathbf{v}^+$, and $\mathbf{v}^-$ instead.

For $\mathbf{v}, \mathbf{w} \in H^1(\Omega, \mathcal{T}_h)$, we define the bilinear form

$$\begin{aligned} B_{\mathrm{DG}}(\mathbf{w}, \mathbf{v}) = & \sum_{\kappa \in \mathcal{T}_h} \int_\kappa \mathbf{w} \cdot \mathcal{L}^* \mathbf{v} \, \mathrm{d}x + \sum_{\kappa \in \mathcal{T}_h} \int_{\partial\kappa\backslash\Gamma} \mathcal{H}(\mathbf{w}^+, \mathbf{w}^-, \boldsymbol{\mu}) \cdot \mathbf{v}^+ \, \mathrm{d}\sigma \\ & + \sum_{\kappa \in \mathcal{T}_h} \int_{\partial\kappa\cap\Gamma} \mathbf{B}_+(\boldsymbol{\mu}) \mathbf{w}^+ \cdot \mathbf{v}^+ \, \mathrm{d}\sigma, \end{aligned}$$

where $\mathcal{L}^*$ is the formal adjoint of $\mathcal{L}$ defined by

$$\mathcal{L}^*\mathbf{v} \equiv -(\mathbf{B}\cdot\nabla)\mathbf{v} + \mathbf{C}^T\mathbf{v}\,,$$

and $\mathcal{H}(\cdot,\cdot,\cdot)$ is a numerical flux function, assumed to be Lipschitz continuous, and such that:

(i) $\mathcal{H}$ is consistent; i.e., $\mathcal{H}(\mathbf{u},\mathbf{u},\boldsymbol{\mu})|_{\partial\kappa\setminus\Gamma} = \mathbf{B}(\boldsymbol{\mu})\mathbf{u}|_{\partial\kappa\setminus\Gamma}$ for all κ in $\mathcal{T}_h$;
(ii) $\mathcal{H}(\cdot,\cdot,\cdot)$ is conservative; i.e., $\mathcal{H}(\mathbf{u}^+,\mathbf{u}^-,\boldsymbol{\mu})|_{\partial\kappa\setminus\Gamma} = -\mathcal{H}(\mathbf{u}^-,\mathbf{u}^+,-\boldsymbol{\mu})|_{\partial\kappa\setminus\Gamma}$.

The choice of the numerical flux function is completely independent of the finite element space employed. For example, one may choose

$$\begin{aligned}\mathcal{H}(\mathbf{u}^+,\mathbf{u}^-,\boldsymbol{\mu}) &= \mathbf{B}_+(\boldsymbol{\mu})\mathbf{u}^+ + \mathbf{B}_-(\boldsymbol{\mu})\mathbf{u}^- \\ &= \frac{1}{2}\left(\mathbf{B}(\boldsymbol{\mu})\mathbf{u}^+ + \mathbf{B}(\boldsymbol{\mu})\mathbf{u}^-\right) - \frac{1}{2}|\mathbf{B}(\boldsymbol{\mu})|(\mathbf{u}^- - \mathbf{u}^+)\,,\end{aligned}$$

where $|\mathbf{B}(\boldsymbol{\mu})| = \mathbf{B}_+(\boldsymbol{\mu}) - \mathbf{B}_-(\boldsymbol{\mu})$. For $\mathbf{v} \in H^1(\Omega,\mathcal{T}_h)$, we consider the linear functional

$$\ell_{\mathrm{DG}}(\mathbf{v}) = \sum_{\kappa\in\mathcal{T}_h}\int_\kappa \mathbf{f}\cdot\mathbf{v}\;\mathrm{d}x - \sum_{\kappa\in\mathcal{T}_h}\int_{\partial\kappa\cap\Gamma} \mathbf{B}_-(\boldsymbol{\mu})\,\mathbf{g}\cdot\mathbf{v}^+\;\mathrm{d}\sigma\,.$$

The *hp*–DGFEM for the numerical solution of the symmetric hyperbolic system (32) is defined as follows: find $\mathbf{u}_{\mathrm{DG}} \in S^{\mathbf{p}}(\Omega,\mathcal{T}_h,\mathbf{F})$ such that

$$B_{\mathrm{DG}}(\mathbf{u}_{\mathrm{DG}},\mathbf{v}) = \ell_{\mathrm{DG}}(\mathbf{v}) \quad \forall \mathbf{v}\in S^{\mathbf{p}}(\Omega,\mathcal{T}_h,\mathbf{F})\,. \tag{34}$$

For the *a priori* error analysis of *hp*–DGFEM, including *hp*–optimal error bounds, we refer to [31,32].

6.2 *A posteriori* error analysis

Suppose that $\mathtt{TOL} > 0$ is a prescribed tolerance and $J(\cdot)$ is a linear functional defined on a linear space which contains $H(\mathcal{L},\Omega) + S^{\mathbf{p}}(\Omega,\mathcal{T}_h,\mathbf{F})$. We shall consider the measurement problem concerned with computing a numerical approximation $\mathbf{u}_{\mathrm{DG}}$ from $S^{\mathbf{p}}(\Omega,\mathcal{T}_h,\mathbf{F})$ such that

$$|J(\mathbf{u}) - J(\mathbf{u}_{\mathrm{DG}})| \le \mathtt{TOL}\,. \tag{35}$$

As in the previous section where we discussed stabilized finite element methods (see, also [30]), we proceed by deriving an *a posteriori* bound on the error between $J(\mathbf{u})$ and $J(\mathbf{u}_{\mathrm{DG}})$ and demanding that this bound does not exceed the prescribed tolerance.

Let us first define the following *dual* problem: find $\mathbf{z}$ in $H(\mathcal{L}^*,\Omega)$ such that

$$B_{\mathrm{DG}}(\mathbf{w},\mathbf{z}) = J(\mathbf{w}) \quad \forall\mathbf{w}\in H(\mathcal{L},\Omega)\,, \tag{36}$$

where $H(\mathcal{L}^*,\Omega)$ denotes the graph space of the adjoint operator $\mathcal{L}^*$ in $L_2(\Omega)$. It will be assumed throughout that (36) possesses a unique solution.

Our *a posteriori* bound on the error between $J(\mathbf{u})$ and $J(\mathbf{u}_{\mathrm{DG}})$ will be expressed in terms of the finite element residual $\mathbf{r}_{h,p}$ defined on $\kappa \in \mathcal{T}_h$ by

$$\mathbf{r}_{h,p}|_\kappa = (\mathbf{f} - \mathcal{L}\mathbf{u}_{\mathrm{DG}})|_\kappa\,,$$

which measures the extent to which $\mathbf{u}_{\mathrm{DG}}$ fails to satisfy the differential equation on the union of the elements κ in the mesh $\mathcal{T}_h$; thus we refer to $\mathbf{r}_{h,p}$ as the *internal residual.* Also, since $\mathbf{u}_{\mathrm{DG}}$ satisfies the boundary conditions only approximately, the vector $\mathbf{B}_-(\boldsymbol{\mu})(\mathbf{g}-\mathbf{u}_{\mathrm{DG}})$ need not be zero on Γ; thus for each element κ with $\partial\kappa \cap \Gamma$ nonempty, we define the *boundary residual* $\boldsymbol{\rho}_{h,p}$ by

$$\boldsymbol{\rho}_{h,p}|_{\partial\kappa\cap\Gamma} = \mathbf{B}_-(\boldsymbol{\mu})(\mathbf{u}^+_{\mathrm{DG}} - \mathbf{g})|_{\partial\kappa\cap\Gamma}\,.$$

If our finite element space consisted of continuous piecewise polynomial functions, as in the streamline–diffusion finite element method with weakly imposed boundary condition, these two residuals would be the only ones we'd encounter. Here, however, due to the fact that a discontinuous piecewise polynomial approximation is used, we need to introduce a third residual which measures the extent to which the numerical solution fails to satisfy continuity of the normal flux across element interfaces. Thus, on $\partial\kappa \setminus \Gamma$, we define the *flux residual*

$$\boldsymbol{\sigma}_{h,p}|_{\partial\kappa\setminus\Gamma} = \left(\mathbf{B}(\boldsymbol{\mu})\mathbf{u}^+_{\mathrm{DG}} - \mathcal{H}(\mathbf{u}^+_{\mathrm{DG}},\mathbf{u}^-_{\mathrm{DG}},\boldsymbol{\mu})\right)|_{\partial\kappa\setminus\Gamma}\,.$$

With these definitions we deduce from (34) the following *Galerkin orthogonality property* of the *hp*–DGFEM:

$$\begin{aligned} B_{\mathrm{DG}}(\mathbf{u}-\mathbf{u}_{\mathrm{DG}},\mathbf{v}) &= \sum_{\kappa\in\mathcal{T}_h}(\mathbf{r}_{h,p},\mathbf{v})_\kappa + \sum_{\kappa\in\mathcal{T}_h}(\boldsymbol{\sigma}_{h,p},\mathbf{v}^+)_{\partial\kappa\setminus\Gamma} \\ &\quad + \sum_{\kappa\in\mathcal{T}_h}(\boldsymbol{\rho}_{h,p},\mathbf{v}^+)_{\partial\kappa\cap\Gamma} = 0 \end{aligned} \tag{37}$$

for all $\mathbf{v}$ in $S^{\mathbf{p}}(\Omega,\mathcal{T}_h,\mathbf{F})$. Upon selecting $\mathbf{w}=\mathbf{u}-\mathbf{u}_{\mathrm{DG}}$ in (36), the linearity of $J(\cdot)$ and (37) give rise to the following (primal) error representation formula:

$$\begin{aligned} J(\mathbf{u}) - J(\mathbf{u}_{\mathrm{DG}}) &= J(\mathbf{u}-\mathbf{u}_{\mathrm{DG}}) \\ &= B_{\mathrm{DG}}(\mathbf{u}-\mathbf{u}_{\mathrm{DG}},\mathbf{z}) = B_{\mathrm{DG}}(\mathbf{u}-\mathbf{u}_{\mathrm{DG}},\mathbf{z}-\mathbf{z}_{h,p}) \\ &\equiv \mathcal{E}_\Omega(\mathbf{u}_{\mathrm{DG}},h,\mathbf{p},\mathbf{z}-\mathbf{z}_{hp}) = \sum_{\kappa\in\mathcal{T}_h}\eta_\kappa\,, \end{aligned} \tag{38}$$

where

$$\eta_\kappa = (\mathbf{r}_{h,p},\mathbf{z}-\mathbf{z}_{h,p})_\kappa + (\boldsymbol{\sigma}_{h,p},(\mathbf{z}-\mathbf{z}_{h,p})^+)_{\partial\kappa\setminus\Gamma} + (\boldsymbol{\rho}_{h,p},(\mathbf{z}-\mathbf{z}_{h,p})^+)_{\partial\kappa\cap\Gamma} \tag{39}$$

for all $\mathbf{z}_{h,p}$ in $S^{\mathbf{p}}(\Omega,\mathcal{T}_h,\mathbf{F})$.

The measurement problem (35) is equivalent to demanding that the *stopping criterion*

$$|\mathcal{E}_{\Omega}(\mathbf{u}_{\mathrm{DG}}, h, \mathbf{p}, \mathbf{z} - \mathbf{z}_{h,p})| \leq \mathtt{TOL} \tag{40}$$

is satisfied, cf. equation (38).

As it stands, the left-hand side of (40) is not 'computable' since it involves the unknown analytical solution $\mathbf{z}$ to the dual problem (36). Let us therefore suppose that problem (36) has been solved numerically, using an *hp*–adaptive DGFEM for example, on a sequence of suitable 'dual' finite element spaces $S^{\tilde{\mathbf{p}}}(\Omega, \tilde{\mathcal{T}}_h, \tilde{\mathbf{F}})$, based on a 'dual' partition $\tilde{\mathcal{T}}_h$ and 'dual' polynomial vector $\tilde{\mathbf{p}}$, which may differ from the 'primal' partition $\mathcal{T}_h$ and 'primal' polynomial vector $\mathbf{p}$, respectively. We shall denote by $\tilde{\mathbf{z}}_{\mathrm{DG}}$ the corresponding approximation to the analytical dual solution $\mathbf{z}$ from the finite element space $S^{\tilde{\mathbf{p}}}(\Omega, \tilde{\mathcal{T}}_h, \tilde{\mathbf{F}})$. Further, $\mathbf{z}_{h,p}$ will denote the $L_2(\Omega)$ orthogonal projection of the numerically computed dual solution $\tilde{\mathbf{z}}_{\mathrm{DG}}$ onto the primal finite element space $S^{\mathbf{p}}(\Omega, \mathcal{T}_h, \mathbf{F})$ defined over the primal subdivision $\mathcal{T}_h$.

Let us now replace $\mathbf{z}$ by $\tilde{\mathbf{z}}_{\mathrm{DG}}$ in η_κ and decompose the error representation formula (38) into terms which are computable, namely those involving the numerical approximation $\tilde{\mathbf{z}}_{\mathrm{DG}}$ to the dual solution, and those that require the analytical dual solution $\mathbf{z}$: on noting that

$$\mathcal{E}_{\Omega}(\mathbf{u}_{\mathrm{DG}}, h, \mathbf{p}, \mathbf{z} - \mathbf{z}_{h,p}) = \mathcal{E}_{\Omega}(\mathbf{u}_{\mathrm{DG}}, h, \mathbf{p}, \tilde{\mathbf{z}}_{\mathrm{DG}} - \mathbf{z}_{h,p}) + \mathcal{E}_{\Omega}(\mathbf{u}_{\mathrm{DG}}, h, \mathbf{p}, \mathbf{z} - \tilde{\mathbf{z}}_{\mathrm{DG}}),$$

we deduce that

$$\begin{aligned} |J(\mathbf{u}) - J(\mathbf{u}_{\mathrm{DG}})| &\leq \mathcal{E}_{\mathrm{P}}^{\mathrm{loc}} + \mathcal{E}_{\mathrm{D}} \\ &\equiv \sum_{\kappa \in \mathcal{T}_h} |\tilde{\eta}_\kappa| + |\mathcal{E}_{\Omega}(\mathbf{u}_{\mathrm{DG}}, h, \mathbf{p}, \mathbf{z} - \tilde{\mathbf{z}}_{\mathrm{DG}})| \,, \end{aligned} \tag{41}$$

where $\tilde{\eta}_\kappa$ is defined analogously to η_κ, cf. (39), with $\mathbf{z}$ replaced by $\tilde{\mathbf{z}}_{\mathrm{DG}}$. Hence a sufficient condition for the validity of our original (noncomputable) stopping criterion (40) is that

$$\mathcal{E}_{\mathrm{P}}^{\mathrm{loc}} + \mathcal{E}_{\mathrm{D}} \leq \mathtt{TOL} \,. \tag{42}$$

We emphasize here that the fundamental difference between the terms $\mathcal{E}_{\mathrm{P}}^{\mathrm{loc}}$ and $\mathcal{E}_{\mathrm{D}}$ is that in the former the absolute value signs appear under the summation over the elements $\kappa \in \mathcal{T}_h$, while in the latter the absolute value sign is outside the sum. It has been shown through numerical experiments in [35] (cf. also [25–27]) that, with a sensible choice of the dual finite element space $S^{\tilde{\mathbf{p}}}(\Omega, \tilde{\mathcal{T}}_h, \tilde{\mathbf{F}})$, $\mathcal{E}_{\mathrm{D}}$ is typically an order of magnitude smaller than $\mathcal{E}_{\mathrm{P}}^{\mathrm{loc}}$. Therefore, $\mathcal{E}_{\mathrm{D}}$ can be safely absorbed into $\mathcal{E}_{\mathrm{P}}^{\mathrm{loc}}$ without compromising the reliability of the adaptive algorithm when the stopping criterion (40) is replaced by

$$\mathcal{E}_{\mathrm{P}}^{\mathrm{loc}} \leq \mathtt{TOL} \,. \tag{43}$$

A bound of the form

$$|J(\mathbf{u}) - J(\mathbf{u}_{\mathrm{DG}})| \lesssim \mathcal{E}_{\mathrm{P}}^{\mathrm{loc}} \equiv \sum_{\kappa \in \mathcal{T}_h} |\tilde{\eta}_\kappa| \tag{44}$$

which explicitly involves the numerical solution $\tilde{\mathbf{z}}_{\mathrm{DG}}$ to the dual problem (through $\tilde{\eta}_\kappa$ defined above) is referred to as a *Type I* (cf. [35]) *a posteriori* error bound.

A possible alternative to using a Type I *a posteriori* error bound would be to completely eliminate the dual solution from the *a posteriori* error bound, through the use of the Cauchy–Schwarz inequality on each of the terms appearing in η_κ, choosing $\mathbf{z}_{h,p}$ as the finite element interpolant (or quasi-interpolant) of $\mathbf{z}$ from $S^{\mathbf{p}}(\Omega, \mathcal{T}_h, \mathbf{F})$, applying interpolation error estimates to bound $\|\mathbf{z} - \mathbf{z}_{h,p}\|_{L_2(\kappa)}$ and $\|\mathbf{z} - \mathbf{z}_{h,p}\|_{L_2(\partial\kappa)}$ in terms of powers of h and p and Sobolev seminorms of $\mathbf{z}$, and finally employing strong stability estimates of Sobolev seminorms of the dual solution in terms of Sobolev norms of the data for the dual problem. Hence, the resulting so-called *Type II a posteriori* error bound will not involve the dual solution. However, such an approach is not without practical drawbacks: quite apart from the fact that the derivation of a Type II *a posteriori* error bound requires a considerable amount of analytical work in order to establish strong stability results for the dual problem in various Sobolev norms and to determine the associated strong stability constants, typically, Type II bounds give a pessimistic over-estimate of the error. Consequently, when employed in an adaptive algorithm to approximate $J(\mathbf{u})$ by $J(\mathbf{u}_{\mathrm{DG}})$ to within a prescribed tolerance, a Type II *a posteriori* error bound may lead to uneconomical meshes; see, [30]. Furthermore, as indicated in Section 2, for hyperbolic problems at least, the presence of the dual solution (or its numerical approximation) as local weight function in the *a posteriori* error bound is highly desirable.

On the other hand, for the Type I error bound (44) to be an accurate approximation of (41) it is essential that $\mathcal{E}_{\mathrm{D}} \ll \mathcal{E}_{\mathrm{P}}^{\mathrm{loc}}$, and for this to be true it is necessary to ensure that the dual finite element space $S^{\tilde{\mathbf{p}}}(\Omega, \tilde{\mathcal{T}}_h, \tilde{\mathbf{F}})$ is sufficiently different from the primal finite element space $S^{\mathbf{p}}(\Omega, \mathcal{T}_h, \mathbf{F})$; for example, if $S^{\tilde{\mathbf{p}}}(\Omega, \tilde{\mathcal{T}}_h, \tilde{\mathbf{F}})$ is chosen to coincide with $S^{\mathbf{p}}(\Omega, \mathcal{T}_h, \mathbf{F})$ then $\mathbf{z}_{h,p} = \tilde{\mathbf{z}}_{\mathrm{DG}}$ and thereby $\mathcal{E}_{\mathrm{P}}^{\mathrm{loc}} = 0$, so $0 < \mathcal{E}_{\mathrm{D}} \ll \mathcal{E}_{\mathrm{P}}^{\mathrm{loc}}$ cannot hold. How can then one select an appropriate dual finite element space $S^{\tilde{\mathbf{p}}}(\Omega, \tilde{\mathcal{T}}_h, \tilde{\mathbf{F}})$?

To answer this question, we begin by noting that

$$N(\cdot) = \mathcal{E}_\Omega(\mathbf{u}_{\mathrm{DG}}, h, \mathbf{p}, \cdot)$$

is a linear functional. Thereby,

$$\mathcal{E}_\Omega(\mathbf{u}_{\mathrm{DG}}, h, \mathbf{p}, \mathbf{z} - \tilde{\mathbf{z}}_{\mathrm{DG}}) = N(\mathbf{z}) - N(\tilde{\mathbf{z}}_{\mathrm{DG}}) .$$

Now, given the tolerance $\mathtt{TOL}$ we select θ, $0 < \theta \ll 1$, split the tolerance

$$\mathtt{TOL} = (1 - \theta)\, \mathtt{TOL} + \theta\, \mathtt{TOL} ,$$

demand that

$$\mathcal{E}_{\mathrm{P}}^{\mathrm{loc}} \leq (1-\theta)\,\mathtt{TOL} \qquad \text{and} \qquad \mathcal{E}_{\mathrm{D}} \leq \theta\,\mathtt{TOL}\,,$$

and aim to satisfy the first of these inequalities with near-equality; for then, both $0 < \mathcal{E}_{\mathrm{D}} \ll \mathcal{E}_{\mathrm{P}}^{\mathrm{loc}}$ and (42) will follow. Thus, it now remains to ensure that $\mathcal{E}_{\mathrm{D}} \leq \theta\,\mathtt{TOL}$; this in turn is equivalent to requiring that

$$|N(\mathbf{z}) - N(\tilde{\mathbf{z}}_{\mathrm{DG}})| \leq \theta\,\mathtt{TOL}\,, \tag{45}$$

which is a measurement problem, analogous to (35), — the problem that we had set out to solve, — except that (45) concerns the dual solution $\mathbf{z}$ while (35) involves the primal solution $\mathbf{u}$. To ensure the validity of (45) we may proceed by deriving an *a posteriori* error bound on $|N(\mathbf{z}) - N(\tilde{\mathbf{z}}_{\mathrm{DG}})|$; the corresponding error representation formula will involve the residuals associated with the numerical solution of the dual problem (36) and the analytical solution of the dual to the dual problem. As a Type I *a posteriori* error bound on $\mathcal{E}_{\mathrm{D}}$ based on such an error representation formula would contain the solution of the dual to the dual problem which would then in turn have to be approximated numerically, in [35] a cruder Type II *a posteriori* error bound on $\mathcal{E}_{\mathrm{D}}$ was employed so as to terminate the potentially infinite succession of mutually dual problems which would otherwise arise. The crudeness of the Type II bound on $\mathcal{E}_{\mathrm{D}}$ is of no concern: from the practical point of view there appears to be little advantage in performing reliable error control for $\mathcal{E}_{\mathrm{D}}$; the aim, when using the Type II bound on $\mathcal{E}_{\mathrm{D}}$, is merely to generate an adequate sequence of finite element approximations $\tilde{\mathbf{z}}_{\mathrm{DG}}$ to the dual solution $\mathbf{z}$ which can then be used to compute $\mathcal{E}_{\mathrm{P}}^{\mathrm{loc}}$.

Thus, given a user–defined tolerance TOL, our approach to the problem of designing an *hp*–finite element space $S^{\mathbf{p}}(\Omega, \mathcal{T}_h, \mathbf{F})$ such that

$$|J(\mathbf{u}) - J(\mathbf{u}_{\mathrm{DG}})| \leq \mathtt{TOL}\,, \tag{46}$$

is as follows. We use the Type I *a posteriori* error bound (42) to construct $S^{\mathbf{p}}(\Omega, \mathcal{T}_h, \mathbf{F})$ such that

$$\mathcal{E}_{\mathrm{P}}^{\mathrm{loc}} \leq \mathtt{TOL} \tag{47}$$

and simply absorb $\mathcal{E}_{\mathrm{D}}$ into $\mathcal{E}_{\mathrm{P}}^{\mathrm{loc}}$. If the stopping criterion (47) is not satisfied, then the elements are marked for refinement/derefinement; to this end, we employ the fixed fraction mesh refinement algorithm (cf. Section 5.2), with refinement and derefinement fractions set to 20% and 10%, respectively.

In the numerical experiments that follow, for the sake of computational simplicity, the dual finite element space $\tilde{S}^{\tilde{\mathbf{p}}}(\Omega, \tilde{\mathcal{T}}_h, \tilde{\mathbf{F}})$ that is used to compute the discontinuous Galerkin approximation $\tilde{\mathbf{z}}_{\mathrm{DG}}$ to $\mathbf{z}$ will be constructed using the same mesh as the one employed for $\mathbf{u}_{\mathrm{DG}}$, i.e., $\tilde{\mathcal{T}}_h \equiv \mathcal{T}_h$, with $\tilde{\mathbf{p}} = \mathbf{p} + \mathbf{1}$ (see [35], for a more general algorithm with $\tilde{\mathcal{T}}_h \neq \mathcal{T}_h$ and $\tilde{\mathbf{p}}$ unrelated to $\mathbf{p}$).

Once an element $\kappa \in \mathcal{T}_h$ has been flagged for refinement or derefinement, a decision must be made whether the local mesh size h_κ or the local degree p_κ of the approximating polynomial should be adjusted accordingly. The choice to perform either h–refinement/derefinement or p–refinement/derefinement is dependent on the local smoothness of the primal and dual solutions $\mathbf{u}$ and $\mathbf{z}$, respectively; cf. [2] and [35].

Let us first consider the case when an element has been flagged for refinement. Clearly, if $\mathbf{u}$ or $\mathbf{z}$ are locally smooth, then p–refinement will be more effective than h–refinement, since the error will be expected to decay quickly within the current element κ as p_κ is increased. On the other hand, if $\mathbf{u}$ and $\mathbf{z}$ have low regularity within the element κ, then h–refinement will be performed. Thus, regions in the computational domain where the primal or dual solution are locally non-smooth are isolated from smooth regions, thereby reducing the influence of singularities/discontinuities as well as making p–refinement more effective. To ensure that the desired level of accuracy is achieved efficiently, in [35] an automatic procedure was developed for deciding when to h– or p–refine, based on the smoothness–estimation strategy proposed by Ainsworth & Senior [3] in the context of norm control for second–order elliptic problems. For a review of various hp-adaptive strategies as well as a description of a new algorithm based on Sobolev index estimation *via* local Legendre expansions, we refer to [33].

If an element has been flagged for derefinement, then the strategy implemented here is to coarsen the mesh in low–error–regions where either the primal or dual solutions $\mathbf{u}$ and $\mathbf{z}$, respectively, are smooth and decrease the degree of the approximating polynomial in low–error–regions when both $\mathbf{u}$ and $\mathbf{z}$ are insufficiently regular, cf. [2] and [35].

6.3 Numerical experiments

The purpose of this section is to assess the performance of our hp–adaptive algorithm for the discontinuous Galerkin finite element method on a simple model problem.

Wave equation. Let us consider the one–dimensional wave equation

$$\frac{\partial^2 \phi}{\partial t^2} - c^2 \frac{\partial^2 \phi}{\partial x^2} = 0\,. \tag{48}$$

We note that by defining $x_1 \equiv x$ and $x_2 \equiv t$, the second–order partial differential equation (48) can be rewritten as the symmetric first–order system

$$\begin{bmatrix} 0 & -c \\ -c & 0 \end{bmatrix} \begin{bmatrix} u_1 \\ u_2 \end{bmatrix}_{x_1} + \begin{bmatrix} 1 & 0 \\ 0 & 1 \end{bmatrix} \begin{bmatrix} u_1 \\ u_2 \end{bmatrix}_{x_2} = \begin{bmatrix} 0 \\ 0 \end{bmatrix}, \tag{49}$$

Table 8. Adaptive algorithm for the wave equation using h–refinement

Elements	DOF	$J(\mathbf{u}-\mathbf{u}_{\mathrm{DG}})$	$\sum_\kappa \tilde{\eta}_\kappa$	θ_1	$\sum_\kappa \|\tilde{\eta}_\kappa\|$	θ_2
32	256	0.1917e-02	0.2462e-02	1.28	0.3139e-02	1.64
47	376	0.1191e-02	0.1392e-02	1.17	0.2108e-02	1.77
74	592	-0.1384e-04	-0.1734e-04	1.25	0.4570e-03	33.01
116	928	-0.6897e-05	-0.6779e-05	0.98	0.2610e-03	37.84
200	1600	-0.3506e-05	-0.3407e-05	0.97	0.6041e-04	17.23
371	2968	-0.8052e-06	-0.9045e-06	1.12	0.2481e-04	30.81
611	4888	-0.7429e-07	-0.7133e-07	0.96	0.6982e-05	93.98
926	7408	-0.1920e-06	-0.1913e-06	1.00	0.4202e-05	21.89
1607	12856	0.3443e-07	0.3382e-07	0.98	0.1218e-05	35.39
2624	20992	-0.1646e-07	-0.1649e-07	1.00	0.6956e-06	42.25
4382	35056	0.4606e-08	0.4562e-08	0.99	0.2853e-06	61.95
7247	57976	0.4882e-09	0.4800e-09	0.98	0.1173e-06	240.29

where $u_1 = \partial\phi/\partial x_2$ and $u_2 = c\partial\phi/\partial x_1$. Here we let $\Omega = (0,1)\times(0,1/2)$ and $c = 1/2$, with

$$u_1(x_1,0) = \begin{cases} 1, & \text{for } 0.45 < x_1 \le 0.55\,, \\ 0, & \text{otherwise}\,, \end{cases}$$

and

$$u_2(x_1,0) = c\,e^{-100(x_1-1/2)^2}\,,$$

for $0 \le x_1 \le 1$; this is a variant of the test problem considered in [29].

Our aim is to compute the weighted line integral of the first component of $\mathbf{u}$ along the upper side of the boundary Γ corresponding to $0 \le x_1 \le 1$, $x_2 = 1/2$, defined as

$$J(\mathbf{u}) = \int_0^1 \psi(x_1)\,u_1(x_1,1/2)\,\mathrm{d}x_1\,,$$

where the weight function ψ is defined by the Gaussian

$$\psi(x_1) = \mathrm{e}^{-\sigma^2(x_1-1/2)^2}\,, \qquad \text{for } 0 \le x_1 \le 1\,,$$

with $\sigma = 40$. The true value of the functional is $J(\mathbf{u}) \approx 0$.

In order to further highlight the advantages of designing an adaptive finite element algorithm based on the weighted refinement indicators $|\tilde{\eta}_\kappa|$ over traditional refinement strategies which do not require the solution of the dual problem (36), cf. Example 2 in Section 5.4, we first study the performance

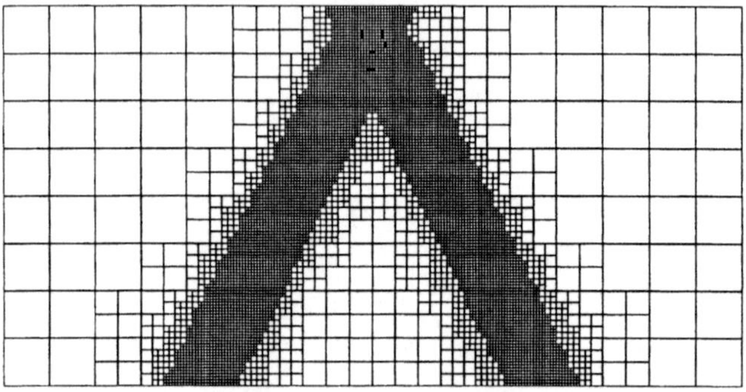

(a)

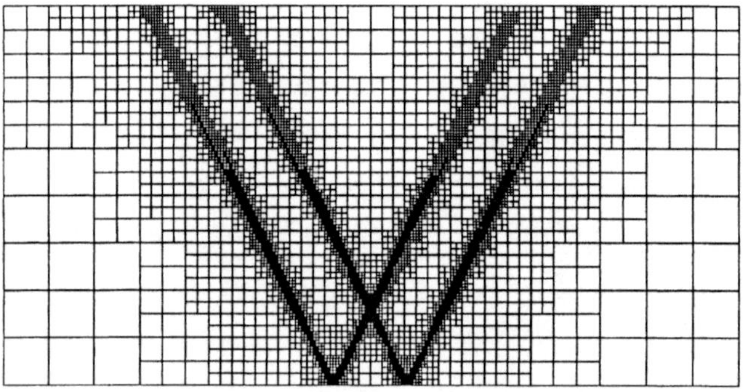

(b)

Fig. 11. Wave equation. (a) Mesh constructed using dual refinement indicator with 7247 elements and 57976 degrees of freedom ($|J(\mathbf{u}) - J(\mathbf{u}_{\mathrm{DG}})| = 4.882 \times 10^{-10}$); (b) Mesh constructed using the empirical refinement indicator with 9293 elements and 74344 degrees of freedom ($|J(\mathbf{u}) - J(\mathbf{u}_{\mathrm{DG}})| = 2.050 \times 10^{-8}$)

of our adaptive strategy with h–refinement. Here, we consider the empirical refinement indicator

$$\eta_\kappa^{\mathrm{empirical}} = \|h\mathbf{r}_{h,p}\|_{L_2(\kappa)} + \|h^{1/2}\boldsymbol{\sigma}_{h,p}\|_{L_2(\partial\kappa\setminus\Gamma)} + \|h^{1/2}\boldsymbol{\rho}_{h,p}\|_{L_2(\partial\kappa\cap\Gamma)} \,,$$

which stems from a Type II *a posteriori* error analysis, cf. [35].

In Table 8 we illustrate the performance of the adaptive algorithm with $\mathbf{p} = \mathbf{1}$; here, we show the number of elements and degrees of freedom (DOF) in $S^{\mathbf{1}}(\Omega, \mathcal{T}_h, \mathbf{F})$, the true error in the functional $J(\mathbf{u} - \mathbf{u}_{\mathrm{DG}})$, the computed error representation formula $\tilde{\mathcal{E}}_\Omega \equiv \sum_{\kappa\in\mathcal{T}_h} \tilde{\eta}_\kappa$, the *a posteriori* error bound (44) and their respective effectivity indices θ_1 and θ_2. We see that the quality of the computed error representation formula $\tilde{\mathcal{E}}_\Omega$ is extremely good, in the sense that $\theta_1 = \tilde{\mathcal{E}}_\Omega / J(u - u_h)$ is close to one, even on relatively coarse meshes. Furthermore, we observe that the Type I *a posteriori* error bound is reliable, in the sense that the second effectivity index θ_2 overestimates the true error

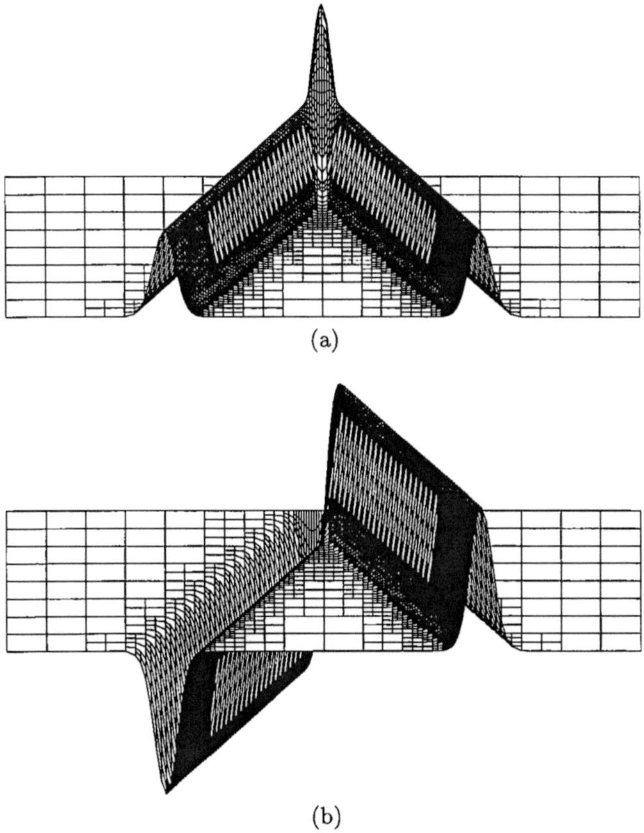

Fig. 12. Dual solution for the wave equation: (a) $\mathbf{z}_1$; (b) $\mathbf{z}_2$

in the computed functional by a consistent factor as the finite element space $S^1(\Omega, \mathcal{T}_h, \mathbf{F})$ is enriched.

In Fig. 11 we show the meshes generated using both the weighted refinement indicator $|\tilde{\eta}_\kappa|$ and the empirical refinement indicator $\eta_\kappa^{\text{empirical}}$. From Fig. 11(a), we see that there is virtually no refinement in the regions of the computational domain where the discontinuities in $\mathbf{u}$ are located when $|\tilde{\eta}_\kappa|$ is employed. Indeed, most of the mesh refinement is concentrated in two regions upstream of the Gaussian weight function ψ. This is due to the presence of the weighting terms in $|\tilde{\eta}_\kappa|$, involving the difference between the approximate dual solution $\tilde{\mathbf{z}}_{\mathrm{DG}}$ and its projection $\mathbf{z}_{h,p}$ onto the primal finite element space, which multiply the computable residual terms involving the numerical solution $\mathbf{u}_{\mathrm{DG}}$, cf. (39) with $\mathbf{z}$ replaced by $\tilde{\mathbf{z}}_{\mathrm{DG}}$. We see from Fig. 12 that each component of the dual solution consists of two 'hills' transported backwards along the characteristics $x_1 \pm cx_2$. In contrast with this, when the weight-

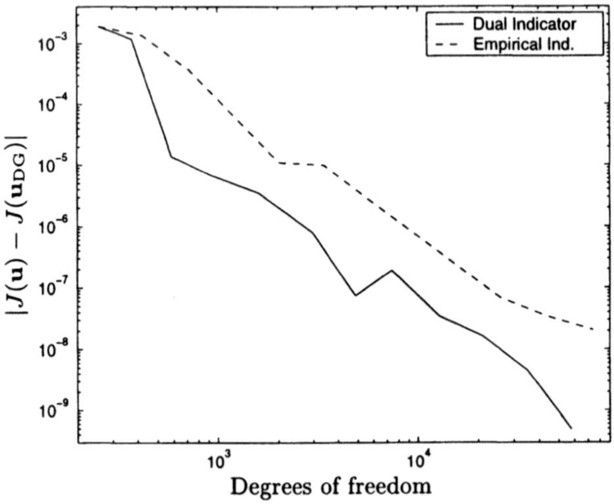

Fig. 13. Convergence of $|J(\mathbf{u}) - J(\mathbf{u}_{\mathrm{DG}})|$ using the dual and empirical refinement indicators for the wave equation

ing terms involving the dual solution are neglected, as in $\eta_\kappa^{\mathrm{empirical}}$, then the resulting mesh is concentrated in the vicinity of the discontinuities in the primal solution $\mathbf{u}$. In Fig. 13, we compare the true error in the computed functional $J(\cdot)$ using the two mesh refinement strategies. Here, we clearly observe the superiority of the Type I *a posteriori* error bound based on the weighted refinement indicators $|\tilde{\eta}_\kappa|$; on the final mesh the true error in the linear functional is almost two orders of magnitude smaller than $|J(\mathbf{u})-J(\mathbf{u}_{\mathrm{DG}})|$ computed on the sequence of meshes produced using $\eta_\kappa^{\mathrm{empirical}}$. This clearly indicates that a good numerical resolution of the discontinuities in this example is irrelevant for the accurate approximation of the functional of interest.

Let us now turn our attention to *hp*–adaptivity; in Table 9 we show the performance of the proposed adaptive finite element algorithm employing *hp*–refinement. Here, we again see that the quality of the computed error representation formula is extremely good, with $\theta_1 \approx 1$ even on very coarse meshes. Furthermore, the Type I *a posteriori* error bound (44) overestimates the true error in the computed functional by about an order of magnitude.

In Fig. 14 we plot $|J(\mathbf{u}) - J(\mathbf{u}_{\mathrm{DG}})|$, using both h– and *hp*–refinement against the square–root of the number of degrees of freedom on a linear–log scale. We see that after the initial transient, the error in the computed functional using *hp*–refinement becomes (on average) a straight line, thereby indicating exponential convergence of $J(\mathbf{u}_{\mathrm{DG}})$ to $J(\mathbf{u})$, despite the fact that $\mathbf{u}$ is only piecewise continuous; this occurs since $\mathbf{z}$ is a real analytic function on $\bar{\Omega}$. Fig. 14 also demonstrates the superiority of the adaptive *hp*–refinement strategy over the standard adaptive h–refinement algorithm. On the final mesh the true error between $J(\mathbf{u})$ and $J(\mathbf{u}_{\mathrm{DG}})$ using *hp*–refinement is al-

Table 9. Adaptive algorithm for the wave equation using *hp*–refinement

Elements	DOF	$J(\mathbf{u}-\mathbf{u}_{\mathrm{DG}})$	$\sum_\kappa \tilde{\eta}_\kappa$	θ_1	$\sum_\kappa \lvert\tilde{\eta}_\kappa\rvert$	θ_2
32	576	-0.3768e-03	-0.3727e-03	0.99	0.6615e-03	1.76
32	646	-0.1917e-03	-0.2558e-03	1.33	0.3495e-03	1.82
32	688	-0.3753e-04	-0.8638e-04	2.30	0.1490e-03	3.97
53	1248	0.1563e-04	0.1869e-04	1.20	0.2657e-04	1.70
86	2024	0.3341e-06	0.4492e-06	1.34	0.4501e-05	13.47
107	2868	0.1061e-06	0.9900e-07	0.93	0.7144e-06	6.73
119	3918	0.6372e-08	0.1900e-08	0.30	0.1401e-06	21.99
140	5486	0.3203e-08	0.3051e-08	0.95	0.2866e-07	8.95
203	9152	0.4293e-08	0.3655e-08	0.85	0.1476e-07	3.44
272	12648	-0.3332e-09	-0.3691e-09	1.11	0.1627e-08	4.88
344	17512	-0.3862e-11	-0.3862e-11	1.00	0.6570e-10	17.01
401	21616	0.6639e-12	0.6508e-12	0.98	0.9235e-11	13.91
464	27202	0.3406e-13	0.3477e-13	1.02	0.1799e-11	52.81

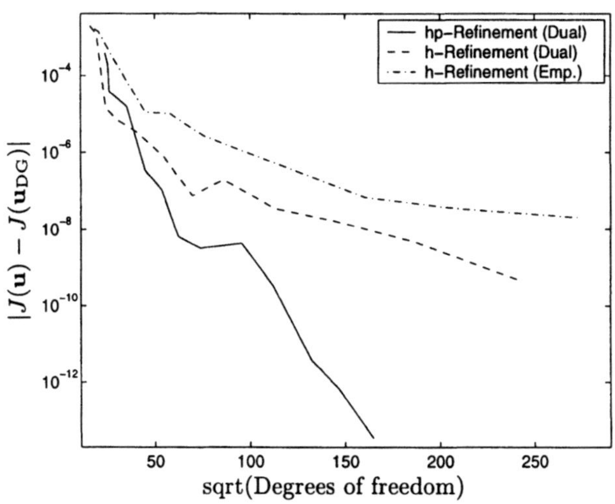

Fig. 14. Comparison between *h*– and *hp*–adaptive mesh refinement for the wave equation

most 6 orders of magnitude smaller than the corresponding quantity when *h*–refinement is employed alone.

Figs. 15 and 16 depict the primal mesh after 7 and 11 adaptive mesh refinement steps, respectively. For clarity, we show the *h*–mesh alone, as well

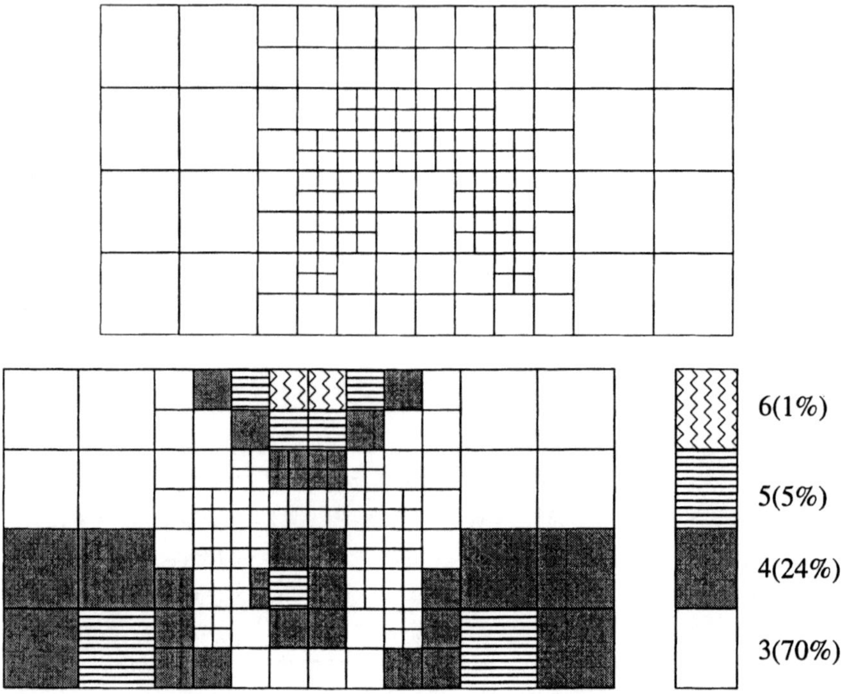

Fig. 15. Wave equation. h– and hp–meshes after 7 refinements, with 140 elements and 5486 degrees of freedom; here, $|J(\mathbf{u}) - J(\mathbf{u}_{\mathrm{DG}})| = 3.203 \times 10^{-9}$

as the corresponding distribution of the polynomial degree on this mesh and the percentage of elements with that degree. We see that some h–refinement of the primal mesh has occurred in the region of the computational domain where the two 'hills' in the dual solution $\mathbf{z}$, cf. Fig. 12, are located. In particular, the h–mesh is slightly finer in the regions where the discontinuities in the primal solution $\mathbf{u}$ intersect with these localized structures in $\mathbf{z}$. Once the h–mesh has adequately captured the structure of the primal and dual solutions, the hp–adaptive algorithm performed p–refinement to accurately resolve the Gaussians present in the dual solution.

6.4 Nonlinear hyperbolic problems

The purpose of this section is to develop the *a posteriori* error analysis of the discontinuous Galerkin finite element method for systems of nonlinear hyperbolic conservation laws.

Given a bounded open polyhedral domain Ω in $\mathbb{R}^n$, $n \geq 1$, with boundary $\partial\Omega$, we consider the following problem: find $\mathbf{u} : \Omega \to \mathbb{R}^m$, $m \geq 1$, such that

$$\operatorname{div} \mathcal{F}(\mathbf{u}) = 0 \quad \text{in } \Omega\,, \tag{50}$$

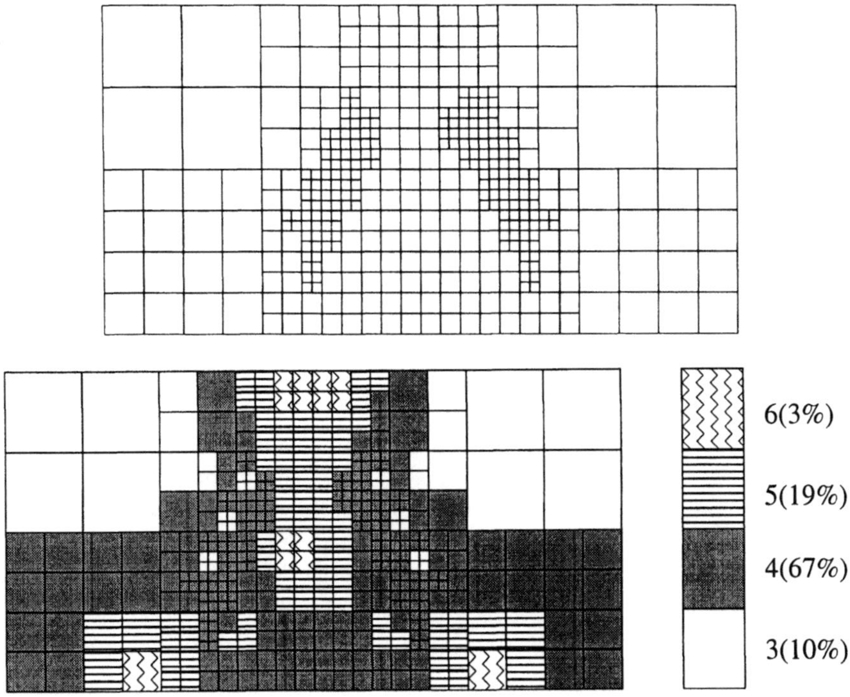

Fig. 16. Wave equation. h– and hp–meshes after 11 refinements, with 401 elements and 21616 degrees of freedom; here, $|J(\mathbf{u}) - J(\mathbf{u}_{\mathrm{DG}})| = 6.639 \times 10^{-13}$

where, $\mathcal{F} : \mathbb{R}^m \to \mathbb{R}^{m\times n}$ is continuously differentiable. We assume that the system of conservation laws (50) may be supplemented by appropriate initial/boundary conditions. For example, assuming that

$$B(\mathbf{u},\boldsymbol{\nu}) := \sum_{i=1}^{n} \nu_i \nabla_{\mathbf{u}} \mathcal{F}_i(\mathbf{u})$$

has m real eigenvalues and a complete set of linearly independent eigenvectors for all $\boldsymbol{\nu} = (\nu_1, \ldots, \nu_n) \in \mathbb{R}^n$; then at inflow/outflow boundaries, we require that

$$B^-(\mathbf{u},\boldsymbol{\mu})(\mathbf{u}-\mathbf{g}) = \mathbf{0}\,,$$

where $\boldsymbol{\mu}$ denotes the unit outward normal vector to $\partial\Omega$, $B^-(\mathbf{u},\boldsymbol{\mu})$ is the negative part of $B(\mathbf{u},\boldsymbol{\mu})$ and $\mathbf{g}$ is a (given) real–valued vector function.

To formulate the discontinuous Galerkin finite element approximation of (50), with $p_\kappa \in \mathbb{N}$ for $\kappa \in \mathcal{T}_h$ we define the finite element space

$$\begin{aligned} S^{\mathbf{p}}(\Omega,\mathcal{T}_h,\mathbf{F}) = \{\mathbf{v} \in [L_2(\Omega)]^m \; : \; & \mathbf{v}|_\kappa \circ F_\kappa \in [\mathcal{Q}_{p_\kappa}]^m \text{ if } F_\kappa^{-1}(\kappa) \text{ is} \\ & \text{the open unit hypercube and} \quad \mathbf{v}|_\kappa \circ F_\kappa \in [\mathcal{P}_{p_\kappa}]^m \\ & \text{if } F_\kappa^{-1}(\kappa) \text{ is the open unit simplex; } \kappa \in \mathcal{T}_h\}\,, \end{aligned}$$

where

$$\mathcal{Q}_p = \mathrm{span}\,\{\hat{x}^\alpha : 0 \le \alpha_i \le p,\ \ 1 \le i \le n\}\,, \quad \mathcal{P}_p = \mathrm{span}\,\{\hat{x}^\alpha : 0 \le |\alpha| \le p\}\,,$$

cf. (33). Given that $\mathbf{v} \in \left[H^1(\kappa)\right]^m$ for each $\kappa \in \mathcal{T}_h$, we denote by $\mathbf{v}^+$ (respectively, $\mathbf{v}^-$) the interior (respectively, exterior) trace of $\mathbf{v}$ on $\partial\kappa$.

The *hp*–DGFEM approximation of (50) is defined as follows: find $\mathbf{u}_{\mathrm{DG}}$ in $S^{\mathbf{p}}(\Omega, \mathcal{T}_h, \mathbf{F})$ such that

$$\sum_{\kappa\in\mathcal{T}_h} \left\{ -\int_\kappa \mathcal{F}(\mathbf{u}_{\mathrm{DG}}) \cdot \nabla \mathbf{v}\,\mathrm{d}x + \int_{\partial\kappa} \mathcal{H}(\mathbf{u}_{\mathrm{DG}}^+, \mathbf{u}_{\mathrm{DG}}^-, \boldsymbol{\mu}_\kappa)\,\mathbf{v}^+\,\mathrm{d}\sigma \right. \\ \left. + \int_\kappa \varepsilon \nabla \mathbf{u}_{\mathrm{DG}} \cdot \nabla \mathbf{v}\,\mathrm{d}x \right\} = 0 \qquad (51)$$

for all $\mathbf{v} \in S^{\mathbf{p}}(\Omega, \mathcal{T}_h, \mathbf{F})$; cf. [26,27,37], for example. Here, $\mathcal{H}(\cdot,\cdot,\cdot)$ denotes a *numerical flux* function, assumed to be Lipschitz continuous, consistent and conservative. As in the linear case, the choice of the numerical flux function is completely independent of the finite element space employed; in the numerical experiments we employ the (local) Lax–Friedrichs flux function. In (51), ε denotes the coefficient of artificial viscosity defined by

$$\varepsilon|_\kappa = C_\varepsilon \left(\frac{h_\kappa}{p_\kappa}\right)^{2-\beta} |\,\mathrm{div}\mathcal{F}(\mathbf{u}_{\mathrm{DG}}|_\kappa)|\,, \qquad (52)$$

where C_ε is a positive constant and $0 < \beta < 1/2$; see [37]. For elements $\kappa \in \mathcal{T}_h$ whose boundary intersects $\partial\Omega$, we replace $\mathbf{u}_{\mathrm{DG}}^-$ by appropriate boundary/initial conditions on $\partial\kappa \cap \partial\Omega$.

In this section, we focus on a measurement problem concerned with controlling the error in the approximation of a given linear or nonlinear target functional $J(\cdot)$. Quantities of real or physical interest include the drag and lift coefficients of a body (such as an aircraft wing) immersed into an inviscid fluid governed by the compressible Euler equations of gas dynamics, cf. Section 4.2. Writing S to denote the surface of the body (or, rather, the contour of its two-dimensional cross-section) and $\bar{l}$ its chord length, these target functionals are defined by

$$J_{\mathrm{drag}}(\mathbf{u}) = \frac{2}{\bar{l}\bar{\rho}|\bar{\mathbf{v}}|^2} \int_S (\psi_d \cdot \boldsymbol{\mu})\,p\,\mathrm{d}s\,, \qquad J_{\mathrm{lift}}(\mathbf{u}) = \frac{2}{\bar{l}\bar{\rho}|\bar{\mathbf{v}}|^2} \int_S (\psi_l \cdot \boldsymbol{\mu})\,p\,\mathrm{d}s\,,$$

respectively. Here, $\boldsymbol{\mu}$ is the unit outward normal to S; $\bar{\mathbf{v}}$ and $\bar{\rho}$ denote the reference (or free-stream) velocity and density, respectively, and

$$\psi_d = \begin{pmatrix} \cos\alpha & -\sin\alpha \\ \sin\alpha & \cos\alpha \end{pmatrix} \begin{pmatrix} 1 \\ 0 \end{pmatrix}, \qquad \psi_l = \begin{pmatrix} \cos\alpha & -\sin\alpha \\ \sin\alpha & \cos\alpha \end{pmatrix} \begin{pmatrix} 0 \\ 1 \end{pmatrix},$$

where α denotes the angle of attack. We note that since the pressure p is derived from the conserved variables $(\rho, \rho u, \rho v, \rho E)$, both target functionals

$J_{\rm drag}(\cdot)$ and $J_{\rm lift}(\cdot)$ are nonlinear. Other examples of $J(\cdot)$ include the local mean value of the field or its flux through the outflow boundary of the computational domain Ω, and the point evaluation of a component of $\mathbf{u}$ in Ω.

Before we embark on our *a posteriori* error analysis, we shall introduce some notation. Assuming that the functional of interest $J(\cdot)$ is differentiable, we write $\bar{J}(\cdot;\cdot)$ to denote the mean value linearization of $J(\cdot)$ defined by

$$\begin{aligned}\bar{J}(\mathbf{u},\mathbf{u}_{\rm DG};\mathbf{u}-\mathbf{u}_{\rm DG}) &= J(\mathbf{u}) - J(\mathbf{u}_{\rm DG}) \\ &= \int_0^1 J'[\theta\mathbf{u}+(1-\theta)\mathbf{u}_{\rm DG}](\mathbf{u}-\mathbf{u}_{\rm DG})\,{\rm d}\theta\,, \end{aligned} \tag{53}$$

where $J'[\mathbf{w}](\cdot)$ denotes the Gateaux derivative of $J(\cdot)$ evaluated at some $\mathbf{w}$ in V. Here, V is some suitably chosen function space such that $S^{\mathbf{p}}(\Omega,\mathcal{T}_h,\mathbf{F}) \subset V$. Analogously, letting $\mathcal{N}(\mathbf{u}_{\rm DG},\mathbf{v})$ denote the left–hand side of (51), we denote by $\mathcal{M}(\mathbf{u},\mathbf{u}_{\rm DG};\cdot,\cdot)$ the mean value linearization of $\mathcal{N}(\cdot,\cdot)$ defined by

$$\begin{aligned}\mathcal{M}(\mathbf{u},\mathbf{u}_{\rm DG};\mathbf{u}-\mathbf{u}_{\rm DG},\mathbf{v}) &= \mathcal{N}(\mathbf{u},\mathbf{v}) - \mathcal{N}(\mathbf{u}_{\rm DG},\mathbf{v}) \\ &= \int_0^1 \mathcal{N}'_{\mathbf{u}}[\theta\mathbf{u}+(1-\theta)\mathbf{u}_{\rm DG}](\mathbf{u}-\mathbf{u}_{\rm DG},\mathbf{v})\,{\rm d}\theta \end{aligned} \tag{54}$$

for all $\mathbf{v}$ in V. Here, $\mathcal{N}'_{\mathbf{u}}[\mathbf{w}](\cdot,\mathbf{v})$ denotes the Gateaux derivative of $\mathbf{u} \mapsto \mathcal{N}(\mathbf{u},\mathbf{v})$, for $\mathbf{v} \in V$ fixed, at some $\mathbf{w}$ in V. The linearization defined in (54) is only a *formal* calculation, in the sense that $\mathcal{N}'_{\mathbf{u}}[\mathbf{w}](\cdot,\cdot)$ may not in general exist. Instead, a suitable approximation to $\mathcal{N}'_{\mathbf{u}}[\mathbf{w}](\cdot,\cdot)$ must be determined, for example, by computing appropriate finite difference quotients of $\mathcal{N}(\cdot,\cdot)$, cf. [26]. For the proceeding analysis, we *assume* that the linearization (54) is well–defined. Under this hypothesis, we introduce the following *dual* problem: find $\mathbf{z} \in V$ such that

$$\mathcal{M}(\mathbf{u},\mathbf{u}_{\rm DG};\mathbf{w},\mathbf{z}) = \bar{J}(\mathbf{u},\mathbf{u}_{\rm DG};\mathbf{w}) \quad \forall \mathbf{w} \in V\,. \tag{55}$$

We assume that (55) possesses a unique solution; of course, the validity of this assumption depends on both the definition of $\mathcal{M}(\mathbf{u},\mathbf{u}_{\rm DG};\cdot,\cdot)$ and the choice of the target functional under consideration, cf. [26]. For the proceeding error analysis, we must therefore *assume* that the dual problem (55) is well–posed. We then have the following result.

Theorem 3. *Let* $\mathbf{u}$ *and* $\mathbf{u}_{\rm DG}$ *denote the solutions to (50) and (51), respectively, and suppose that the dual problem (55) is well–posed. Then,*

$$J(\mathbf{u}) - J(\mathbf{u}_{\rm DG}) = \mathcal{E}_\Omega \equiv \sum_{\kappa\in\mathcal{T}_h} \eta_\kappa\,, \tag{56}$$

where

$$\begin{aligned}\eta_\kappa = \int_\kappa \mathbf{r}_{h,p}\cdot(\mathbf{z}-\mathbf{z}_{h,p})\,{\rm d}x + \int_{\partial\kappa} \boldsymbol{\sigma}_{h,p}\cdot(\mathbf{z}-\mathbf{z}_{h,p})^+\,{\rm d}\sigma \\ - \int_\kappa \varepsilon\nabla\mathbf{u}_{\rm DG}\cdot\nabla(\mathbf{z}-\mathbf{z}_{h,p})\,{\rm d}x\end{aligned}$$

for all $\mathbf{z}_{h,p}$ *in* $S^{\mathbf{p}}(\Omega, \mathcal{T}_h, \mathbf{F})$. *Here,*

$$\mathbf{r}_{h,p}|_\kappa = -\mathrm{div}\mathcal{F}(\mathbf{u}_{\mathrm{DG}}) \quad \textit{and} \quad \boldsymbol{\sigma}_{h,p}|_\kappa = \mathcal{F}(\mathbf{u}_{\mathrm{DG}}) \cdot \mathbf{n}_\kappa - \mathcal{H}(\mathbf{u}_{\mathrm{DG}}^+, \mathbf{u}_{\mathrm{DG}}^-, \mathbf{n}_\kappa)$$

denote the internal and boundary finite element residuals, respectively, defined on each $\kappa \in \mathcal{T}_h$.

Proof. Choosing $\mathbf{w} = \mathbf{u} - \mathbf{u}_{\mathrm{DG}}$ in (55), recalling the linearization performed in (53), and exploiting the Galerkin orthogonality property $\mathcal{N}(\mathbf{u}, \mathbf{v}) - \mathcal{N}(\mathbf{u}_{\mathrm{DG}}, \mathbf{v}) = 0$ for all $\mathbf{v}$ in $S^{\mathbf{p}}(\Omega, \mathcal{T}_h, \mathbf{F})$, we get

$$\begin{aligned} J(\mathbf{u}) - J(\mathbf{u}_{\mathrm{DG}}) &= \bar{J}(\mathbf{u}, \mathbf{u}_{\mathrm{DG}}; \mathbf{u} - \mathbf{u}_{\mathrm{DG}}) = \mathcal{M}(\mathbf{u}, \mathbf{u}_{\mathrm{DG}}; \mathbf{u} - \mathbf{u}_{\mathrm{DG}}, \mathbf{z}) \\ &= \mathcal{M}(\mathbf{u}, \mathbf{u}_{\mathrm{DG}}; \mathbf{u} - \mathbf{u}_{\mathrm{DG}}, \mathbf{z} - \mathbf{z}_{h,p}) = -\mathcal{N}(\mathbf{u}_{\mathrm{DG}}, \mathbf{z} - \mathbf{z}_{h,p}) \end{aligned}$$

for all $\mathbf{z}_{h,p}$ in $S^{\mathbf{p}}(\Omega, \mathcal{T}_h, \mathbf{F})$. Equation (56) now follows by employing the divergence theorem.

Based on the general error representation formula derived in Theorem 3, *a posteriori* error estimates bounding the error in the computed functional $J(\cdot)$ may be deduced. Again, we shall confine ourselves to considering a Type I *a posteriori* error bound; it is a straightforward consequence of the error representation formula stated in the previous theorem.

Corollary 1. *Under the assumptions of Theorem 3, we have*

$$|J(\mathbf{u}) - J(\mathbf{u}_{\mathrm{DG}})| \leq \mathcal{E}_{|\Omega|} \equiv \sum_{\kappa \in \mathcal{T}_h} |\eta_\kappa| \,. \tag{57}$$

Proof. The error bound (57) follows from (56) by application of the triangle inequality.

We end this section by noting that for nonlinear hyperbolic conservation laws and/or nonlinear target functionals $J(\cdot)$, both the error representation formula (56) and the Type I *a posteriori* error bound (57) depend on the unknown analytical solution to the primal and dual problems. Thus, in order to render these quantities computable, both $\mathbf{u}$ and $\mathbf{z}$ must be replaced by suitable approximations. To this end, the linearizations leading to $\mathcal{M}(\mathbf{u}, \mathbf{u}_{\mathrm{DG}}; \cdot, \cdot)$ and $\bar{J}(\mathbf{u}, \mathbf{u}_{\mathrm{DG}}; \cdot)$ are performed about $\mathbf{u}_{\mathrm{DG}}$ and the dual solution $\mathbf{z}$ is replaced by a discontinuous Galerkin approximation $\hat{\mathbf{z}}_{\mathrm{DG}}$ computed on the same mesh $\mathcal{T}_h$ as for the primal solution $\mathbf{u}_{\mathrm{DG}}$, but with a higher degree polynomial, i.e., $\hat{\mathbf{z}}_{\mathrm{DG}} \in S^{\hat{\mathbf{p}}}(\Omega, \mathcal{T}_h, \mathbf{F})$, $\hat{\mathbf{p}} = \mathbf{p} + \mathbf{p}_{\mathrm{INC}}$; in Section 6.5, we set $\mathbf{p}_{\mathrm{INC}} = \mathbf{1}$.

6.5 Numerical experiments

In this section we present some numerical examples to assess the performance of our *hp*–adaptive algorithm for the discontinuous Galerkin finite element method.

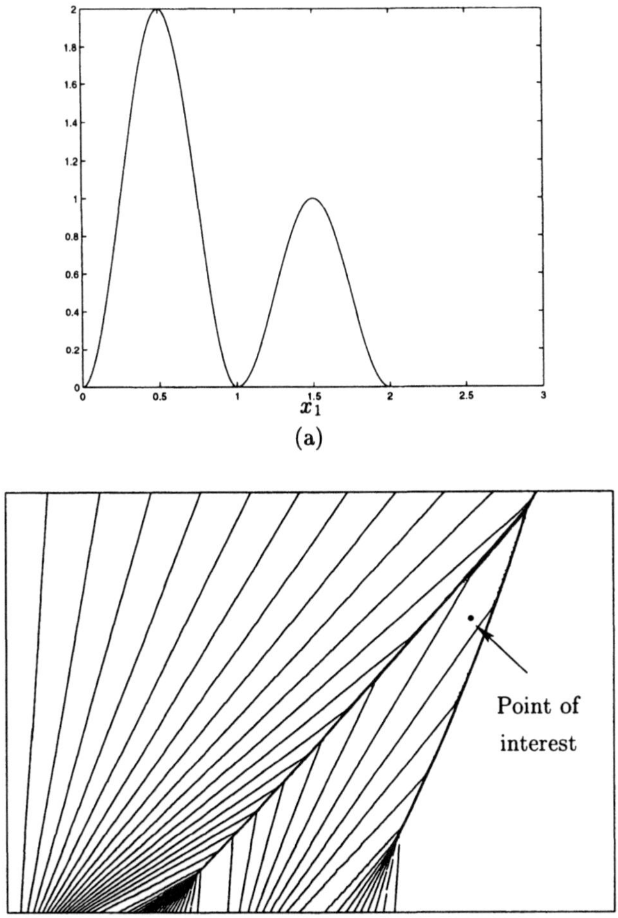

(a)

(b)

Fig. 17. Burgers' problem: (a) Initial condition; (b) Solution isolines

Burgers' equation. In this example, we consider the one–dimensional unsteady inviscid Burgers' equation for the scalar variable $\mathbf{u} \equiv u$; i.e., writing x_2 to denote time, we have

$$u_{x_2} + \left((1/2)\,u^2\right)_{x_1} = 0\,, \tag{58}$$

on the (space–time) domain $\Omega = (0,3) \times (0,2)$, subject to the initial condition

$$u(x,0) = \begin{cases} 2\,\sin^2(\pi x)\,, & 0 \le x \le 1\,, \\ \sin^2(\pi x)\,, & 1 \le x \le 2\,, \\ 0\,, & 2 \le x \le 3\,, \end{cases} \tag{59}$$

Table 10. Adaptive algorithm for Burgers' problem

Elements	DOF	$J(u - u_{DG})$	$\sum_\kappa \tilde{\eta}_\kappa$	θ_1	$\sum_\kappa \lvert\tilde{\eta}_\kappa\rvert$	θ_2
24	216	0.5010e-01	0.1033	2.06	0.1913	3.82
36	324	-0.9031e-01	0.1096	-1.21	0.2058	2.28
81	736	0.1175e-01	-0.4193e-01	-3.57	0.1614	13.73
120	1150	-0.1961e-01	-0.1740e-01	0.89	0.9974e-01	5.09
207	2096	-0.9130e-02	-0.6216e-02	0.68	0.4372e-01	4.79
291	3211	0.7391e-03	0.8754e-03	1.18	0.8192e-02	11.08
336	4279	-0.3173e-03	-0.5243e-03	1.65	0.3101e-02	9.77
420	6227	-0.6514e-04	-0.4742e-04	0.73	0.8027e-03	12.32
525	9008	0.1030e-04	0.8515e-05	0.83	0.1232e-03	11.96
660	13045	-0.5401e-06	-0.1246e-05	2.31	0.3717e-04	68.82
792	18070	-0.5414e-06	-0.5697e-06	1.05	0.1095e-04	20.23
999	26020	0.1010e-06	0.5856e-07	0.58	0.1112e-05	11.01
1269	37181	0.1455e-08	0.1085e-08	0.75	0.8654e-07	59.49
1725	57850	-0.6196e-10	-0.4914e-10	0.79	0.2841e-08	45.86
2541	94649	0.3163e-11	0.2822e-11	0.89	0.9505e-10	30.05

and boundary condition $u(0, x_2) = 0$, for $x_2 \in [0, 2]$, cf. Fig. 17(a). The analytical solution to (58), (59) consists of two smooth 'hills' which form shock waves as time increases. These shocks will eventually merge to form a single line of discontinuity in the (x_1, x_2)–plane. Here, we select the functional of interest $J(\cdot)$ to be the value of the solution before these two shocks interact with one another. More precisely, we choose

$$J(u) = u(2.3, 1.5)\,,$$

cf. Fig. 17(b); thereby $J(u) = 0.664442403975254670$.

In Table 10 we show the performance of our *hp*–adaptive finite element algorithm with $C_\varepsilon = 1/20$ and $\beta = 1/10$. As for the wave equation presented in Section 6.3, we see that the quality of the computed error representation formula is quite poor on coarse meshes; though, as the finite element space $S^{\mathbf{p}}(\Omega, \mathcal{T}_h, \mathbf{F})$ is enriched the effectivity index θ_1 is very close to one. Furthermore, we again observe that the Type I *a posteriori* error bound (57) is reliable and asymptotically sharp; indeed, $\tilde{\mathcal{E}}_{|\Omega|} = \sum_\kappa |\tilde{\eta}_\kappa|$ overestimates $|J(u - u_{DG})|$ by just over an order of magnitude.

In Fig. 18(a) we compare the performance of the h– and *hp*–mesh refinement algorithms for this problem. Again, we observe exponential convergence of the error in the computed functional using *hp*–refinement; on the linear–

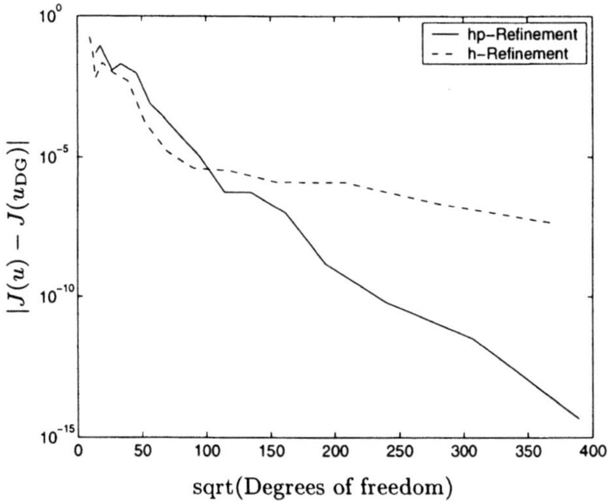

(a)

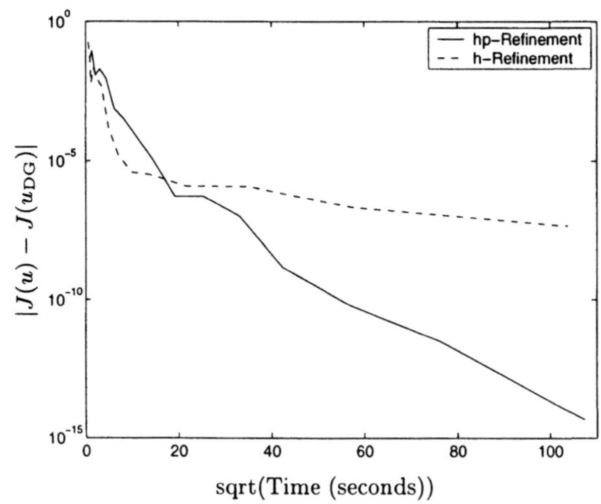

(b)

Fig. 18. Burgers' equation. Comparison between h– and hp–adaptive mesh refinement: (a) $|J(u)-J(u_{\mathrm{DG}})|$ versus number of degrees of freedom; (b) $|J(u)-J(u_{\mathrm{DG}})|$ versus computational time

log scale, the convergence line is straight. On the final mesh the true error between $J(u)$ and $J(u_{\mathrm{DG}})$ using hp–refinement is almost 5 orders of magnitude smaller than the corresponding quantity when h–refinement is employed alone. Furthermore, from Fig. 18(b) we observe that the hp–refinement algo-

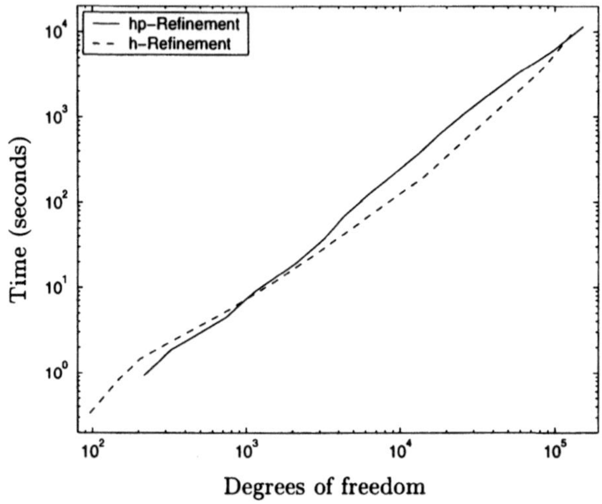

Fig. 19. Burgers' equation. Computational time versus number of degrees of freedom

rithm also outperforms the h–refinement strategy, when comparing the error in the computed target functional with respect to the computational cost. Indeed, Fig. 19 clearly shows that for the hp–DGFEM the cost per degree of freedom when hp–refinement is employed is comparable to that of using h–refinement.

Finally, in Fig. 20 we show the primal mesh after 11 adaptive hp–mesh refinements. Here, we see that the h–mesh has been refined in the region upstream of the point of interest, thereby isolating the smooth region of u from the two interacting shock waves; this renders the subsequent p–refinement in this region much more effective.

Subsonic flow around a NACA0012 airfoil. In this second example, we consider the subsonic flow around a NACA0012 airfoil; here, the upper and lower surfaces of the airfoil geometry are specified by the function $g^{\pm}$, respectively, where

$$g^{\pm}(s) = \pm 5 \times 0.12 \times (0.2969 s^{1/2} - 0.126 s - 0.3516 s^2 + 0.2843 s^3 - 0.1015 s^4),$$

cf. Fig. 21. We note that the chord length l of the airfoil is $l \approx 1.00893$; thereby, we write $\hat{g}$ to denote the rescaling of g to yield an airfoil of unit (chord) length. The computational domain Ω is now defined to be an annulus with inner boundary $\hat{g}$ and outer boundary consisting of a circle of radius 10 units. On the outer boundary, we prescribe a Mach 0.5 flow at a zero angle of attack, with farfield density $\rho = 1$ and pressure $p = 1$. The solution to this problem consists of a strictly subsonic flow, symmetric about the x–axis,

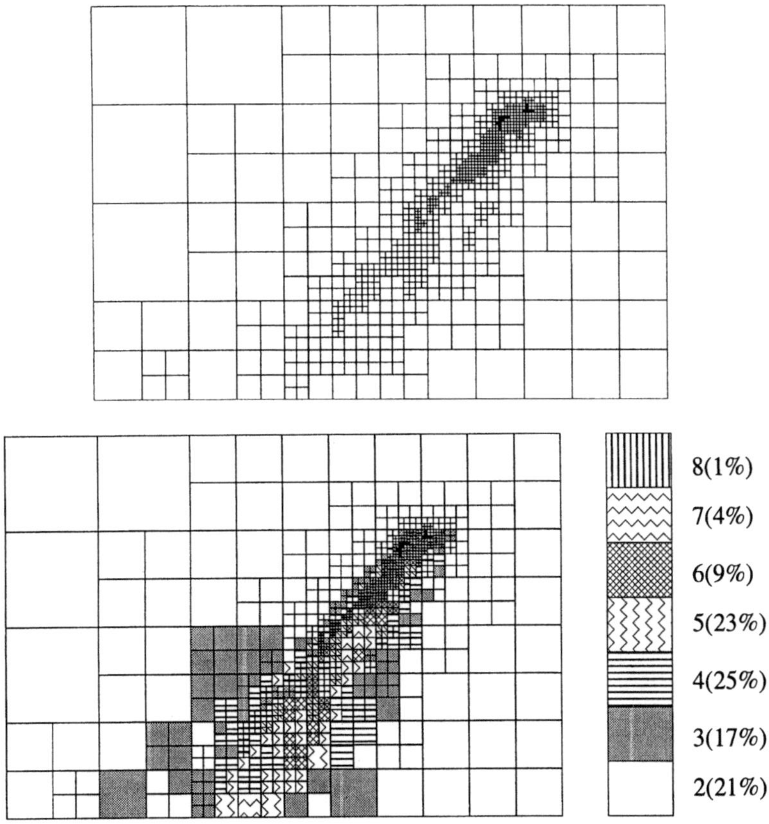

Fig. 20. Burgers' equation. h– and hp–meshes after 11 refinements, with 999 elements and 26020 degrees of freedom; here, $|J(u) - J(u_{\mathrm{DG}})| = 1.010 \times 10^{-7}$

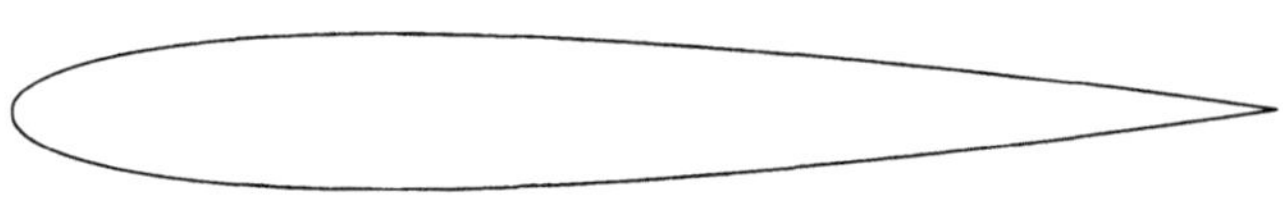

Fig. 21. Profile of the NACA0012 airfoil

cf. Fig. 22. Given that this flow is smooth, no artificial viscosity is required; thereby, we set $C_\varepsilon = 0$.

Here, we consider the evaluation of the drag coefficient on the surface of the airfoil, i.e. $J(\cdot) \equiv J_{\mathrm{drag}}(\cdot)$. Since that the angle of attack $\alpha = 0$, the true value of the target functional $J_{\mathrm{drag}}(\mathbf{u})$ is zero. In Fig. 23 we show the performance of the adaptive algorithm using both h– and hp–mesh refinement. Since the primal solution is smooth we again observe exponential convergence

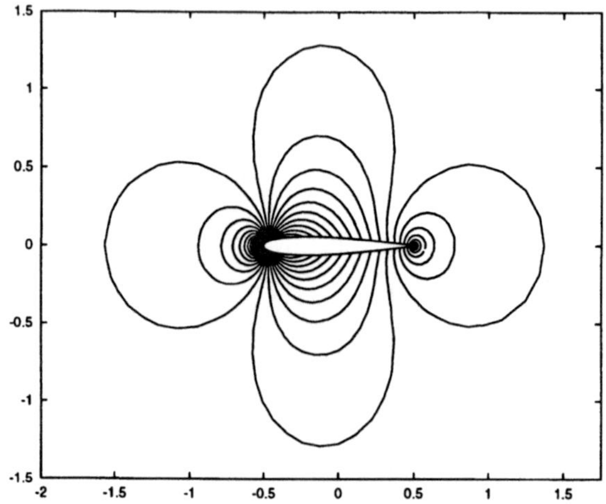

Fig. 22. Mach isolines for the subsonic NACA0012 flow

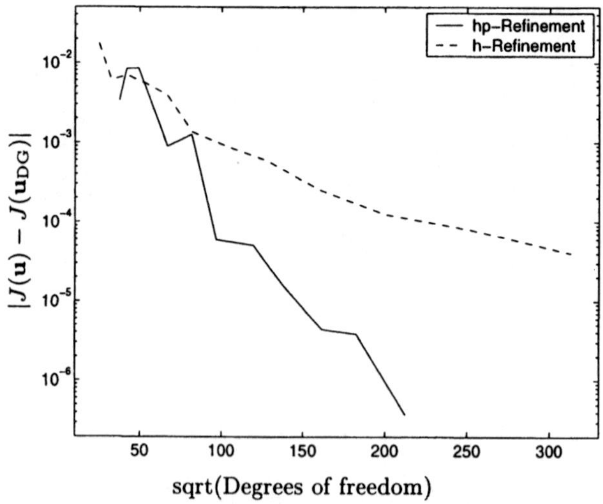

Fig. 23. Subsonic NACA0012 flow. Comparison between h– and hp–adaptive mesh refinement

of $J(\mathbf{u}_{\mathrm{DG}})$ to $J(\mathbf{u})$ when the hp–refinement is employed; on a linear–log scale the convergence line is (on average) straight. Fig. 23 also demonstrates the superiority of the adaptive hp–refinement strategy over the standard adaptive h–refinement algorithm. On the final mesh the true error between $J(\mathbf{u})$ and $J(\mathbf{u}_{\mathrm{DG}})$ using hp–refinement is over 2 orders of magnitude smaller than the corresponding quantity when h–refinement is employed alone.

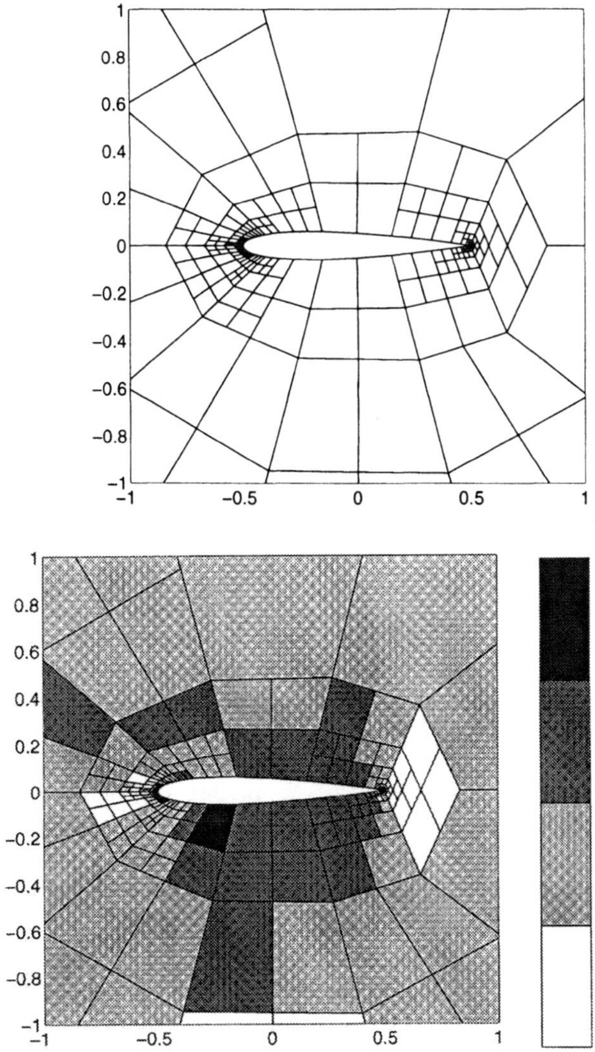

Fig. 24. Subsonic NACA0012 flow. h– and hp–meshes after 10 refinements, with 325 elements and 45008 degrees of freedom; here, $|J(\mathbf{u}) - J(\mathbf{u}_{\mathrm{DG}})| = 3.756 \times 10^{-7}$

Finally, in Fig. 24 we show the primal mesh after 10 adaptive hp–mesh refinements. Here, we see that the h–mesh has been refined in the neighborhood of the leading and trailing edges of the airfoil, cf. [27]. Moreover, elements located on the upper and lower surfaces of the airfoil have been p–enriched where the primal solution is relatively smooth.

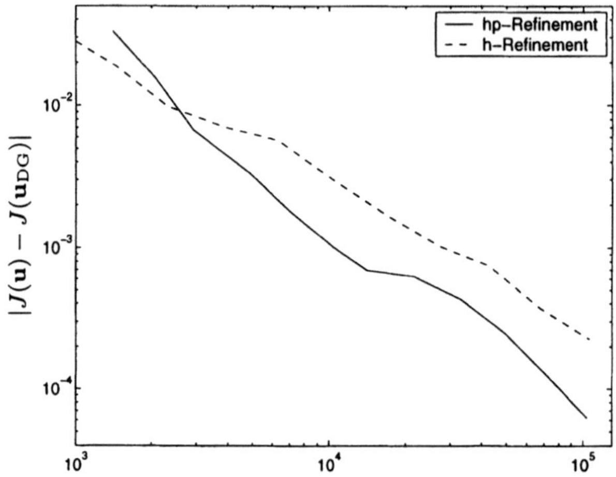

Fig. 25. Transonic NACA0012 flow. Comparison between h– and hp–adaptive mesh refinement

Transonic flow around a NACA0012 airfoil. In this final example, we again compute the flow around a NACA0012 airfoil; here, we consider a Mach 0.8 flow with angle of attack $\alpha = 1.25°$, and farfield density $\rho = 1$ and pressure $p = 1$. The structure of the solution in this example is now quite different from the subsonic flow studied in the previous section. Here, as the impinging subsonic flow passes the airfoil, it first accelerates to become supersonic, before slowing to be subsonic again; this results in the development of shocks on both the upper and lower surfaces of the airfoil, cf. [27]. Throughout this section we set the artificial viscosity parameters appearing in (52) as follows: $C_\varepsilon = 3/100$ and $\beta = 1/10$.

Here, we again consider the evaluation of the drag coefficient on the surface of the airfoil, i.e. $J(\cdot) \equiv J_{\text{drag}}(\cdot)$. On the basis of a fine grid computation, the true value of the functional is given by $J(\mathbf{u}) \approx 0.0201$ (see, [27]). In Fig. 25 we show the performance of the adaptive algorithm using both h– and hp–mesh refinement. In this example both the primal and dual solutions are nonsmooth, so we no longer expect to observe exponential convergence of the error in the functional $J(\cdot)$; thus, here we plot the same quantities as in Fig. 23 against the number of degrees of freedom in $S^{\mathbf{p}}(\Omega, \mathcal{T}, \mathbf{F})$ on a log–log scale. As in our previous examples, we observe that the true error in $J(\cdot)$ is (almost) always smaller if hp–refinement is employed. Indeed, on the final mesh the true error in the drag coefficient is over 70% smaller if hp–refinement is employed as opposed to h–refinement only.

Finally, in Fig. 26 we show the primal mesh after 10 adaptive hp–mesh refinements. As in the previous example, we see that the h–mesh has been refined in the vicinity of the leading and trailing edges of the airfoil; addi-

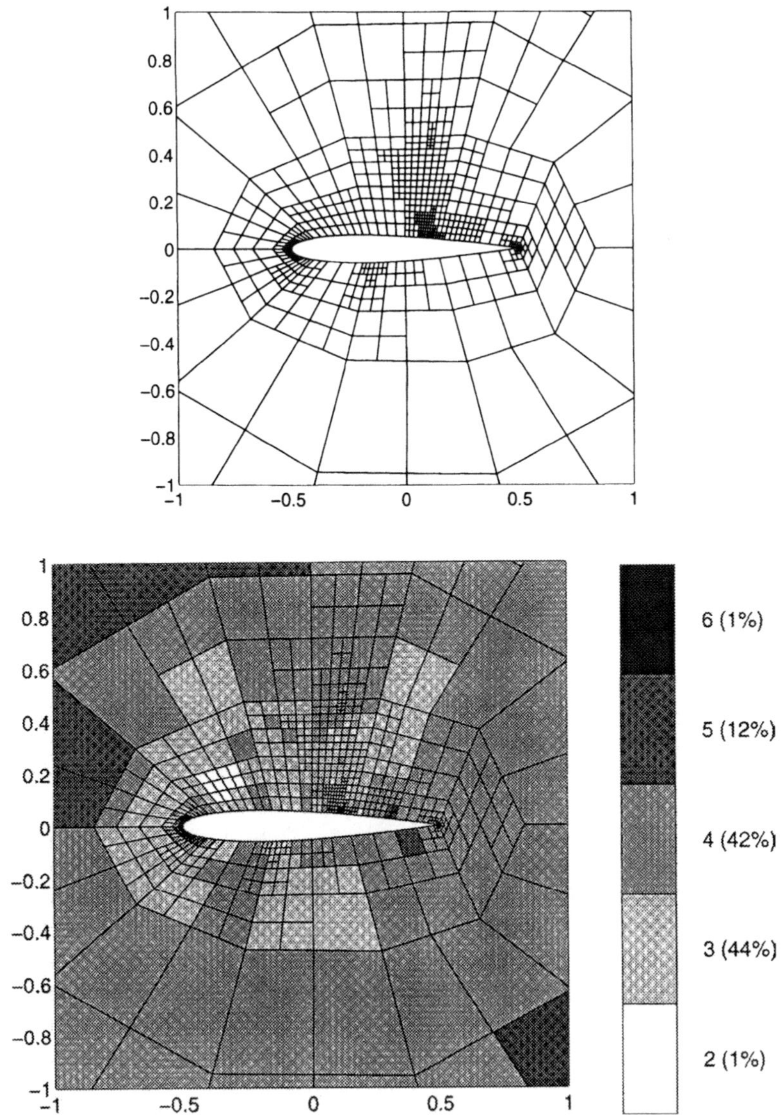

Fig. 26. Transonic NACA0012 flow. h– and hp–meshes after 10 refinements, with 783 elements and 69956 degrees of freedom; here, $|J(\mathbf{u}) - J(\mathbf{u}_{\mathrm{DG}})| = 1.311 \times 10^{-4}$

tional h–refinement has also taken place in the neighbourhood of the shocks emanating from the upper and lower surfaces of the airfoil, cf. [27]. This renders the subsequent p–refinement in the regions of the computational domain where either the primal or dual solution are locally smooth much more effective.

7 Conclusions

Using duality arguments, in this paper we developed Type I *a posteriori* error bounds for the h–version of the streamline–diffusion finite element method and the hp–version of the discontinuous Galerkin finite element method. The bounds have been implemented into adaptive finite element algorithms that are capable of delivering approximations to general output functionals of the solution for both linear and nonlinear hyperbolic problems, accurate to within a prescribed tolerance.

Unlike Type II *a posteriori* error bounds, Type I bounds explicitly involve the dual solution as local weight. Indeed, a Type I bound is always sharper than the corresponding Type II bound, although the extra sharpness does not come for free since a Type I bound requires the numerical solution of the dual problem. As the dual problem is always linear even when the primal problem is nonlinear, for nonlinear problems the additional cost of solving a single dual problem is negligible, cf. [26]. Of course if the primal problem is linear, then the computational cost is roughly doubled.

While the systematic comparative study of Type I and Type II *a posteriori* error bounds on representative model problems will require further work, it is perhaps helpful to point out some potential advantages and disadvantages of the two approaches. Although a Type II *a posteriori* error bound does not explicitly involve the dual solution as local weight, it does, nevertheless, include the strong stability constant C_{stab} of the dual problem. Analytical estimates of C_{stab} are inevitably pessimistic since they are obtained by admitting very large classes of functions as data (e.g., any possible element of $L_2(\Omega)$ or $H^1(\Omega)$ as forcing function). In addition, the analytical proof of strong stability may demand a nontrivial amount of human effort. Computational estimation of C_{stab} involves the solution of the dual problem for a large set of 'typical' data, — a task that is computationally much more demanding than solving a single dual problem for a Type I error bound. On the other hand, once a library of representative stability constants is available [15], they do not need to be recomputed; also, storing representative stability constants is clearly less memory-intensive than storing the dual solution, particularly for three-dimensional problems. Having said this, the storage of the dual solution, for use in a Type I error bound, may pay off if the resulting computational mesh is more economical than the one that stems from a Type II *a posteriori* error bound.

At any rate, on the basis of our computational experience we would argue that the need to perform a single additional computation for the dual solution involved in a Type I *a posteriori* error bound is a worthwhile price to pay for the availability of reliable error control based on a rigorous, systematic, simple and general mathematical framework. Cheap, heuristic alternatives to Type I or Type II *a posteriori* error bounds offer no control on the error; for goal-oriented computations (such as the approximation of output functionals), the optimality of the meshes that result from such heuristic devices is highly questionable.

Acknowledgements

The authors are grateful to Rémi Abgrall, Mark Ainsworth, Ivo Babuška, Tim Barth, Bernardo Cockburn, Leszek Demkowicz, Joe Flaherty, Mike Giles, Ralf Hartmann, Kathryn Harriman, Claes Johnson, Mats Larson, John Mackenzie, Peter Monk, Niles Pierce, Rolf Rannacher, Christoph Schwab, Bill Senior, Thomas Sonar, and Gerald Warnecke for helpful discussions on the subject of this paper. Part of this work was performed using the University of Leicester Mathematical Modelling Centre's supercomputer which was purchased through the EPSRC Strategic Equipment Initiative.

References

1. Adams, R.A. (1975) Sobolev Spaces. Academic Press, New York
2. Adjerid, S., Aiffa, M., and Flaherty, J.E. (1998) Computational methods for singularly perturbed systems. In Cronin, J. and O'Malley, R.E. (eds.), Singular Perturbation Concepts of Differential Equations, AMS, Providence
3. Ainsworth, M. and Senior, B. (1998) An adaptive refinement strategy for *hp*–finite element computations. Appl. Numer. Maths. **26**, 165–178
4. Ainsworth, M. and Oden, J.T. (1996) A Posteriori Error Estimation in Finite Element Analysis. Series in Computational and Applied Mathematics. Elsevier
5. Bardos, C. (1970) Problèmes aux limites pour les équations aux dérivées partielles du premier ordre à coefficients réels; théorèmes d'approximation; application à l'équation de transport. Ann. Sci. École Norm. Sup. **4**, 185–233
6. Barth, T.J. (1998) Numerical methods for gas dynamics systems on unstructured meshes. In: Kröner, D., Ohlberger, M., Rohde, C. (eds.) An Introduction to Recent Developments in Theory and Numerics of Conservation Laws. Lecture Notes in Computational Science and Engineering **5**, pp. 195–285. Springer, Berlin Heidelberg
7. Becker, R. and Rannacher, R. (1996) Weighted a posteriori error control in FE methods. Interdisziplinäres Zentrum für Wissenschaftliches Rechnen, Universität Heidelberg, Preprint No. 1
8. Becker, R. and Rannacher, R. (2001) An optimal control approach to a-posteriori error estimation in finite element methods. Iserles, A. (ed.), Acta Numerica (to appear)
9. Bey, K.S. and Oden, J.T. (1996) *hp*-Version discontinuous Galerkin methods for hyperbolic conservation laws. Comput. Methods Appl. Mech. Engrg. **133**, 259–286
10. Braess, D. (1997) Finite Elements. Theory, Fast Solvers, and Applications in Solid Mechanics. Cambridge University Press, Cambridge
11. Cockburn, B. and Gau, H. (1995) A posteriori error estimates for general numerical methods for scalar conservation laws. Comput. Appl. Math. **14**, 37–47

12. Cockburn, B., Karniadakis, G.E., and Shu, C.-W. (2000) The development of discontinuous Galerkin methods. In: Cockburn, B., Karniadakis, G.E., Shu C.-W. (eds.) Discontinuous Galerkin Finite Element Methods. Lecture Notes in Computational Science and Engineering, **11**. Springer, Berlin Heidelberg
13. Dautray, R. and Lions, J.-L. (1993) Mathematical Analysis and Numerical Methods for Science and Technology. **6**, Evolution Problems II. Springer, Berlin Heidelberg
14. Eriksson, K., Estep, D., Hansbo, P., and Johnson, C. (1995) Introduction to adaptive methods for differential equations. In: Iserles, A. (ed.), Acta Numerica, 105–158
15. Estep, D, Larson, M., and Williams, R. (2002) Estimating the error of numerical solutions of systems of reaction-diffusion equations. Memoirs of the American Mathematical Society (to appear)
16. Friedrichs, K.O. (1958) Symmetric positive linear differential equations. Commun. Pure Appl. Math. **11**, 333–418
17. Giles, M.B., Larsson, M., Levenstam, M., and Süli, E. (1997) Adaptive error control for finite element approximations of the lift and drag coefficients in viscous flow. Numerical Analysis Group Research Report NA-97/06, University of Oxford.
`http://web.comlab.ox.ac.uk/oucl/publications/natr/na-97-06.html`
18. Giles, M.B. and Pierce, N.A. (1997) Adjoint equation in CFD: duality, boundary conditions and solution behaviour. AIAA Paper 97–1850
19. Giles, M.B. (1998) On adjoint equations for error analysis and optimal grid adaptation in CFD. In: Caughey, D. and E. Hafez, E. (eds.), Frontiers of Computational Fluid Dynamics 1998. World Scientific
20. Giles, M.B. (1999) Improved lift and drag estimates using adjoint Euler equations. AIAA Paper 99–3293
21. Giles, M.B. (2000) An introduction to the adjoint design approach and analysis. Numerical Analysis Group Research Report NA-00/04, University of Oxford.
`http://web.comlab.ox.ac.uk/oucl/publications/natr/na-00-04.html`
22. Giles, M.B and Süli, E. (2002) Adjoint methods for PDEs: a posteriori error analysis and postprocessing by duality. Iserles, A. (ed.), Acta Numerica 2002 (to appear)
23. Gui, W. and Babuška, I. (1986) The h, p and h–p versions of the finite element method in 1 Dimension. Part III. The adaptive h–p version. Numer. Math. **49**, 659–683
24. Hansbo, P. and Johnson, C. (1991) Adaptive streamline diffusion finite element methods for compressible flow using conservative variables. Comput. Methods Appl. Mech. Engrg. **87**, 267–280
25. Hartmann, R. (2000) Adaptive FE-methods for conservation equations. In: Warnecke, G. (ed.), Eighth International Conference on Hyperbolic Problems. Theory, Numerics, Applications (HYP2000) (to appear)
26. Hartmann, R. and Houston, P. (2001) Adaptive discontinuous Galerkin finite element methods for nonlinear hyperbolic conservation laws. (Submitted for publication)
27. Hartmann, R. and Houston, P. (2001) Adaptive discontinuous Galerkin finite element methods for the compressible Euler equations. (Submitted for publication)

28. Hartmann, R., Houston, P., and Süli, E. (2001) Adaptive discontinuous Galerkin finite element method for nonlinear hyperbolic problems. In M. Baines, editor, Numerical Methods for Fluid Dynamics VII, ICFD, pp. 347–353
29. Houston, P., Mackenzie, J.A., Süli, E, and Warnecke, G. (1999) A posteriori error analysis for numerical approximations of Friedrichs systems. Numer. Math. **82**, 433–470
30. Houston, P., Rannacher, R. and Süli, E. (2000) A posteriori error analysis for stabilized finite element approximations of transport problems. Comput. Meth. Appl. Mech. Engrg. **190**, 1483–1508
31. Houston, P., Schwab, Ch. and Süli, E. (2000) Stabilized *hp*–finite element methods for first–order hyperbolic problems. SIAM J. Numer. Anal. **37**, 1618–1643
32. Houston, P., Schwab, Ch., and Süli, E. (2002) Discontinuous *hp*–finite element methods for advection–diffusion–reaction problems. SIAM J. Numer. Anal. (to appear)
33. Houston, P., Senior, B. and Süli, E. (2002) Sobolev regularity estimation for *hp*–adaptive finite element methods. (Submitted for publication)
34. Houston, P. and Süli, E. (2001) Stabilized *hp*–finite element approximation of partial differential equations with non-negative characteristic form. Computing **66**, 99–119
35. Houston, P. and Süli, E. (2001) *hp*–Adaptive discontinuous Galerkin finite element methods for hyperbolic problems. SIAM J. Sci. Comp., **23**, 1225–1251
36. Hughes, T.J.R., Franca, F.P. and Mallet, M. (1986) A new finite element formulation for computational fluid dynamics: symmetric forms of the compressible Euler and Navier–Stokes equations and the second law of thermodynamics. Comp. Meth. App. Mech. Engrg. **54**, 223–234
37. Jaffre, J., Johnson, C. and Szepessy, A. (1995) Convergence of the discontinuous Galerkin finite element method for hyperbolic conservation laws. Math. Models Methods Appl. Sci. **5**, 367–386
38. Johnson, C., Nävert, U. and Pitkäranta, J. (1984) Finite Element Methods for linear hyperbolic problems. Comp. Meth. Appl. Mech. Engrg. **45**, 285–312
39. Kröner, D. (1997) Numerical Schemes for Conservation Laws. Wiley-Teubner
40. Kröner, D. and Ohlberger, M. (2000) A-posteriori error estimates for upwind finite volume schemes for nonlinear conservation laws in multi dimensions. Math. Comput. **69**, 25–39
41. Lax, P.D. and Phillips, R.S. (1960) Local boundary conditions for dissipative symmetric linear differential operators. Comm. Pure Appl. Math., **13**, 427–455
42. Larson, M. and Barth, T.J. (2000) In: Cockburn, B., Karniadakis, G.E., Shu C.-W. (eds.) Discontinuous Galerkin Finite Element Methods. Lecture Notes in Computational Science and Engineering, **11**, pp. 363–368. Springer, Berlin Heidelberg
43. Mock, M.S. (1980) Systems of conservation laws of mixed type. J. Diff. Equ. **37**, 70–88
44. Monk, P. and Süli, E. (1998) The adaptive computation of far field patterns by a posteriori error estimates of linear functionals. SIAM J. Numer. Anal. **36**, 251–274
45. Oden, J.T. and Prudhomme, S. (1999) On goal-oriented error estimation for elliptic problems: Application to control of pointwise errors. Comput. Meth. Appl. Mech. Engrg. **176**, 313–331

46. Paraschivoiu, M., Peraire, J. and Patera, A. (1997) A posteriori finite element bounds for linear functional outputs of elliptic partial differential equations. Comput. Meth. Appl. Mech. Engrg. **150**, 289–312
47. Peraire, J. and Patera, A.T. (1997) Bounds for linear functional outputs of coercive partial differential equations: local indicators and adaptive refinement. In: Ladeveze, P. and Oden, J.T. (eds.), New Advances in Adaptive Computational Methods in Mechanics. Elsevier
48. Pierce, N.A. and Giles, M.B. (1998) Adjoint recovery of superconvergent functionals from approximate solutions of partial differential equations. Numerical Analysis Group Research Report NA-98/18, University of Oxford. `http://web.comlab.ox.ac.uk/oucl/publications/natr/na-98-18.html`
49. Rannacher, R. (1998) Adaptive finite element methods. In: Bulgak, H. and Zenger, C. (eds.), Proc. NATO-Summer School on Error Control and Adaptivity in Scientific Computing, Kluwer Academic Publishers, 247–278
50. Rauch, J. (1972) L_2 is a continuable initial condition for Kreiss' mixed problems. Comm. Pure Appl. Math. **25**, 265–285
51. Schwab, Ch. (1998) *p*- and *hp*-Finite Element Methods. Theory and Applications to Solid and Fluid Mechanics. Oxford University Press, Oxford
52. Sonar, T. and Süli, E. (1998) A dual graph-norm refinement indicator for finite volume approximations of the Euler equations. Numer. Math. **78**, 619–658
53. Süli, E. (1998) A posteriori error analysis and adaptivity for finite element approximations of hyperbolic problems. In: Kröner, D., Ohlberger, M., Rohde, C. (eds.) An Introduction to Recent Developments in Theory and Numerics of Conservation Laws. Lecture Notes in Computational Science and Engineering **5**, pp. 123–194. Springer, Berlin Heidelberg
54. Süli, E., Houston, P. and Schwab, Ch. (1999) *hp*–Finite element methods for hyperbolic problems. In: Whiteman, J.R. (ed.), The Mathematics of Finite Elements and Applications X. MAFELAP 1999. Elsevier, 143–162
55. Süli, E., Houston, P. and Senior, B. (2001) *hp*–Discontinuous Galerkin finite element methods for nonlinear hyperbolic problems. Int. J. Numer. Meth. Fluids (to appear)
56. Szabó, B. and Babuška, I. (1991) Finite Element Analysis. J. Wiley & Sons, New York
57. Verfürth, R. (1996) A Review of a Posteriori Error Estimation and Adaptive Mesh-Refinement Techniques. B.G. Teubner, Stuttgart

Editorial Policy

§1. Volumes in the following three categories will be published in LNCSE:

i) Research monographs
ii) Lecture and seminar notes
iii) Conference proceedings

Those considering a book which might be suitable for the series are strongly advised to contact the publisher or the series editors at an early stage.

§2. Categories i) and ii). These categories will be emphasized by Lecture Notes in Computational Science and Engineering. **Submissions by interdisciplinary teams of authors are encouraged.** The goal is to report new developments - quickly, informally, and in a way that will make them accessible to non-specialists. In the evaluation of submissions timeliness of the work is an important criterion. Texts should be well-rounded, well-written and reasonably self-contained. In most cases the work will contain results of others as well as those of the author(s). In each case the author(s) should provide sufficient motivation, examples, and applications. In this respect, Ph.D. theses will usually be deemed unsuitable for the Lecture Notes series. Proposals for volumes in these categories should be submitted either to one of the series editors or to Springer-Verlag, Heidelberg, and will be refereed. A provisional judgment on the acceptability of a project can be based on partial information about the work: a detailed outline describing the contents of each chapter, the estimated length, a bibliography, and one or two sample chapters - or a first draft. A final decision whether to accept will rest on an evaluation of the completed work which should include

- at least 100 pages of text;
- a table of contents;
- an informative introduction perhaps with some historical remarks which should be accessible to readers unfamiliar with the topic treated;
- a subject index.

§3. Category iii). Conference proceedings will be considered for publication provided that they are both of exceptional interest and devoted to a single topic. One (or more) expert participants will act as the scientific editor(s) of the volume. They select the papers which are suitable for inclusion and have them individually refereed as for a journal. Papers not closely related to the central topic are to be excluded. Organizers should contact Lecture Notes in Computational Science and Engineering at the planning stage.

In exceptional cases some other multi-author-volumes may be considered in this category.

§4. Format. Only works in English are considered. They should be submitted in camera-ready form according to Springer-Verlag's specifications. Electronic material can be included if appropriate. Please contact the publisher. Technical instructions and/or TEX macros are available via http://www.springer.de/author/tex/help-tex.html; the name of the macro package is "LNCSE - LaTEX2e class for Lecture Notes in Computational Science and Engineering". The macros can also be sent on request.

General Remarks

Lecture Notes are printed by photo-offset from the master-copy delivered in camera-ready form by the authors. For this purpose Springer-Verlag provides technical instructions for the preparation of manuscripts. See also *Editorial Policy.*

Careful preparation of manuscripts will help keep production time short and ensure a satisfactory appearance of the finished book.

The following terms and conditions hold:

Categories i), ii), and iii):
Authors receive 50 free copies of their book. No royalty is paid. Commitment to publish is made by letter of intent rather than by signing a formal contract. Springer-Verlag secures the copyright for each volume.

For conference proceedings, editors receive a total of 50 free copies of their volume for distribution to the contributing authors.

All categories:
Authors are entitled to purchase further copies of their book and other Springer mathematics books for their personal use, at a discount of 33,3 % directly from Springer-Verlag.

Addresses:

Timothy J. Barth
NASA Ames Research Center
NAS Division
Moffett Field, CA 94035, USA
e-mail: barth@nas.nasa.gov

Michael Griebel
Institut für Angewandte Mathematik
der Universität Bonn
Wegelerstr. 6
D-53115 Bonn, Germany
e-mail: griebel@iam.uni-bonn.de

David E. Keyes
Computer Science Department
Old Dominion University
Norfolk, VA 23529-0162, USA
e-mail: keyes@cs.odu.edu

Risto M. Nieminen
Laboratory of Physics
Helsinki University of Technology
02150 Espoo, Finland
e-mail: rni@fyslab.hut.fi

Dirk Roose
Department of Computer Science
Katholieke Universiteit Leuven
Celestijnenlaan 200A
3001 Leuven-Heverlee, Belgium
e-mail: dirk.roose@cs.kuleuven.ac.be

Tamar Schlick
Department of Chemistry
Courant Institute of Mathematical
Sciences
New York University
and Howard Hughes Medical Institute
251 Mercer Street
New York, NY 10012, USA
e-mail: schlick@nyu.edu

Springer-Verlag, Mathematics Editorial IV
Tiergartenstrasse 17
D-69121 Heidelberg, Germany
Tel.: *49 (6221) 487-8185
e-mail: peters@springer.de
http://www.springer.de/math/
peters.html

Lecture Notes in Computational Science and Engineering

Vol. 1 D. Funaro, *Spectral Elements for Transport-Dominated Equations.* 1997. X, 211 pp. Softcover. ISBN 3-540-62649-2

Vol. 2 H. P. Langtangen, *Computational Partial Differential Equations.* Numerical Methods and Diffpack Programming. 1999. XXIII, 682 pp. Hardcover. ISBN 3-540-65274-4

Vol. 3 W. Hackbusch, G. Wittum (eds.), *Multigrid Methods V.* Proceedings of the Fifth European Multigrid Conference held in Stuttgart, Germany, October 1-4, 1996. 1998. VIII, 334 pp. Softcover. ISBN 3-540-63133-X

Vol. 4 P. Deuflhard, J. Hermans, B. Leimkuhler, A. E. Mark, S. Reich, R. D. Skeel (eds.), *Computational Molecular Dynamics: Challenges, Methods, Ideas.* Proceedings of the 2nd International Symposium on Algorithms for Macromolecular Modelling, Berlin, May 21-24, 1997. 1998. XI, 489 pp. Softcover. ISBN 3-540-63242-5

Vol. 5 D. Kröner, M. Ohlberger, C. Rohde (eds.), *An Introduction to Recent Developments in Theory and Numerics for Conservation Laws.* Proceedings of the International School on Theory and Numerics for Conservation Laws, Freiburg / Littenweiler, October 20-24, 1997. 1998. VII, 285 pp. Softcover. ISBN 3-540-65081-4

Vol. 6 S. Turek, *Efficient Solvers for Incompressible Flow Problems.* An Algorithmic and Computational Approach. 1999. XVII, 352 pp, with CD-ROM. Hardcover. ISBN 3-540-65433-X

Vol. 7 R. von Schwerin, *Multi Body System SIMulation.* Numerical Methods, Algorithms, and Software. 1999. XX, 338 pp. Softcover. ISBN 3-540-65662-6

Vol. 8 H.-J. Bungartz, F. Durst, C. Zenger (eds.), *High Performance Scientific and Engineering Computing.* Proceedings of the International FORTWIHR Conference on HPSEC, Munich, March 16-18, 1998. 1999. X, 471 pp. Softcover. 3-540-65730-4

Vol. 9 T. J. Barth, H. Deconinck (eds.), *High-Order Methods for Computational Physics.* 1999. VII, 582 pp. Hardcover. 3-540-65893-9

Vol. 10 H. P. Langtangen, A. M. Bruaset, E. Quak (eds.), *Advances in Software Tools for Scientific Computing.* 2000. X, 357 pp. Softcover. 3-540-66557-9

Vol. 11 B. Cockburn, G. E. Karniadakis, C.-W. Shu (eds.), *Discontinuous Galerkin Methods.* Theory, Computation and Applications. 2000. XI, 470 pp. Hardcover. 3-540-66787-3

Vol. 12 U. van Rienen, *Numerical Methods in Computational Electrodynamics.* Linear Systems in Practical Applications. 2000. XIII, 375 pp. Softcover. 3-540-67629-5

Vol. 13 B. Engquist, L. Johnsson, M. Hammill, F. Short (eds.), *Simulation and Visualization on the Grid.* Parallelldatorcentrum Seventh Annual Conference, Stockholm, December 1999, Proceedings. 2000. XIII, 301 pp. Softcover. 3-540-67264-8

Vol. 14 E. Dick, K. Riemslagh, J. Vierendeels (eds.), *Multigrid Methods VI.* Proceedings of the Sixth European Multigrid Conference Held in Gent, Belgium, September 27-30, 1999. 2000. IX, 293 pp. Softcover. 3-540-67157-9

Vol. 15 A. Frommer, T. Lippert, B. Medeke, K. Schilling (eds.), *Numerical Challenges in Lattice Quantum Chromodynamics.* Joint Interdisciplinary Workshop of John von Neumann Institute for Computing, Jülich and Institute of Applied Computer Science, Wuppertal University, August 1999. 2000. VIII, 184 pp. Softcover. 3-540-67732-1

Vol. 16 J. Lang, *Adaptive Multilevel Solution of Nonlinear Parabolic PDE Systems.* Theory, Algorithm, and Applications. 2001. XII, 157 pp. Softcover. 3-540-67900-6

Vol. 17 B. I. Wohlmuth, *Discretization Methods and Iterative Solvers Based on Domain Decomposition.* 2001. X, 197 pp. Softcover. 3-540-41083-X

Vol. 18 U. van Rienen, M. Günther, D. Hecht (eds.), *Scientific Computing in Electrical Engineering.* Proceedings of the 3rd International Workshop, August 20-23, 2000, Warnemünde, Germany. 2001. XII, 428 pp. Softcover. 3-540-42173-4

Vol. 19 I. Babuška, P. G. Ciarlet, T. Miyoshi (eds.), *Mathematical Modeling and Numerical Simulation in Continuum Mechanics.* Proceedings of the International Symposium on Mathematical Modeling and Numerical Simulation in Continuum Mechanics, September 29 - October 3, 2000, Yamaguchi, Japan. 2002. VIII, 301 pp. Softcover. 3-540-42399-0

Vol. 20 T. J. Barth, T. Chan, R. Haimes (eds.), *Multiscale and Multiresolution Methods.* Theory and Applications. 2002. X, 389 pp. Softcover. 3-540-42420-2

Vol. 21 M. Breuer, F. Durst, C. Zenger (eds.), *High Performance Scientific and Engineering Computing.* Proceedings of the 3rd International FORTWIHR Conference on HPSEC, Erlangen, March 12-14, 2001. 2002. XIII, 408 pp. Softcover. 3-540-42946-8

Vol. 22 K. Urban, *Wavelets in Numerical Simulation.* Problem Adapted Construction and Applications. 2002. XV, 181 pp. Softcover. 3-540-43055-5

Vol. 23 L. F. Pavarino, A. Toselli (eds.), *Recent Developments in Domain Decomposition Methods.* 2002. XII, 243 pp. Softcover. 3-540-43413-5

Vol. 24 T. Schlick, H. H. Gan (eds.), *Computational Methods for Macromolecules: Challenges and Applications.* Proceedings of the 3rd International Workshop on Algorithms for Macromolecular Modeling, New York, October 12-14, 2000. 2002. IX, 504 pp. Softcover. 3-540-43756-8

Vol. 25 T. J. Barth, H. Deconinck (eds.), *Error Estimation and Adaptive Discretization Methods in Computational Fluid Dynamics.* 2003. VII, 344 pp. Hardcover. 3-540-43758-4

Vol. 26 M. Griebel, M. A. Schweitzer (eds.), *Meshfree Methods for Partial Differential Equations.* 2003. IX, 466 pp. Softcover. 3-540-43891-2

For further information on these books please have a look at our mathematics catalogue at the following URL: `http://www.springer.de/math/index.html`

Druck: Strauss Offsetdruck, Mörlenbach
Verarbeitung: Schäffer, Grünstadt

Printed in the United States
61565LVS00001B/187-189

9 783540 437581